Biogeochemistry of Trace Elements in Arid Environments

ENVIRONMENTAL POLLUTION

VOLUME 13

Editors

Brian J. Alloway, *Department of Soil Science, The University of Reading, U.K.*
Jack T. Trevors, *Department of Environmental Biology, University of Guelph, Ontario, Canada*

Editorial Board

T. Anderson, *The Institute of Environmental and Human Health, Texas Tech University, Lubbock, U.S.A.*
T.H. Christensen, *Department of Environmental Science and Engineering, Danish Technical University, Lyngby, Denmark*
I. Colbeck, *Institute for Environmental Research, Department of Biological Sciences, University of Essex, Colchester, U.K.*
K.C. Jones, *Institute of Environmental and Natural Sciences, Lancaster University, U.K.*
W. Salomons, *GKSS Research Center, Geesthacht, Germany*

The titles published in this series are listed at the end of this volume.

Biogeochemistry of Trace Elements in Arid Environments

By

Fengxiang X. Han

Mississippi State University
Starkville, MS, USA

With a Review Chapter on Arid Zone Soil

by

Arieh Singer

Hebrew University of Jerusalem
Rehovot, Israel

 Springer

A C.I.P. Catalogue record for this book is available from the Library of Congress.

ISBN 978-1-4020-6023-6 (HB)
ISBN 978-1-4020-6024-3 (e-book)

Published by Springer,
P.O. Box 17, 3300 AA Dordrecht, The Netherlands.

www.springer.com

Printed on acid-free paper

Cover Image © 2007 JupiterImages Corporation

All Rights Reserved
© 2007 Springer
No part of this work may be reproduced, stored in a retrieval system, or transmitted
in any form or by any means, electronic, mechanical, photocopying, microfilming, recording
or otherwise, without written permission from the Publisher, with the exception
of any material supplied specifically for the purpose of being entered
and executed on a computer system, for exclusive use by the purchaser of the work.

To my wife, Chunxia Michelle, and children Bowen, Kayla, and Samuel for their understanding, patience, and support.

To my mother, Liangying Hu, for her heartfelt love, diligence, and hard work in raising, educating, and supporting us, never giving up hope.

and

To Professor Amos Banin at The Hebrew University of Jerusalem, Israel for his dedicated training, guidance, and encouragement.

Contents

List of Figures... ix
List of Tables... xxi
Preface... xxv

SECTION I
Introduction on arid zone soil
Arieh Singer... 1

Chapter 1
Arid zone soils: Nature and properties................................... 3

SECTION II
Biogeochemistry of trace elements in arid environments
Fengxiang X. Han.. 45

Chapter 2
Trace element distribution in arid zone soils...........................47

Chapter 3
Solution chemistry of trace elements in arid zone soils................ 69

Chapter 4
Selective sequential dissolution for trace elements in arid zone
soils... 107

Chapter 5
Binding and distribution of trace elements among solid-phase
components in arid zone soils... 131

Chapter 6
Transfer fluxes of trace elements in arid zone soils – a case
study: Israeli arid soils 169

Chapter 7
Bioavailability of trace elements in arid zone soils....................221

Chapter 8
Trace element pollution in arid zone soils.............................267

Chapter 9
Global perspectives of anthropogenic interferences
in the natural trace element distribution: Industrial age inputs
of trace elements into the pedosphere 303

References 329

Index 357

List of Figures

Figure 1.1. (a) Climate categories based on mean annual precipitation; (b) Desert and steppe boundaries as a function of both mean annual precipitation and temperature according to the Koeppen system; arrows show how boundaries shift according to whether precipitation falls mainly in in the summer or winter (after Monger ct al., 2004. Reprinted from Encyclopedia of Soils in the Environment, D. Hillel, Tropical soils: Arid and Semi-arid, p183, Copyright (2004), with permission from Elsevier).

Figure 1.2. Global land area by aridity zone (%) (after UNEP, 1993. Reprinted from World Atlas of Desertification, Figure of global land area by aridity zones, p5, Copyright (1993), with permission from UNEP).

Figure 1.3. Distributions of arid and semi-arid areas (Reprinted from Soils of Arid Regions, Dregne H.E., p 6, Copyright (1976), with permission from Elsevier).

Figure 1.4. Phase diagram for the system K_2O-Al_2O_3-SiO_2-H_2O at 25 °C and 1 atmosphere. Open circles are analytical data for water from springs or seeps in a granitic terrain in the Sierra Nevada (from Feth and others) (%) (from Birkeland, 1984, after Feth and others, 1964. Reprinted from Soils and Geomorphology, p111, Copyright (1984), with permission from Oxford University Press, New York, Oxford).

Figure 1.5. Schematic Reg soil pedon: the gravel is chalk, underneath the gravel cover is the porous vesicular A horizon (%) (from Singer, 2007, Reprinted from The Soils of the Land of Israel, Copyright (2007), with permission from Springer).

Figure 1.6. The transport of fine particles by duststorm (after Singer, 2007. Reprinted from The Soils of the Land of Israel, Copyright (2007), with permission from Springer).

Figure 1.7. Distribution of salts in chloride-sulfate Solonchak during the various seasons of the year (after Ravikovitch and Bidner-Bar Hava, 1948) (from Singer, 2007, after Ravikovitch and Bidner-BarHava, 1948. Reprinted from The Soils of the Land of Israel, Copyright (2007), with permission from Springer).

Figure 1.8. Cracks, surface mulch and soil structure in a Vertisol during the dry season (from FAO, 2001. Reprinted from World Soil Resources Reports 94, P. Driessen, J. Deckers, O. Spaargaren, F. Nachtergaele, eds., p80, Copyright (2001), with permission from FAO).

Figure 1.9. Illustration of soil taxonomy suborders and great groups that have aridic moisture regimes (shaded) and their moisture counterparts that have ustic and xeric moisture regimes (after Monger, 2002. Reprinted from Encyclopedia of Soil Science, H. Tan, ed., Arid Soils, Monger, p86, Copyright (2002), with permission from Marcel Dekker).

Figure 2.1. Distribution of chemical elements in the earth's crust (mg/kg).

Figure 2.2. Elemental enrichment factors in soils, plotted on a log scale against the ionic potential of the elements.

Figure 2.3. Elemental enrichment factors in baterial and fungi, plotted on a log scale against the ionic potential of the elements (after Banin and Navrot, 1975. Reprinted from Science, 189, Banin A. and Navrot J., Origin of Life: Clues from relations between chemical compositions of living organisms and natural environments, pp 550-551, Copyright (1975), with permission from AAAS).

Figure 2.4. Elemental enrichment factors in plants and land animals, plotted on a log scale against the ionic potential of the elements (after Banin and Navrot, 1975. Reprinted from Science, 189, Banin A. and Navrot J., Origin of Life: Clues from relations between chemical compositions of living organisms and natural environments, pp 550-551, Copyright (1975), with permission from AAAS).

Figure 2.5. Elemental enrichment factors in seawater, related to the ionic potential of the elements (after Banin and Navrot, 1975. Reprinted from Science, 189, Banin A. and Navrot J., Origin of Life: Clues from relations between chemical compositions of living organisms and natural environments, pp 550-551, Copyright (1975), with permission from AAAS).

Figure 2.6. Effects of soil pH on total Cu concentrations in Israeli arid and semi-arid soils (R: 0.50, N: 39).

Figure 3.1. Changes of Zn activity in 10 Colorado soils with pH (data extracted from Ma and Lindsay, 1990).

Figure 3.2. Changes of Cd speciation in soil solutions of a typical Israeli calcareous soil with pH 4 - 9 (after Hirsch and Banin, 1990, with permission from Soil Sci. Soc. Am).

Figure 3.3. Changes of Cd speciation in soil solutions of Cd nitrate-treated Domino soil from California (data extracted from Candelaria and Chang, 1997).

Figures

Figure 3.4. Effects of phosphate levels on Cd and Zn solution speciation in California soils that received sludge application (data extracted from Villarroel et al., 1993).

Figure 3.5. Effects of Cl concentrations (NaCl or $CaCl_2$) on Cd solution speciation in a California soil (data extracted from Bingham et al., 1983).

Figure 4.1. Removal of carbonate from Israeli arid soils as indicated by the X-ray diffractograms after extraction of the carbonate fraction by NaOAc-HOAc solutions at various pHs for 16 hours. C: calcite: d = 3.04 Å, and D: dolomite, d = 2.89 Å. Number 1, 2, 3, 4, 5, and 6 indicate non-treated soil (No. 1), treatments (No. 2-6) with NaOAc-HOAc solutions at pH 7.0, 6.0, 5.5, 5.0 and 4.0, respectively (after Han and Banin, 1995. Reprinted from Commun Soil Sci Plant Anal, 26, Han and Banin A., Selective sequential dissolution techniques for trace metals in arid-zone soils: The carbonate dissolution step, p 563, Copyright (1995), with permission from Taylor & Francis US).

Figure 4.2. Dissolution of Ca from Israeli arid soils by NaOAc-HOAc solutions at various pHs after the extraction of the exchangeable fraction (after Han and Banin, 1995. Reprinted from Commun Soil Sci Plant Anal, 26, Han and Banin A., Selective sequential dissolution techniques for trace metals in arid-zone soils: The carbonate dissolution step, p 568, Copyright (1995), with permission from Taylor & Francis US).

Figure 4.3. Dissolution kinetics of $CaCO_3$ by NaOAc-HOAc solutions at varying pHs. a: Dissolution of pure $CaCO_3$ by NaOAc-HOAc (0.1 g $CaCO_3$, 25 mL buffer solution), b and c: Dissolution of soil $CaCO_3$ by NaOAc-HOAc at pH 5.0 (1.0 g soil, 25 ml buffer solution) (after Han and Banin, 1995. Reprinted from Commun Soil Sci Plant Anal, 26, Han and Banin A., Selective sequential dissolution techniques for trace metals in arid-zone soils: The carbonate dissolution step, p 565, Copyright (1995), with permission from Taylor & Francis US).

Figure 4.4. Effects of pH of NaOAc-HOAc solutions for the CARB fraction on the subsequent fractions (Soil $CaCO_3$: 33.3%) (after Han and Banin, 1995. Reprinted from Commun Soil Sci Plant Anal, 26, Han and Banin A., Selective sequential dissolution techniques for trace metals in arid-zone soils: The carbonate dissolution step, p 572, Copyright (1995), with permission from Taylor & Francis US).

Figure 4.5. Cumulative sums of extractable elements in soil J3 ($CaCO_3$: 33.3%) in consecutive steps of the selective sequential dissolution procedure as a function of the cumulative steps. 1: CARB (carbonate); 2: CARB (carbonate) + ERO (easily reducible oxide); 3: CARB + ERO + OM (organic matter); 4: CARB + ERO + OM + RO (reducible oxide) (after Han and Banin, 1995. Reprinted from Commun Soil Sci Plant Anal, 26, Han and Banin A., Selective sequential dissolution techniques for trace metals in arid-zone soils: The carbonate dissolution step, p 573, Copyright (1995), with permission from Taylor & Francis US).

Figure 4.6. Cumulative sums of extractable Ca, Mg, and Mn in arid soil J3, B10, and S1 in consecutive steps of two sequential dissolution procedures.

Figure 5.1. Correlation between Co and Fe contents in Israeli arid soils.

Figure 5.2. Correlation between Cu and MnO contents in Israeli arid soils.

Figure 5.3. Effects of soil pH on the Zn amount bound to the Fe oxide fraction (amorphous/crystalline Fe oxide and overall Fe oxide bound fractions) in soils from China with pH 3.73 - 8.1 and 0 - 14.7% $CaCO_3$ (after Han et al., 1995. Reprinted from Geoderma, 66, Han F.X., Hu A.T., Qi H.Y., Transformation and distribution of forms of zinc in acid, neutral and calcareous soils of China, p 128, Copyright (1995), with permission from Elsevier).

Figure 5.4. Effects of pH on Zn adsorption on an untreated and treated soils with peroxide and dithionite.

Figure 5.5. Kinetics of Zn adsorption on a soil and soil with removals of organic matter and Fe oxides (1 g soil reacted with 20 ml of 10 mg/L Zn solution concentration).

Figure 5.6. Relationship between Co and Mn contents extracted from solid-phase of six Israeli arid-zone soils with sequential dissolution procedures (after Han et al., 2002b. Reprinted from J Environ Sci Health, Part A, 137, Han F.X., Banin A., Kingery W.L., Li Z.P., Pathways and kinetics of transformation of cobalt among solid-phase components in arid-zone soils, p 184, Copyright (2002), with permission from Taylor & Francis).

Figure 6.1. The fractional loading isotherms of Cu in a contaminated Israeli loessial soil at an initial (one hour) period and after 48 weeks. The soil was treated with increasing levels of metal nitrates and was incubated under the field capacity regime. Horizontal solid line represents the native content of Cu in the nonamended soil (*Figure 6.1 – Figure 6.4,* after Han and Banin, 2001. Reprinted from Commun Soil Sci Plant Anal, 32, Han F.X and Banin A.,The

Figures xiii

Figure 6.2. fractional loading isotherm of heavy metals in an arid-zone soil, pp 2700-2703, Copyright (2001), with permission from Taylor & Francis)

Figure 6.2. The fractional loading isotherms of Cr in a contaminated Israeli loessial soil at an initial (one hour) period and after 48 weeks.

Figure 6.3. The fractional loading isotherms of Ni in a contaminated Israeli loessial soil at an initial (one hour) period and after 48 weeks.

Figure 6.4. The fractional loading isotherms of Zn in a contaminated Israeli loessial soil at an initial (one hour) period and after 48 weeks.

Figure 6.5. The initial reduced partition index, I_R, of six metals in two Israeli arid soils. Two soils were treated with metal nitrates at various loading levels. Soils were incubated under the field capacity moisture regime (modified after Han and Banin, 1999, with permission from Springer Science and Business Media).

Figure 6.6. Long-term changes of U_{ts} of metals in a loessial soil from Israel. The soil received metal nitrates under the field capacity moisture regime (after Han and Banin, 1999. Reprinted from Water Air Soil Pollut, 114, Han F.X and Banin A., Long-term transformations and redistribution of potentially toxic heavy metals in arid-zone soils. II: Incubation under field capacity conditions, p 245, Copyright (1999), with permission from Springer Science and Business Media).

Figure 6.7. Ranges (arithmetic mean and standard deviation) of the reduced partition index (I_R) of Cd, Cu, Cr, Ni and Zn in 45 Israeli arid-zone soils (after Han et al., 2003a. Reprinted from Adv Environ Res, 8, Han F.X., Banin A., Kingery W.L., Triplett G.B., Zhou L.X., Zheng S.J., Ding W.X., New approach to studies of redistribution of heavy metals in soils, p 118, Copyright (2003), with permission from Elsevier).

Figure 6.8. Changes of Cd fractions in two Israeli soils during 336 days of incubation at the field capacity moisture regime (after Han and Banin, 1999. Reprinted from Water Air Soil Pollut, 114, Han F.X and Banin A., Long-term transformations and redistribution of potentially toxic heavy metals in arid-zone soils. II: Incubation under field capacity conditions, p 238, Copyright (1999), with permission from Springer Science and Business Media).

Figure 6.9. Changes of Cr fractions in two Israeli soils during 336 days of incubation at the field capacity moisture regime (after Han and Banin, 1999. Reprinted from Water Air Soil Pollut, 114, Han F.X and Banin A., Long-term transformations and redistribution of potentially toxic heavy metals in arid-zone soils. II: Incubation under field capacity conditions, p 239, Copyright (1999), with permission from Springer Science and Business Media).

Figure 6.10. Changes of Cu fractions in two Israeli soils during 336 days of incubation at the field capacity moisture regime (after Han and Banin, 1999. Reprinted from Water Air Soil Pollut, 114, Han F.X and Banin A., Long-term transformations and redistribution of potentially toxic heavy metals in arid-zone soils. II: Incubation under field capacity conditions, p 240, Copyright (1999), with permission from Springer Science and Business Media).

Figure 6.11. Changes of Ni fractions in two Israeli soils during 336 days of incubation at the field capacity moisture regime (after Han and Banin, 1999. Reprinted from Water Air Soil Pollut, 114, Han F.X and Banin A., Long-term transformations and redistribution of potentially toxic heavy metals in arid-zone soils. II: Incubation under field capacity conditions, p 241, Copyright (1999), with permission from Springer Science and Business Media).

Figure 6.12. Changes of Zn fractions in two Israeli soils during 336 days of incubation at the field capacity moisture regime (after Han and Banin, 1999. Reprinted from Water Air Soil Pollut, 114, Han F.X and Banin A., Long-term transformations and redistribution of potentially toxic heavy metals in arid-zone soils. II: Incubation under field capacity conditions, p 243, Copyright (1999), with permission from Springer Science and Business Media).

Figure 6.13. Changes of Pb fractions in two Israeli soils during 336 days of incubation at the field capacity moisture regime (after Han and Banin, 1999. Reprinted from Water Air Soil Pollut, 114, Han F.X and Banin A., Long-term transformations and redistribution of potentially toxic heavy metals in arid-zone soils. II: Incubation under field capacity conditions, p 242, Copyright (1999), with permission from Springer Science and Business Media).

Figures   xv

Figure 6.14. Changes of U_{ts} of Cd, Cu and Zn in an Israeli sandy soil after 8 months of sludge application (T is the total metal content in the non-amended soil) (after Han et al., 2003a. Reprinted from Adv Environ Res, 8, Han F.X., Banin A., Kingery W.L., Triplett G.B., Zhou L.X., Zheng S.J., Ding W.X., New approach to studies of redistribution of heavy metals in soils, p 117, Copyright (2003), with permission from Elsevier).

Figure 6.15. The average transfer fluxes of Cr and Cu among solid-phase components in Israeli soils. The soils received metal nitrates at the 3T level under the field capacity moisture regime (T: total metal content in non-amended soils).

Figure 6.16. The average transfer fluxes of Ni and Zn among solid-phase components in Israeli soils. The soils received metal nitrates at the 3T level under the field capacity moisture regime (T: total metal content in non-amended soils).

Figure 6.17. Relationships between the initial (the first day) fluxes of metals in the soluble plus exchangeable fraction and metal loading levels in Israeli soils. Soils received metal nitrates and were incubated in the saturated-paste (SP) and field capacity (FC) moisture regimes.

Figure 6.18. Changes of the annual transfer fluxes (from one day to one year) among solid-phase components with metal loading levels in Israeli soils. Soils received metal nitrates and were incubated in the saturated-paste (SP) and field capacity (FC) moisture regimes.

Figure 6.19. Distribution of Mn in solid-phase fractions in Israeli soils. Soils were incubated at the saturated paste regime (after Han and Banin, 1996. Reprinted from Soil Sci Soc Am J, 60, Han F.X., Banin A., Solid-phase manganese fractionation changes in saturated arid-zone soils: Pathways and kinetics, p 1075, Copyright (1996), with permission from Soil Sci Soc Am).

Figure 6.20. Distribution of Mn in solid-phase fractions in Israeli soils. Soils were incubated at the field capacity regime.

Figure 6.21. Initial and annual distribution and transformations of Mn in the major solid-phase fractions in two Israeli soils. Soils were incubated at the saturated-paste regime (after Han and Banin, 1996. Reprinted from Soil Sci Soc Am J, 60, Han F.X., Banin A., Solid-phase manganese fractionation changes in saturated arid-zone soils: Pathways and kinetics, p 1076, Copyright (1996), with permission from Soil Sci Soc Am).

Figure 6.22. Distribution and transformations of Mn in the major solid-phase fractions in two Israeli soils. Soils were incubated at the field capacity regime.

Figure 6.23. Changes of the redox parameter (pe + pH) in two Israeli arid soils during the saturated paste incubation (after Han and Banin, 1996. Reprinted from Soil Sci Soc Am J, 60, Han F.X., Banin A., Solid-phase manganese fractionation changes in saturated arid-zone soils: Pathways and kinetics, p 1076, Copyright (1996), with permission from Soil Sci Soc Am).

Figure 6.24. Comparisons of Mn changes in the ERO (a) and the EXC and the CARB fractions (b) in an Israeli sandy soil during one year of saturation incubation and subsequent drying processes.

Figure 6.25. Changes of R_{tf} of Mn fractions and U_{ts} in Israeli sandy (a) and loessial (b) soils during one year of incubation at the saturated paste regime (after Han and Banin, 1996. Reprinted from Soil Sci Soc Am J, 60, Han F.X., Banin A., Solid-phase manganese fractionation changes in saturated arid-zone soils: Pathways and kinetics, pp 1077-1078, Copyright (1996), with permission from Soil Sci Soc Am).

Figure 6.26. Kinetics of Mn transformations (described by the parabolic diffusion equation) among three main labile solid-phase fractions of Israeli arid soils. Soils were incubated at the saturated paste regime (after Han and Banin, 1996. Reprinted from Soil Sci Soc Am J, 60, Han F.X., Banin A., Solid-phase manganese fractionation changes in saturated arid-zone soils: Pathways and kinetics, p 1079, Copyright (1996), with permission from Soil Sci Soc Am).

Figure 6.27. Distribution of Co in the solid-phase components in two Israeli arid soils incubated at the saturated paste regime (after Han et al., 2002b. Reprinted from J Environ Sci Health, Part A, 137, Han F.X., Banin A., Kingery W.L., Li Z.P., Pathways and kinetics of transformation of cobalt among solid-phase components in arid-zone soils, p 187, Copyright (2003), with permission from Taylor & Francis).

Figure 6.28. Initial changes of Co concentrations in the main solid-phase fractions of two Israeli arid soils. Soils were incubated under the saturated paste regime (after Han et al., 2002b. Reprinted from J Environ Sci Health, Part A, 137, Han F.X., Banin A., Kingery W.L., Li Z.P., Pathways and kinetics of transformation of cobalt among solid-phase components in arid-zone soils, p 188, Copyright (2003), with permission from Taylor & Francis).

Figures

Figure 6.29. Relationships between the Co and Mn contents in the CARB and the ERO fractions in Israeli soils during the saturated paste incubation.

Figure 6.30. Kinetics of transformation of Co in the ERO fraction in two Israeli soils according to the two-constant rate model (a) and the simple Elovich model (b), respectively. Soils were incubated under the saturation paste regime (modified after Han et al., 2002b. Reprinted from J Environ Sci Health, Part A, 137, Han F.X., Banin A., Kingery W.L., Li Z.P., Pathways and kinetics of transformation of cobalt among solid-phase components in arid-zone soils, p 192, Copyright (2003), with permission from Taylor & Francis).

Figure 7.1. Trace element concentrations in plants on California Donimo soil (pH 7.5) amended with metal sulfate-enriched sludge (Data from Mitchell et al., 1978).

Figure 7.2. Ratios of DPTA-extractable trace elements in the rhizosphere over bulk Egyptian soils irrigated with sewage water for 10, 40, and 80 years (Data from El-Motaium and Badawy, 1999).

Figure 7.3. Changes of Cd, Pb and Zn concentrations in soil solutions with pH from ten Colorado arid soils (data from Ma and Lindsay, 1990).

Figure 7.4. Comparisons of decreases in NH_4NO_3-extractable Zn, Ni and Cu in an Israeli loessial soil receiving metal nitrates and incubated under saturated paste, field capacity, and wetting/drying cycle moisture regimes (Han and Banin, 1997, 1999, and Han et al., 2001a).

Figure 7.5. The AB-DTPA-extractable Cu and Zn in a semi-arid soil from Colorado after termination of five excessive biosolids applications. Open circle (fine, smectitic, mesic Aridic Argiustolls) and open triangle (fine, smectitic, mesic Aridic Paleustolls) represent control soil without biosolid application, while solid circle and solid triangle indicate biosolid applications (data extracted from Barbarick ad Ippolito, 2003).

Figure 7.6. Decreases in bioavailability (as NH_4NO_3-extractable) of Cu, Ni and Zn in an Israeli loess soil receiving metal salts and incubated under field capacity regime for one year (data from Han and Banin, 1999).

Figure 7.7. Changes of Cu and Zn concentrations in soil solution with DOC from South Australia (Data from Fotovat et al., 1997).

Figure 7.8. Swiss chard Cd concentrations as affected by chloride levels in soils (modified from Weggler-Beaton et al., 2000).

Figure 7.9. Effects of N rates on Cd concentrations in wheat shoots and grain (data from Mitchell et al., 2000).

Figure 8.1. The accumulation of trace elements in two Californian soils. The soils received four-year sludge applications at the rate of 0, 22.5, 45, and 90 tons ha^{-1} year^{-1}. Domino loam soil, fine-loamy, mixed, thermic Xerollic Calciorthid; Greenfield sandy loam soil, coarse-loamy, mixed, thermic Typic Haploxeralf (Data from Sposito et al., 1982).

Figure 9.1. Measured and fitted global annual industrial age As production. (a) by mining; (b) released by coal burning; (c) released by petroleum burning; and (d) gross annual release into the environment (See explanations in the text) (after Han et al., 2003b. Reprinted from Naturwissenschaften, 90, Han F.X., Su Y., Monts D.L., Plodinec M.J., Banin A., Triplett G.B., Assessment of global industrial-age anthropogenic arsenic contamination, pp 396-397, Copyright (2003), with kind permission of Springer Science and Business Media).

Figure 9.2. (a) The percentages of annual and cumulative As generated from the world-wide coal and petroleum industries over the global gross As production. (b) The percentages of annual and cumulative As generated from the world-wide petroleum industry over As generated from both coal and petroleum industries (after Han et al., 2003b. Reprinted from Naturwissenschaften, 90, Han F.X., Su Y., Monts D.L., Plodinec M.J., Banin A., Triplett G.B., Assessment of global industrial-age anthropogenic arsenic contamination, pp 396-398, Copyright (2003), with kind permission of Springer Science and Business Media).

Figure 9.3. Measured and fitted global annual industrial age Hg mine production and Hg production from world coal industry (after Han et al., 2002a. Reprinted from Naturwissenschaften, 89, Han F.X., Banin A., Su Y., Monts D.L., Plodinec M.J., Kingery W.L., Triplett G.B., Industrial age anthropogenic inputs of heavy metals into the pedosphere, p 500, Copyright (2002), with kind permission of Springer Science and Business Media).

Figures   xix

Figure 9.4. Measured and fitted world annual metal production. The measured world annual metal production data were collected from U.S. Geological Survey-Minerals Information, 1997; Adriano, 1986; Woytinsky and Woytinsky, 1953. Chromium production was calculated from chromite production assuming an average of 27.0% Cr. The fitted Cd data before 1963, were calculated from the world annual Zn production (Cd mainly as a by-product) (after Han et al., 2002a. Reprinted from Naturwissenschaften, 89, Han F.X., Banin A., Su Y., Monts D.L., Plodinec M.J., Kingery W.L., Triplett G.B., Industrial age anthropogenic inputs of heavy metals into the pedosphere, p 499, Copyright (2002), with kind permission of Springer Science and Business Media).

Figure 9.5. World cumulative metal productions since 1860 based on fitted world annual metal productions (after Han et al., 2002a. Reprinted from Naturwissenschaften, 89, Han F.X., Banin A., Su Y., Monts D.L., Plodinec M.J., Kingery W.L., Triplett G.B., Industrial age anthropogenic inputs of heavy metals into the pedosphere, p 499, Copyright (2002), with kind permission of Springer Science and Business Media).

Figure 9.6. Global cumulative industrial age Hg mine production, Hg production from coal, and total global Hg production since 1860 based on estimated world annual Hg mine and coal production (inset: The percentage of Hg production from coal over the total global Hg production) (after Han et al., 2002a. Reprinted from Naturwissenschaften, 89, Han F.X., Banin A., Su Y., Monts D.L., Plodinec M.J., Kingery W.L., Triplett G.B., Industrial age anthropogenic inputs of heavy metals into the pedosphere, p 500, Copyright (2002), with kind permission of Springer Science and Business Media).

Figure 9.7. Global cumulative industrial age As production from mining, As generated from coal and petroleum, and cumulative gross As production since 1850 (after Han et al., 2003b. Reprinted from Naturwissenschaften, 90, Han F.X., Su Y., Monts D.L., Plodinec M.J., Banin A., Triplett G.B., Assessment of global industrial-age anthropogenic arsenic contamination, p 397, Copyright (2003), with kind permission of Springer Science and Business Media).

List of Tables

Table 1.1. Aridity zones by region (x10^6 ha) (after UNEP, 1993).
Table 1.2. Land area by aridity zones in Africa (x10^6 ha) (after UNEP, 1993).
Table 1.3. Global extent of arid and semi-arid soils (km^2) based on the Soil Taxonomy system) (after Monger et al., 2004).
Table 2.1. Concentrations of trace elements in the earth's crust, rocks and world soils[a].
Table 2.2. Concentrations of selected trace elements in representative arid and semi-arid soils of North China[a] (from Liu, 1996).
Table 2.3. Geometric means for selected trace elements in arid and semi-arid soils of the U.S.
Table 2.4. Averages and ranges of concentrations of selected trace elements in arid and semi-arid soils of Israel[a].
Table 3.1. The major ion composition of dispersed soil solutions and saturation extracts from selected soils in California, U.S.
Table 3.2. Concentrations of trace elements in soil solutions of the California soils that received sludge applications.
Table 3.3. Concentrations of trace elements in soil solutions extracted by saturated paste from two metal salt-spiked Israeli soils incubated at saturated regime.
Table 3.4. Solution complexes of calcium and magnesium at $25^0 C$[a].
Table 3.5. Solution complexes of selected trace elements at $25^0 C$[a].
Table 3.6. The first hydrolysis constants of selected elements in soil chemistry[a].
Table 3.7. Summary of major representative species of selected trace elements in arid soil solutions.
Table 3.8. Possible soil solution solubility controls for selected trace elements[a].
Table 3.9. Solubility product constants of common compounds and minerals of selected major and trace elements at 25 °C[a].
Table 4.1. Some selective sequential dissolution procedures employed to fractionate trace elements in soils[a, b].
Table 4.2. Dissolution of some major elements extracted by NaOAc-HOAc solutions at various pHs after removing the exchangeable fraction[a].
Table 4.3. Dissolution of some trace elements extracted by NaOAc-HOAc at various pHs after removing the exchangeable fraction (mg/Kg)[a].

Table 4.4. Comparisons of two selective sequential dissolution procedures[a].
Table 4.5. Extractants for trace elements bound to various solid-phase components in soils.
Table 5.1. Percentages of trace metals bound in soil organic matter in two Israeli arid soils treated with metal nitrates[a].
Table 5.2. Percentages of trace metals bound in the Mn and Fe oxides in two Israeli arid soils treated with metal nitrates[a].
Table 5.3. Percentages of trace metals bound in carbonates in two Israeli arid soils treated with metal nitrates.
Table 5.4. Distribution of trace elements among solid-phase components[a] in 45 Israeli arid soils
Table 5.5. Distribution of Zn among solid-phase components in some Chinese arid soils[a].
Table 5.6. Some basic atomic properties of studied and related elements[a].
Table 5.7. Solubility equilibria of selected minerals of related elements[a]
Table 6.1. The relevant soil properties.
Table 6.2. The original metal concentration (HNO_3-extractable) in native soils and levels of addition to the two Israeli arid soils incubated under the saturated paste regime condition.
Table 6.3. Treatment levels of trace metals in the two Israeli arid soils incubated under the field capacity and wetting-drying cycle moisture regimes
Table 6.4. Protocol for the selective sequential dissolution procedure.
Table 6.5. Reduced partition index (I_R) of trace metals in arid-zone soils incubated under the saturated paste regime (after Han and Banin, 1997. Reprinted from Water Air Soil Pollut, 95, Han F.X., Banin A., Long-term transformations and redistribution of potentially toxic heavy metals in arid-zone soils. I: Incubation under saturated conditions, p 411, Copyright (1997), with permission from Springer Science and Business Media).
Table 6.6. Comparisons of the redistribution of metals in two Israeli soils at $3T^a$ treatment after one year of incubation under the saturated paste (SP), field capacity (FC), and wetting-drying cycle (Cycle) moisture regimes (% of the sum of fractions, as means of two replicates) (from Han et al., 2001a, with permission from Lippincott Williams & Wilkins).
Table 6.7. The average residence time (hour) of heavy metals in Israeli soil solutions.

Tables   xxiii

Table 6.8. The kinetics of the initial changes of Mn concentration of three fractions in incubated Israeli soils under saturated regime (C: mg kg-1; t: sec) (Han and Banin, 1996, with permission from Soil Sci Soc Am).
Table 7.1. Concentrations of selected trace elements in plants grown on noncontaminated arid and semi-arid soils[a].
Table 7.2. Extractants for bioavailable trace elements in arid and semi-arid soils.
Table 7.3. Bioavailability of trace elements in arid and semi-arid soils.
Table 7.4. Zinc sources and application rates for correcting zinc deficiency on crops[a].
Table 7.5. Solubility product constants of common compounds/minerals of selected major and trace elements at 25 $^{\circ}C$[a].
Table 7.6. Concentrations of bioavailable trace elements in arid and semi-arid soils of selected countries[a].
Table 7.7. Concentrations of bioavailabile Zn (DTPA-extractable Zn), Cu (DTPA-extractable Cu), and B (water-soluble B) in arid and semi-arid soils of North China[a] (from Liu, 1996).
Table 7.8. Plants/animals and countries susceptible for micronutrient deficiency.
Table 7.9. Diagnostic soil and plant tissue Cd criteria for a 25% and 50% yield decrements[a].
Table 8.1. Agricultural irrigation water standards.
Table 8.2. Trace element composition of sewage treatment plant effluents from the U.S., southern California, and Israel.
Table 8.3. Concentrations of trace elements in Israeli arid soils (0-30 cm) irrigated with secondary sewage effluents (from Banin et al., 1981).
Table 8.4. Trace element composition of sewage sludge from the U.S. and Israel.
Table 8.5. Cumulative loading rates of trace elements for land applications of sewage sludge in selected countries[a].
Table 8.6. Maximum annual loading rates of trace elements for land applications of sewage sludge in selected countries[a].
Table 8.7. Maximum permissible trace element levels of sewage sludge for land applications in the U.S. and China[a].
Table 8.8. Average concentrations of trace elements in phosphate rocks (PR)[a].
Table 8.9. Concentrations of trace elements in vegetation and soils at Tharsis mined areas of semi-arid South West Spain[a].
Table 8.10. Summary of the effects of Cd on microbial processes and enzyme activities in arid and semi-arid soils.

Table 8.11. Chemicals and minerals for the remediation of trace element contaminated arid and semi-arid soils.

Table 8.12. Plant species for the phytoremediation of trace element contaminated soils.

Table 9.1. Global industrial age annual production of selected trace elements and heavy metals in 1880, 1900, 1950, 1990 and 2000 (million tons) (Data extracted from Han et al., 2002a, 2003b).

Table 9.2. Global industrial age cumulative production of selected trace elements and heavy metals in 1880, 1900, 1950, 1990 and 2000 (million tons)[a].

Table 9.3. Potential anthropogenic inputs in world soils and global metal burden per capita[a].

Table 9.4. Trace element contents of the lithosphere and world soils (mg/kg) (Data from Han et al. 2002a, with kind permission of Springer Science and Business Media).

Table 9.5. Arsenic contents of the lithosphere and world soils (Data from Han et al., 2003b, with kind permission of Springer Science and Business Media).

Preface

The arid regions of the world are the sites of major and rapid development in recent decades. Their share in global food production is increasing as a result of large irrigation development projects. The fragile nature of these ecosystems and their limited water resources are threatened by Global Warming processes, risking the livelihood of hundreds of millions of inhabitants of the arid belts spanning Africa, Asia and America. Management of arid zone resources to produce more food in a sustainable manner in order to feed the world growing population has become increasingly important. A key issue is to understand the risks of toxic trace and heavy elements released into the soils by increased human activities in arid environments.

Years ago, the author and Dr. Amos Banin from the Hebrew University of Jerusalem, Israel started to plan such a timely book to meet the current needs of the world's arid environments. Much research on biogeochemistry of trace elements in arid soils was done by the author and Dr. Banin during the period from 1992 to 1997 when the author was at the Hebrew University. Therefore, a majority of the chapters of this book were formed at that time. Of course, many materials and first-hand data are from the previous publications by the author and Dr. Banin's laboratory. Unfortunately, due to the health problems of Dr. Banin, he was unable to get this project finished. Thus this book is a product that resulted from the collaborative efforts and coauthoring between the author and Dr. Banin.

The book focuses on the biogeochemistry of trace elements in arid and semiarid zone soils and includes an introductory chapter on the nature and properties of arid zone soils. It presents an updated overview and a comprehensive coverage of the major aspects of trace elements and heavy metals that are of most concern in the world's arid and semi-arid soils. These include the content and distribution of trace elements in arid soils, their solution chemistry, their solid-phase chemistry, selective sequential dissolution techniques for trace elements in arid soils, the bioavailability of trace elements, and the pollution and remediation of contaminated arid soils. A comprehensive and focused case study on transfer fluxes of trace elements in Israeli arid and semi-arid soils is presented. The book concludes with a discussion of a quantitative global perspective on anthropogenic interferences in the natural trace elements distributions. The elements discussed in this book include Cd, Cu, Cr, Ni, Pb, Zn, Hg, As, Se, Co, B, Mo and others. This book is an excellent reference for students and professionals in the environmental, ecological, agricultural and geological sciences.

The author wishes to sincerely thank Dr. A. Banin for his initiation, support, and encouragement and Dr. Arieh Singer (the Hebrew University of Jerusalem, Israel) for his contribution of the introductory review chapter on the nature and properties of arid zone soils. I am most grateful to my colleagues, Dr. Yi Su, Dr. David L. Monts, Dr. Brian Kauffman, Dr. Perry Norton, Dr. Susan Scherrer and Mr. Dean Patterson (Institute for Clean Energy Technology, Mississippi State University) for their time and reviewing the chapters. My thanks are also due to Dr. Patricia Dill (Mississippi State University) for her editing. I also express my thanks to Dr. Paul Roos and Mrs. Betty van Herk and the Springer publishing staff for their help, support, and encouragement in the publication of this book.

Fengxiang X. Han
Starkville, Mississippi
October, 2006

Section I
Introduction on Arid Zone Soil

By

Arieh Singer

**Department of Soil and Water Sciences
Hebrew University of Jerusalem
Rehovot 76100, Israel**

Chapter 1

ARID ZONE SOILS: NATURE AND PROPERTIES

1.1 CLIMATIC DEFINITION OF ARID AND SEMI-ARID ZONES

Arid and semi-arid zones may be defined in several ways, but the climatic is the one most commonly accepted. According to that definition, aridity represents a lack of moisture in average climatic conditions (UNEP, 1993). This situation may be caused by one of four climatic conditions, which may interact in the case of specific arid/semi-arid zones: continentality, topography, anticyclonic subsidence and oceanic cold currents.

The rainfall regime in arid areas is characterized by low, irregular and unpredictable precipitation, often concentrated in a few rainstorms, creating humid conditions in the soil for a short period and over a limited area. In many arid areas, several years may elapse between successive rainfalls. The moisture supplied to the soil from rain is offset by evaporation, that is related to air temperature, air humidity and intensity of solar radiation. Because of the irregular rainfall distribution, mean precipitation values have little meaning, if not also the range of variation is indicated.

Arid and semi-arid regions are distinguished on the basis of their annual precipitation sums and include:
1) Deserts with an annual precipitation of <50 mm/year and devoid of vegetation (Fig. 1.1).
2) Arid regions with 50–250 mm/year precipitation and sparse vegetation, and
3) Semi-arid regions with a precipitation sum of 250 to 500 mm/year and a steppe savannah/prairie/pampa vegetation.

One expression of the combined influence of both precipitation and temperature is the de Martonne aridity index based on the formula

$$I_a = P_{mm}/(T°C + 10) \qquad (1.1)$$

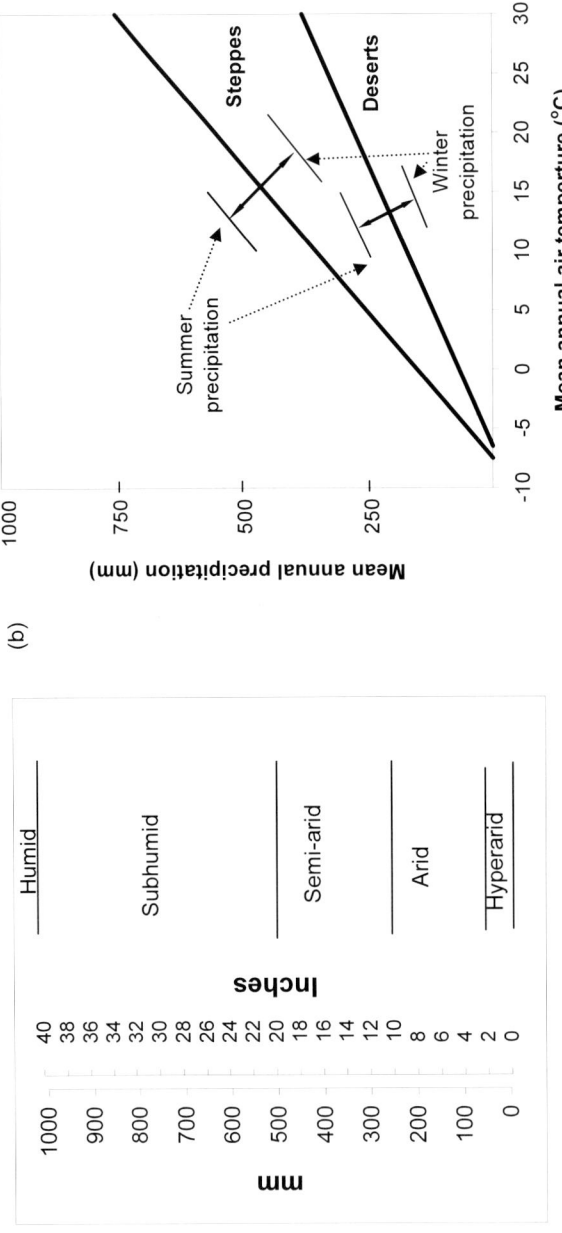

Figure 1.1. (a) Climate categories based on mean annual precipitation; (b) Desert and steppe boundaries as a function of both mean annual precipitation and temperature according to the Koeppen system; arrows show how boundaries shift according to whether precipitation falls mainly in in the summer or winter (after Monger et al., 2004. Reprinted from Encyclopedia of Soils in the Environment, D. Hillel, Tropical soils: Arid and Semi-arid, p183, Copyright (2004), with permission from Elsevier)

Where the aridity index (I_a) is equal to the mean annual precipitation in mm (P_{mm}) divided by the mean annual temperature in °C plus 10. According to this index, values below 5 characterize true deserts, values of approximately 10 define dry steppes, values of about 20 represent prairies, and values above 30 typify forest.

Thus, the Martonne aridity index defines not only climatic parameters but also vegetational ones. Seasonality of precipitation is another climatic factor that affects desert and steppe boundaries. For a given mean annual temperature, the boundary of a steppe will extend into wetter climates if its precipitation falls mainly in the summer. This is because summer evapotranspiration depletes soil moisture more thoroughly than winter evapotranspiration. For similar reasons, the size of a desert will be larger if its precipitation falls in the summer rather than in the winter (Monger et al., 2004).

An aridity definition that takes account of the relation precipitation/potential evaporation is the Aridity Index used in the Atlas of Desertification (UNEP, 1993). Four different degrees of aridity can be recognized. These are best defined using the Aridity Index (AI), calculated as the ratio P/PET, where P stands for precipitation and PET for (potential) evapotranspiration. AI values of <1.0 indicate an average annual moisture deficit. According to this criterion, four subzones can be discussed:

Hyper-arid environments (P/PET < 0.05): cover 7.5% of the global land surface and have very limited and highly variable rainfall amounts both interannually (up to 100%) and on a monthly basis such that there is no seasonal rainfall regime. In virtually all cases where data are available, year-long periods without rainfall have been recorded.

Arid areas ($0.05 \leq$ P/PET < 0.20): mean annual precipitation values up to about 200 mm in winter rainfall areas and 300 mm in summer rainfall areas but more importantly inter-annual variability in the 50–100% range. Pastoralism is possible but without mobility or the use of groundwater resources is highly susceptible to climatic variability.

Semi-arid areas ($0.20 \leq$ P/PET < 0.50): highly seasonal rainfall regimes and mean annual values up to about 800 mm in summer rainfall areas and 500 mm in winter regimes. Inter-annual variability is nonetheless high (25–50%) so despite the apparent suitability for grazing of semi-arid grasslands, this and other sedentary agricultural activities are susceptible to seasonal and inter-annual moisture deficiency.

Dry subhumid areas ($0.50 \leq$ P/PET < 0.65): less than 25% inter-annual rainfall variability and rain-fed agriculture is widely practiced (Fig. 1.2).

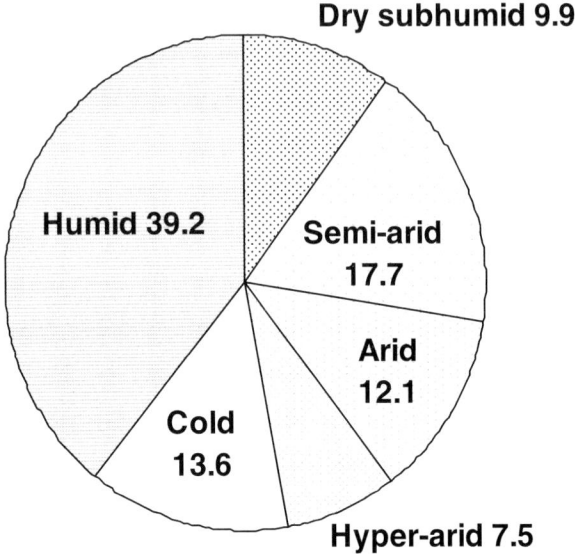

Figure 1.2. Global land area by aridity zone (%) (after UNEP, 1993. Reprinted from World Atlas of Desertification, Figure of global land area by aridity zones, p 5, Copyright (1993), with permission from UNEP)

PET can be determined in three ways, each having its own limitations and advantages. First, it can be calculated from actual evaporation determined by direct measurement using lysimeters or evaporation pans, but the available database is small and the equipment used is not standardized. Second, PET can be calculated using theoretical formulae. The method of H.L. Penman, proposed in the late 1940s, is commonly used but requires a large body of directly measured meteorological data including solar radiation, wind velocity, relative humidity and temperature. Its application at the global scale is therefore again constrained by data availability.

A more useful approach relies on knowledge of the empirical relationship between measured PET and certain variables, allowing PET calculation from just mean monthly temperature data and the average number of daylight hours by month. This method, established by C.W. Thornthwaite in 1948, has a practicality relevant to the scale of this study and to the drylands where environmental data are scarce, and was used by P. Meigs for his 1953 UNESCO map of world aridity. Though a more sophisticated related method that calculates evapotranspiration rates for different crop types (CET) is available, it was not used here because it again requires a further data input that is not generally available.

A global Thornthwaite PET surface was calculated from the temperature surface data for a 0.5° resolution grid. The Thornthwaite method is known systematically to underestimate PET for dry conditions and to overestimate values for moist and cold environments. Consequently, an empirical adjustment factor was derived by CRU and applied to the data to bring the values closely in line with those of the Penman method.

Finally, probably the most useful way to define aridity so that it includes also a reference to the capability of the land to sustain plant growth, is by using soil water potential.

Soil water potential is another way of defining boundaries of arid and semi-arid soils. Soil moisture measured as soil water potential, which includes the influence of particle size and salts, is more important to vegetation than annual precipitation alone. The Soil Taxonomy system (Soil Survey Staff, 1975), for example, uses soil water potential to define moisture regimes as a criterion for classifying soils. The aridic moisture regime, for example, includes soils too dry to support non-irrigated crops, and is defined as soil that is moist (i.e., water held at tensions greater than – 1500 kPa) for no more than 90 consecutive days when the soil temperature at a depth of 50 cm is above 8°C. Soils with the ustic and xeric moisture regimes are transitional between the aridic moisture regime and soils of humid climates that have the udic moisture regime. Semi-arid soils occur within ustic and xeric moisture regimes, their drier subdivisions, and wetter subdivisions of the aridic moisture regime, namely, the aridic ustic, aridic xeric, xeric aridic, and ustic aridic regimes.

1.2 GEOMORPHIC PROCESSES AND LANDFORMS IN ARID AND SEMI-ARID ZONES

Most deserts and (semi-) arid regions occur between 10° and 35° latitude (e.g. Sahara desert, Kalahari desert), in the interior parts of continents (e.g. Australia, Gobi desert) and in rain shadow areas in fold belts (e.g. Peru, Nepal). Large parts of the arctic tundra receive less than 250 mm precipitation per annum and qualify as "arid regions" too (FAO, 2001).

Important geomorphic processes in the dry regions of the world differ from those in more humid environments:
1) streams are intermittent or ephemeral (and have very irregular discharges),
2) mass-wasting processes and unconfined sheet floods are prominent,

3) many rivers do not debauch into the sea but end in inland depressions without outlet,
4) salt lakes are a common landscape feature,
5) aeolian processes play an important role, particularly in areas below the 150 mm/year isohyet, and
6) physical weathering processes are prominent whereas hydrolysis of minerals is subdued.

Polar and subtropical fronts have shifted southwards in the (geologically) recent past and many regions that are arid today once had a more humid climate. Conversely, many of the present humid regions were much drier in glacial periods, especially between 20,000 and 13,000 BP when aeolian processes influenced land formation more than at present. Mass wasting, fluvial processes and aeolian processes are the most important landform-shaping factors in arid and semi-arid regions.

1.2.1 Mass-Wasting Processes

Mass-wasting processes are associated with strongly accidented terrain, e.g. where tectonic uplift has created mountains and in areas with steep fault scarps or incised valleys. Mass wasting often produces erosion landforms, such as residual hills or mountains that remain as isolated features in a low-relief plain. The residual elements consist normally of weathering-resistant rocks (e.g. Uluru sandstone, Australia) or are capped with a layer of resistant rock protecting the underlying softer rock from erosion (e.g. Utah, Great Monument National Park). Such a residual hill or table mountain is called an "inselberg" or "mesa". Mountain foot slopes with a low slope angle and consisting of bedrock covered with a thin blanket of debris are termed "pediments". Contrary to what it is often thought, pediments are erosional landforms because material is moved down the slope. Ultimately, severe erosion may create multiple, deeply incised valleys, in particular in areas with soft sedimentary rocks such as shale or marls, and create a "badland".

The only depositional landform associated with mass wasting is the "talus cone" or "rock debris cone". In barren deserts or mountains, temperature differences between day and night can be considerable and this frequently results in thermal disintegration of rocks. Salt crystals in the fissures may accelerate the process. Detached fragments of rocks and stones accumulate in debris cones at the foot of an inselberg or mountain.

1.2.2 Fluvial Landforms in Arid and Semi-Arid Regions

Fluvial processes in arid regions produce typical landforms. These are different in high-relief and low-relief areas.

1.2.2.1 High-Relief Areas

Where a mountain front borders on a level plain, for instance at a major fault scrap or rift valley boundary, "alluvial fans" are likely to form. These form upon deposition of weathering products at the slope break. Debris flows and sheet floods during occasional heavy downpours are discharged from the hinterland via feeder canyons. Erosion products accumulate at the exit point (at the border between hinterland and plain) in a typical half-circular cone, the alluvial fan. The cone has a steep gradient and a pattern of unconfined channels that shift over the depositional body. As the water velocity of the protruding river becomes less, its sediment load can no longer be carried and much of it is deposited right at the entrance to the fan. This rapidly blocks the channel, which then sweeps left and right to evade the obstacle. The result is a low-angle sediment cone. The sediments tend to be coarser at the "proximal" end of the fan (close to the fan head or apex) than at the "distal" part (the fan toe, far into the plain).

1.2.2.2 Low-Relief Areas

Episodic heavy downpours in low-relief areas are often followed by overland flash floods and debris flows that follow existing depressions in the landscape. Such arid-region fluvial valleys are called "wadis". Many wadis that are now found in desert regions formed during a more humid climatic episode between 13,000 and 8,000 years BP, at the transition from the Last Glacial to the Early Holocene. Wadis in desert regions carry water only after torrential rainstorms that normally occur once in a few years. At the onset of the rains, water can still infiltrate into the soil. As the downpours continue, the supply of water soon exceeds the infiltration capacity of the soil and excess water is discharged as surface run-off: a "flash flood" is set in motion. Slaking and caking of the soil surface enhance surface run-off towards the wadis that become torrential braided streams with high sediment loads. These braided streams have only one channel, but multiple bars. After the downpour, the river will completely dry up again until the next event. Many wadis connect with dry, salty basins where individual floodplains merge into extensive "playas", These are salty lakes with properties that will be outlined in the next paragraph.

Many inland depression areas in deserts are former lake areas in which open water was present during the early Holocene. Former river delta sequences and coastline features (e.g. coastal terraces) may still be visible.

1.2.3 Lacustine Landforms in Arid and Semi-Arid Regions

If a low-lying basin has no outlet, incoming water from (flash) floods evaporates inside the basin where its dissolved salts accumulate in the lowest parts. First, $CaCO_3$ and $MgCO_3$ precipitate as calcite, aragonite or dolomite. As the brine becomes further concentrated, gypsum ($CaSO_4.2H_2O$) segregates, and still later, when the lake is almost dry, halite (NaCl) and other highly soluble salts. Such salt lakes indicate that annual evapotranspiration is greater than the sum of incoming floodwater and precipitation. When such a "playa" dries out, the muddle lake floor shrinks and cracks. Accumulated salts crystallize and form crusts on top of the playa floor and in cracks in the surface soil. Much of the accumulated salts stem from (evaporitic marine sediments) outside the basin; many Mesozoic (Triassic, Jurassic) and Tertiary sediments are very rich in evaporates.

It depends on local hydrographic conditions whether a playa is wet around the year or dries out. A playa may stay (almost) permanently wet if it is part of a closed basin that is under the influence of groundwater. Some playas such as the Dead Sea are fed by perennial rivers and will not dry out either but their water is so salty that salts precipitate. Laminated evaporates of considerable thickness can form in this way, with lamination reflecting the periodicity of the seasons.

The largest evaporate basin formed in recent geological history is the Mediterranean basin. A closed, or almost closed, basin formed when mountain building blocked the Straits of Gibraltar some 6 million years ago. Before the Straits opened again, half a million years later, a layer of 1 kilometer of evaporates had accumulated on the basin floor.

Many lakes in present-day arid regions were freshwater lakes in the wet period between 12,000 and 8,000 BP. Terraces and/or shorelines from that period extend well above the present lake or lacustrine plain. The same lakes were completely dry in the arid Late Pleniglacial period (20,000-13,000 BP). Even a comparatively minor climate change can upset sedimentation regimes in arid lands.

1.2.4 Major Landsforms in Landscapes with Sands

Extensive regions with quartz-rich sands exist on earth. By and large, these can be divided into three broad categories:
1) Residual sands are the result of prolonged weathering of quartz-rich rocks such as granite, sandstone and quartzite. Chemical weathering is particularly active in wet and hot tropical regions where it leads to formation of chemically extremely poor substrates.
2) Aeolian sands are deposited by wind action, either in dunes or in extensive sheets ("cover sand areas"). Wind action is particularly effective in hot and dry regions such as deserts but sand dunes are also common in (sub)humid regions with sparse vegetation, notably in overgrazed areas and along beaches and fluvial 'braid plains'. The (weathering) history of the parent materials in the source area determines whether the sands are rich in quartz and/or carbonates.
3) Alluvial sands are transported by water. In general, these sands are less well-sorted and also less weathered – and therefore "richer" – than aeolian sands. An exception are so called "recycled" fluvial sands deposited by rivers that cut through thoroughly weathered rocks, predominantly in tropical regions (such as several tributaries of the Amazon River). Typical landsforms in regions with fluvial sands will be discussed in more detail in the section on lowland region; areas with alluvial sands are less extensive than residual and aeolian sands.

1.2.4.1 Residual Sands

Extensive, horizontal sandstone plateaus occur in tropical shield areas. Well-known examples are the Precambrian Roraima sandstone formations on the Guiana Shield and the Voltaian sandstone formations in Western Africa. Major occurrences of consolidated sands are found in Northern Africa, in Guyana and Surinam, eastern Peru, northeastern Brazil and in Liberia (western Africa). These sandstone formations have a history of tropical weathering in common; they all have a deep weathering mantle of bleached, white sands that are very rich in quartz, poor in clay and excessively drained. Electrolyte contents differ by region: In arid and semi-arid areas where evaporation exceeds precipitation, salts and carbonates may accumulate at or near the surface of the soil.

1.2.4.2 Aeolian Sands

Sandy parent materials are also abundant in areas where sand accumulates after selective transportation of weathering material by wind or water.

During transport, selection of particles (sorting and winnowing) occurs; the momentary wind speed and the size, shape and density of minerals determine how far a particular grain will be transported. Fine gravel travels by creep and sand-sized particles by saltation. Silt-sized particles can be carried over great distances (Saharan "dust" settles regularly in central Europe and, in the past, loess formations have formed extensive blankets far from the source areas). Fine, plate-shaped clay minerals and micas are blown out and travel even farther (which explains why wind-borne sediments are normally poor in micas). This sorting of grains results in deposits that consist of pure sand with a uniform particle size. Many aeolian sand deposits show characteristic large-scale cross bedding, indicative of sand deposition on the slip faces of dunes.

"Fixed dunes" are formed when transported sand settles in the lee of an obstacle such as a piece of rock. The obstacle thus grows in size and more sand settles: the dune grows. The transport capacity of the wind decreases as it drives the sand grains to the top of the dune, causing an increasing part of the transported sand to settle before reaching the dune crest. This steepens the angle of the slope, particularly near the crest. Vegetation growing on (in particular) the lower part of dunes may eventually keep most of the sand in place. Dunes along coasts are often fixed by vegetation (natural or planted by man).

"Free dunes" have no fixed position, but migrate downwind by erosion on the gently inclined windward side and deposition on the leeward side (slip face) in the same way as described for fixed dunes. The smallest free dunes are common wind ripples that measure only a few centimeters in height. Large dunes are found in extensive dune areas in deserts, in sand seas known as "ergs".

Coastal dunes occur along beaches or sand-flats that form part of a non-erosional sandy or deltaic coast. The source areas of the sand will eventually lose all sand, silt and clay particles; some become wet (groundwater) depressions whereas others acquire a rocky or boulder-strewn surface known as a "desert pavement".

Two main types of free dunes are distinguished, viz. "crescentic" dunes and "linear" dunes.
1) Crescentic dunes are typically wider than long. They assume a crescent shape ("barchan dunes") where winds are predominantly uni-directional. The rate of advancement of the sand is roughly

inversely proportional to the height of the crest. This causes the flanks of a shifting dune to advance more rapidly than the central part until the flanks become sheltered by the main mass of the dune. Coalescing barchans produce "transverse dunes".

2) Linear or longitudinal dunes (also known as "seifs") are straight or slightly sinuous, symmetrical sand ridges that are typically much longer than wide. The surface between individual ridges may be covered either with sand or have a gravel or boulder pavement. More or less stable "star dunes" (also known as "ghourds") occur in regions with varying wind directions. A star dune has one high central part, with several arms radiating outwards. Star dunes tend to grow in height rather than move; they are a common feature in the Sahara "erg".

The area of actual "erg" and dune formation is delimited by the 150 mm/yr isohyet. This precipitation boundary appears to have shifted strongly in the recent past. Between 20,000 and 13,000 yr BP, the southern limit of active dune formation in the Sahara desert was 800 km south of its present position and most of the now sparely vegetated Sahelian zone was an area of active dune formation at that time. These dunes, mostly of the longitudinal type, are now fixed by vegetation, but their aeolian parentage is still obvious from their well-sorted material. A similar story can be told for the Kalahari sands. Overgrazing in recent times has reactivated aeolian transport in many regions with sands.

1.2.5 Major Landforms in Steppe Regions

Steppes and steppic regions (pampas, prairies) receive between 250 and 500 mm of precipitation annually, i.e. more than twice the quantity that falls in true desert areas where rainfall is insufficient to support a vegetation that could protect the land from erosion.

Dunes and sand plains form where strong winds carry sand grains "in saltation" over short distances. Particles finer than sand are transported "in suspension" and over greater distances until they settle as "loess", predominantly in the steppe regions adjacent to the desert zone.

1.2.5.1 Landforms in Regions With Loess

Chinese records make mention of extensive loess deposition between 400 and 600 AD, between 1000 and 1200 AD and between 1500 and 1900 AD (during the "Little Ice Age"). However, the most extensive occurrences of loess on Earth (in the steppe regions of Eastern and Central Europe and the USA) are of Pleistocene age.

During the Late Pleniglacial, between 20,000 and 13,000 BP, some 25% of the land surface became covered with continental ice sheets (versus some 10% today). With so much water stored in ice sheets, the sea level dropped to about 120 meters below the present level and large parts of the world became extremely arid. The Amazon rain forest dwindled to isolated refugia, European forests disappeared but for small sheltered areas, and large parts of the globe turned to tundra, steppe, savannah or desert.

Clearly, aeolian processes were much more important at that time than at present. Large parts of the present temperate zone, from the "cover sands" of the Netherlands to the sand dunes in north-east Siberia are Ice Age (aeolian) sands. South and east of this cover sand belt lies a belt of loess deposits, extending from France, across Belgium, the southern Netherlands, Germany and large parts of Eastern Europe into the vast steppes of Russia, and further east to Siberia and China. A similar east-west loess belt exists in the USA and less extensive areas occur on the Southern Hemisphere, e.g. in the Argentinean pampas.

Loess is a well-sorted, usually calcareous, non-stratified, yellowish-grey, aeolian clastic sediment. It consists predominantly of silt-sized particles (2–50 mm), and contains normally less than 20 percent clay and less than 15 percent sand. It covers the land surface as a blanket, which is less than 8 meters thick in the Netherlands (exceptionally 17 meters) but can reach up to 40 meters in Eastern Europe and 330 meters in China.

Loess is a very porous material and vertical walls remain remarkably stable, but loess slakes easily so that exposed surface areas are prone to water erosion. The loess material itself is produced by abrasion of rock surfaces by glaciers and blown out from glacial outwash plains and alluvium. It is generally difficult to identify the exact source areas of specific loess deposits because the various loess deposits have a surprisingly similar mineralogy. "Typical" loess contains quartz, feldspar, some micas, calcium carbonate and clay minerals. A possible explanation may be that glaciers abrade large surfaces of diverse mineralogy, so that the mineralogical variation between different source areas is averaged out. Further mixing and homogenization of dust particles from various sources occurs during transport. Loess is absent from regions that were covered by glaciers in the last glacial period, nor does it occur in the humid tropics. The vast areas of loess in China may not have a glacial origin: the loess grades into the sandy loess and sands of the Gobi desert in Mongolia. Deposition is still going on today at a rate of several millimeters per year. Long-distance transport of dust particles from the Gobi desert seems to be responsible for the thick Chinese loess deposits.

Loess settles when dust-laden winds slow down to speeds between 7 (on dry surfaces) to 14 meters per second (on moist surfaces). The pore distribution of loess lets it quickly be retained by capillary forces if it lands on a moist surface. The presence of a vegetation cover may also enhance the rate of loess deposition, and many authors maintain that the northern limit of loess deposition coincides with the northernmost extent of grass steppes during arid periods in the Pleistocene.

1.2.5.2 Landforms in Regions Without Loess

Vast undulating till plains occur in North America, between the Canadian shield area and the loess belt. This area is either covered with thick tills or with "deglaciation" sediments, lacustrine sediments in particular. The lake areas are level as such but the till landscape has a typical "hummocky" relief. The main characteristic of hummocky tills (40% of the total area) is the predominance of very local drainage patterns (mainly in depressions). Tills and loess have in common that they are internally uniform and that they all date back to deglaciation periods.

The vast loess and till plains are now colonized by grasses and/or forest. They are the home of some of the best soils of the world: the "black earths". Deep, black Chernozems occupy the central parts of the Eurasian and North American steppe zones. Brown Kastanozems are typical of the drier parts of the steppe zone and border on arid and semi-arid lands. Dusky red Phaeozems occur in slightly more humid areas such as the American prairies and pampas.

1.3   DISTRIBUTION OF ARID AND SEMI-ARID AREAS

The North-American arid and semi-arid zones reach from almost the Canadian border through Mexico (Fig. 1.3) (Nettleton and Peterson, 1983). The Great Plains, on the east, consist of more or less parallel planes of sedimentary rocks such as shale, sandstone and limestone that slope gently eastward from the Rocky Mountains. The topography is gently undulating to rolling. A cap of aeolian material (loess) overlies much of the Great Plains. Shale, sandstone and limestone are the dominant underlying rocks of the intermountain-plateau subregions. Aeolian loess also covers the basalt

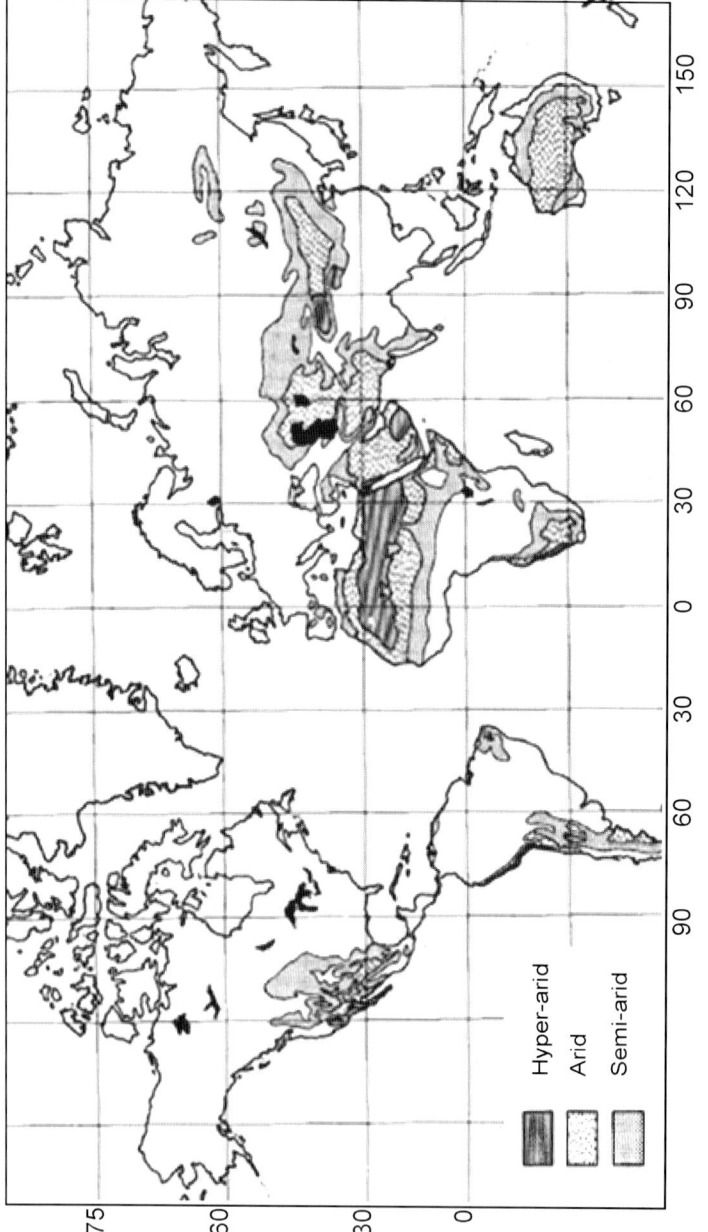

Figure 1.3. Distributions of arid and semi-arid areas (Reprinted from Soils of Arid Regions, Dregne H.E., p 6, Copyright (1976), with permission from Elsevier)

that underlies much of the Columbia Plateau. The ranges that extend between the Rocky Mountains and the Sierra Nevada consist mainly of old sedimentary rocks, with some igneous rocks in between. Intervening basins are of the closed type with no outlet to the Pacific Ocean. The mountain footslopes consist of coarse textured alluvial and colluvial material, while the centers of the basins are occupied by saline playas.

South America has two major areas, separated by the Andes Mountain ridges, where arid and semi-arid zones dominate. In the barren, coastal Atacama desert of Peru and northern Chile, precipitation is extremely limited and comes as winter mists or drizzles. In the rain shadow east of the Andes in Argentina, arid zones are widespread. South America apparently has only a slightly greater proportion of dry zones than North America.

The basement complex for the Patagonian desert and arid northeastern Brazil is formed by metamorphosed Precambrian rocks. Landscapes are characterized by level erosion surfaces of different ages. The landscape is dissected by a large number of valleys. Large depressions are filled with marine and continental beds of sedimentary rocks. Rocks in the Andean system, that stretches the entire length of the west side of the continent, vary greatly. Many depressions are filled with sediments. In addition, many active volcanoes are responsible for periodic lava flows and the deposition of volcanic ash. East of the Andes, the land surface is level and slopes towards the Atlantic Ocean. Broad depressions contain saline or sodic soils.

Sediments of Tertiary and Quaternary age, including volcanic ash and aeolian materials, make up the parent material of the soils. In the more arid parts of the Andean System (the coastal plain of Peru and Chile, and the Altiplano of Bolivia) the topography is level. The Altiplano is a very large closed basin with numerous salt flats. In northwestern Argentina, the planar topography is broken by mountains composed of Precambrian rocks and Quaternary sediments.

Arid and semi-arid lands are quite widespread in Asia, where deserts make up some 11% of this continent (Table 1.1); in Central Asia, north and east of the Caspian to Aral Sea area. In the Turkestan desert, large dunes form huge "sand seas" (Nettleton and Peterson, 1983). West and North of China, the Takla-Makan and Gobi deserts dominate the landscapes. More to the south, the Great Indian Desert extends up to the Himalayan foothills. More to the west are the hyperarid areas of the Dasht-i-Lat and Dasht-i-Kavri deserts in Iran and Syrian deserts. Immense areas in the north-central part of the Arabian Peninsula and in the Rub al Khali deserts in its southeastern part are hyperarid or arid.

Surface deposits of arid and semi-arid lands are diverse, ranging from stony, gravelly and sandy to clayey. These are exposed in a series of great plains broken by mountain ranges and interrupted by closed basins in large and small depressions, having no connection to the sea. Commonly, the uplands are coarse textured and the lowlands are fine textured. In the uplands, stone deserts (hamadas) are common. The landscape consists of vegetation – less rough broken land, undulating to mountainous. Sand deserts, with giant dunes, are more characteristic for the lowland plains. Clay deposits are common in the closed basins. Most of these deposits are slightly to moderately saline. When a high water table is close to the surface, the clay deposits become highly saline, and become covered with salt crusts. Fine-grained deposits that are subject to occasional flooding become covered by fine clay/silt crusts that crack on drying. The largest salt desert is the ancient alluvial plain Dasht-i-Kavir in Iran, in which run-off water from surrounding higher lands had evaporated. Also the highlands of Tibet, parts of Mongolia and Central Asia include depressions with saline, fine-grained material.

Table 1.1. Aridity zones by region ($\times 10^6$ ha) (after UNEP, 1993)

Zone	Region						
	Africa	Asia	Austral-Asia	Europe	North America	South America	Total
Cold	0	1082.5	0	27.9	616.9	37.7	1765
Humid	1007.6	1224.3	218.9	622.9	838.5	1188.1	5100.4
Dry Subhumid	268.7	352.7	51.3	183.5	231.5	207	1294.7
Semi-arid	513.6	693.4	309	105.2	419.4	264.5	2305.3
Arid	503.5	625.7	303	11	81.5	44.5	1569.2
Hyper-arid	672	277.3	0	0	3.1	25.7	978.1
Total	965.6	4256	882.2	950.5	2190.9	1767.5	13012.7

The semi-arid areas receiving somewhat higher rainfall are covered by a steppe-like fairly continuous vegetative cover of Xerophilous shrubs and grasses. In the Central Asian Plains, east of the Caspian Sea, wind erosion and transport are the dominant features. Sand dunes dominate the landscape for hundreds of kilometers.

Although Australia has a smaller desert area than either Africa or Asia, among the continents it has the highest proportion of arid and semi-arid zones. In fact, aside from some humid and semi-humid areas on the eastern, northern and south-western fringes of the continent, the core of the continent is arid or semi-arid.

Most of Australia's arid and semi-arid regions consist of a highly weathered, extremely old land surface. In the western Platform Archaean granite and gneiss are occasionally exposed, but for the most part covered by erosion products (Dregne, 1976). In the eastern Central Basin, Tertiary sands cover marine Mesozoic claystone and soft sandstone. Also in the Central Basin lies the Great Artesian Basin with saline groundwater. Both the platform and Great Basin rocks had undergone intensive weathering particularly during the Middle Tertiary, producing ancient soils with indurated layers – duri-crusts. The crusts were mostly of the iron-laterite type on the platforms and silicrete in the Basins, suggesting humid paleoclimates in the past. In their turn, these crust-containing soils have been eroded away for the most part, preserved only locally. Their residues have given rise to more modern aridic soils containing calcrete. The contemporaneous landscape is that of flat tablelands or plains interrupted by a few mountain ranges.

Antarctica too is very dry and could classify as arid or semi-arid, but in contrast to the other zones it is a very cold dryness.

The second most arid continent after Australia is Africa (Table 1.2). The largest desert region in Africa is the Sahara desert. North of the Sahara, large arid areas extend throughout the greater Maghreb, including Morocco, Algeria, Tunisia, Libya, Egypt and Western Sahara; the Nile Valley too is arid, though the climatic effects of aridity are mitigated by the Nile water. South of the Sahara, the arid/semi-arid Sahel extends for approximately 7000 km across a continuous west-east belt that is over 1000 wide for much of its length, and includes Senegal, Mauritania, Mali, Burkina Faso, Niger, Chad and much of Kenya, and north-eastern Uganda. Huge tracts of land in South Africa, Botswana, Namibia, Southern Zambia, central Mozambique and parts of Zimbabwe are arid/semi-arid too.

Table 1.2. Land area by aridity zones in Africa ($\times 10^6$ ha) (after UNEP, 1993)

Aridity zone	Region				
	North	Sahel	South	Others	Total
Hyper-arid	385.4	276.4	8.2	0	670
Arid	98.1	348.6	54.1	2.7	503.5
Semi-arid	37.4	303.7	159.4	13.3	513.8
Dry sub-humid	15.1	150.1	81.5	22	268.7
Humid	9.3	260	127.7	612.6	1009.6
Total	545.3	1338.8	430.9	650.6	2965.6

In Africa, Precambrian crystalline rocks are exposed over about one quarter to one third of the arid/semi-arid land surfaces (Dregne, 1976). While mountain chains are located in the northwest, east and south, much of the rest of the arid and semi-arid land consists of sandstone, limestone, loose sands, alluvium and colluvium. Sand-dune fields ("ergs") cover extensive areas in northwest Africa. Sand-dune fields are also prominent in other areas of northern Africa. A second major feature of northern Africa arid areas are the desert pavements (to be described below). Dry salt lakes (sebkhas) are conspicuous in a broad zone extending from Tunisia to Mauritania. The largest among these is the Quattara Depression in northwest Egypt.

In Europe, a minor area of aridic zones occurs near the Black and Caspian Seas. According to Dregne (1976), the southern parts of Spain, Italy and Greece are semi-arid.

Calcareous and gypsiferous formations are widespread in the arid regions of Spain. Limestone, sandstone and shale are the principal rocks in the northern and southern tablelands and in the old marine basins of the Ebro and Guadalquivir valleys. Igneous rocks are present in the mountain ranges which rise to over 3,500 m in places. The tablelands have an undulating to rolling topography, as do the uplands along the Ebro River. Flood plains of the numerous rivers are narrow but there are broad level areas along some of the major streams. Sand dunes are found along the Mediterranean coast in the south.

Contrary to popular concepts, sands do not always dominate arid and semi-arid zones. Aridisols occur on a wide variety of landforms, lithological formations and are of different ages. They are most common on stable land surfaces of Late Pleistocene or greater age in tectonically active deserts where they comprise alluvial fan, alluvial flats or stream terraces. Arid zones also include mountainous terrain with steep slopes (Nettleton and Peterson, 1983). Many arid or semi-arid zones include fluvial and aeolian materials of Pleistocene age.

1.4 COMMON PROPERTIES OF ARID SOILS

Scarcity of rain is the dominant characteristic of arid and semi-arid soils. Many of the features of these soils are related to their moisture regime, which is xeric and/or ustic, using the terminology of "Soil Taxonomy" (Soil Survey Staff, 1975). Soils in these regions are unique because they are only rarely water saturated for extended periods of time. Relatively little water percolates deep enough to reach groundwater. As a result, the most poignant chemical/mineralogical feature of most of these soils is the accumulation of carbonates, gypsum and more soluble salts in all or some part of their soil profiles. They are rarely acidic, and some are quite alkaline. Yet, besides climate also other soil formation factors such as parent material, relief, time and vegetation affect arid and semi-arid soils. Indeed, even climate is not a constant in the soil formation process because of the relatively long duration of this process. On the same site, different climates (paleoclimates) may have given rise to different soils.

Surface horizons of arid soils exhibit some characteristic features. As a rule, they have much lower amounts of organic matter than soils from more humid areas, commonly not exceeding 1.5%. Many soils are covered by a crust of salt efflorescences formed by shallow groundwater that had risen by capillarity to the surface and evaporated there. Others are covered to various degrees by gravel forming a "desert pavement". Very often a vesicular horizon can be found underneath this stone cover. Others again, mostly of a sandy texture, had developed microbiotic crusts (MBC). These crusts are very important in affecting moisture distribution and erosional processes in these soils. Subsurface horizons in arid soils invariably include accumulations of secondary carbonates, gypsum, and soluble salts. The ochric epipedon is the most common one in aridic soils (Soil Survey Staff, 1975).

Coarse-sized particles dominate the particle size distribution of arid soils. Some soils are also quite gravelly. The subsurface horizons commonly exhibit accumulation zones of carbonates, gypsum and more soluble salts. Many arid zone soils are shallow and gravelly, some are alkaline. Their structure is weak. From most soils, clay accumulation horizons (argillic horizons) are absent, or are only weakly developed, and so are minerals that indicate an advanced degree of weathering.

The moisture regime in semi-arid soils, owing to more humid climatic conditions, is xeric and occasionally even ustic. Owing to a greater wealth in natural vegetation, their surface horizons have a higher organic matter content. Desert pavements, vesicular horizons, and salt crusts are rare. The variety in surface horizons is greater. While the ochric epipedon is still the most common, mollis epipedons do occur also, particularly in the more humid fringes of the semi-arid zone, where often rich vegetations, such as tall-grass prairies, dominate. Also, the subsurface horizons are of a greater variety. Translocation of clay occurs more often, and as a result argillic and even albic horizons are not rare. While the accumulation of less soluble salts, such as carbonates, is still widespread, the occurrence of the more soluble ones such as gypsum and chlorides is rarer and restricted to the drier margins of the semi-arid zones, or to specific physiographic situations. Being transitional between arid and humid soils, semi-arid soils have a large variety of subsurface horizons. As with arid soils, semi-arid soils might have calcic, petrocalcic, duripan, gypsic, nitric and salic horizons (for terminology, see USDA Soil Taxonomy). Yet, as with humid soils, semi-arid soils might have albic, argillic, and kandic horizons, as well as fragipans (Monger et al., 2004).

1.5 SOIL FORMING FACTORS AND PROCESSES

The five accepted soil forming factors act and interact in a variety of pathways to create arid and semi-arid soils. Climate, by determining water supply, is the principal factor. Water – its amount and depth of penetration – is the major force that gives rise to differences between arid and humid soils (Monger et al., 2004). In humid soils >50% of the water entering the soil drains downward through the profile to groundwater. In arid soils, <10% flushes through the soil profile to the groundwater. A humid soil receiving 1300 mm of rain, for example, would have about 650 mm that percolated through its profile in a year. An arid soil receiving 200 mm would have <20 mm that percolated through its profile. Thus, nearly all water in arid soils and much water in semi-arid soils enters and leaves via the soil surface. Notable exceptions are low-lying areas that receive runoff water (e.g., playas). In these topographic lows, soils are non-saline because of deep leaching, if the water table is deep and the soils have high permeability. However, if the water table is shallow or soils have slow permeability, these topographic lows are zones of salt accumulation.

The rainfall regime in arid/semi-arid areas is characterized by low, irregular and unpredictable precipitation, often concentrated in a few rainstorms, creating humid conditions in the soil for a short period and over a limited area. In semi-arid areas, this irregularity gradually disappears. The moisture supplied to the soil from rain is offset by evaporation that is enhanced by low air humidity, high solar radiation and high air temperature.

Because of the irregular rainfall distribution, mean precipitation values have little meaning in the (semi)-arid zone, if not also the range of variation is indicated. This variability refers to both temporal and spatial variability. Temporal variability affects not only the onset and duration of the rains in the year, but plays also a role in year-by-year differences. The variability is highest in the hyper-arid zone, where the mean precipitation value is composed of a few intensive rainstorms. When these fall on a heated barren surface – as is often the case in the arid zone – a part of it is immediately evaporated and lost for soil processes. High rainfall intensity results on the other hand in a rapid saturation of the surface layers and creates lateral runoff and erosion, in particular on sloping land. Many arid and semi-arid soils show therefore features of gully and sheet erosion.

Field studies have shown that the first 5–6 mm of rain falling on a heated desert surface evaporate almost immediately, whilst single storms with more than 20 mm rain lose a major part of it by lateral runoff. Hence, it is estimated that from the already low rainfall in the arid zone an important part is lost for weathering and soil formation. The importance of this runoff depends on a number of site-specific factors such as rainfall intensity, slope, surface sealing, soil texture and vegetation. Low amounts of rain penetrate the soil over a small depth, the latter being determined by the amount of rain and by the permeability of the soil, itself being affected by texture and the occurrence of surface sealing. This moisture permits very weak chemical weathering only such as dissolution of calcium carbonate according to equations (1.2) and (1.3) or of salts according to formula (1.4):

$$CO_2 + H_2O \rightleftharpoons H_2CO_3 \qquad (1.2)$$
$$CaCO_3 + H_2CO_3 \rightleftharpoons Ca + 2HCO_3 \rightleftharpoons Ca(HCO_3)_2 \qquad (1.3)$$
$$Na_2SO_4 + 10H_2O \rightleftharpoons Na_2SO_4 \cdot 10H_2O \qquad (1.4)$$

The low amounts of moisture considerably retard weathering processes, foremost among them the hydrolysis of aluminosilicates. Soil solutions rapidly attain saturation with the dissolved hydrolysis products and further progress is stalled. This is evident from the mineral stability diagram shown in Fig. 1.4. Thus, for example, the desilication process will not progress very much under arid or semi-arid conditions, and soil minerals will be Si-rich (except in situations where the soil parent material consists of strongly weathered minerals or paleosols, such as the semi-arid soils of Western Africa). As a result, SiO_2/Al_2O_3 ratios in arid/semi-arid soils will be high, and they are likely to contain little weathered minerals such as plagiolases and mafic minerals. Among the clay minerals, mica and illite, at the starting point of the clay mineral weathering sequence, will dominate. Clay minerals of relatively low stabilite, such as palygorskite, will be found exclusively in arid and semi-arid soils.

Because the wetting front is limited to the upper part of the soil, horizon differentiation is restricted to the surface layers, and the soils take the form of a shallow A-C-R or A-R profile, occasionally A-Cca-R profile. In semi-arid soils affected by higher rainfall, the infiltrating water reaches a greater depth, and the profile takes an A-(B)-Ca or A-Bca-C form (Verheye, 2006).

Due to the spatial irregularity of the rains and the occurrence of lateral runoff concentrating additional water in (micro) depressions, soil profile development may vary quite considerably from one location to another in arid zones. This phenomenon might, to some extent, explain the occurrence of A-Bt-C profiles, e.g., with clear clay illuviation features, at some isolated locations in desert zones.

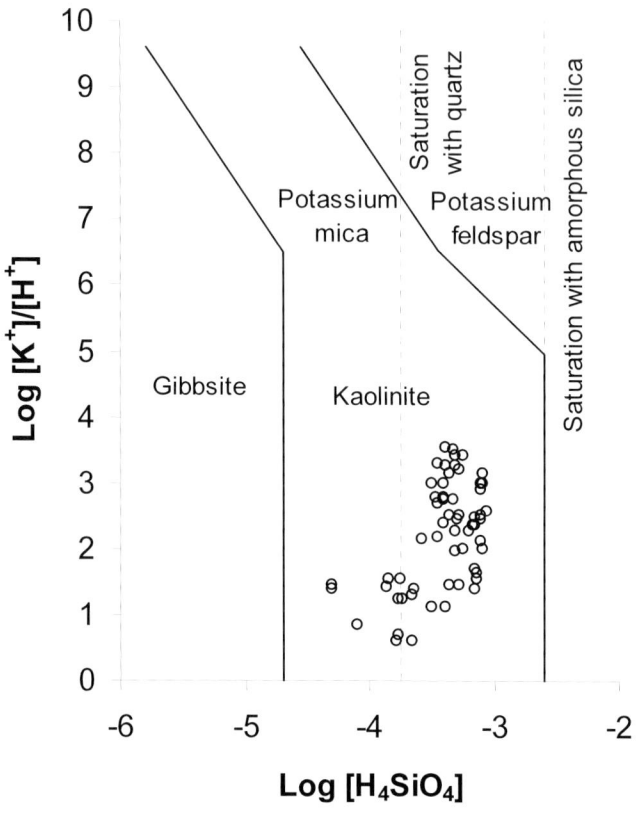

Figure 1.4. Phase diagram for the system K_2O-Al_2O_3-SiO_2-H_2O at 25 °C and 1 atmosphere. Open circles are analytical data for water from springs or seeps in a granitic terrain in the Sierra Nevada (from Feth and others) (%) (from Birkeland, 1984, after Feth and others, 1964. Reprinted from Soils and Geomorphology, p111, Copyright (1984), with permission from Oxford University Press, New York, Oxford)

Air temperatures in arid zones are generally high and show significant daily variations. Because of the absence of a protective vegetative cover and a specific thermal absorption and desorption effects due to surface or rock colors or slope aspect, air temperatures of 35–45°C may reach peaks up to 50–60°C or more on rock surfaces.

Air temperature variations are reasonably well-reflected in the soil surface but disappear rapidly in depth; at 50 cm depth daily variations are hardly significant, and only seasonal temperature fluctuations are registered. Almost similar situations occur in hard consolidated rocks. Those, although having a somewhat higher thermal conductivity than loose materials, do not allow either a high thermal penetration. In rocks, thermal gradients are in the order of 5° to 10 °C per 10 cm.

1.5.1 Soil Processes in Arid (Desert) Areas

1.5.1.1 Desert Pavement Formation – Reg Soils

The high temperature gradients over short distances result in breakdown and mechanical splitting of the larger stones on the desert surface. These physical weathering processes are most important in rocks with a heterogeneous composition because of the uneven thermal expansion of minerals. Even in one and the same mineral, the linear expansion can widely vary depending on the direction within the crystal lattice. In calcite, for example, the linear expansion is approximately 5 times higher parallel to the L3 axis (25.6×10^{-6}) as compared to the direction perpendicular to that same axis (5.5×10^{-6}) (Verheye, 2006).

Reg soils are closely associated with desertic regions. They have developed on stable surfaces where coarse, gravelly desert alluvium is exposed, and are characterized by a well-developed desert pavement and exhibit some well-defined soil horizons. They occur mostly on depositional surfaces where stones and gravels have been deposited since Neogene times. The surfaces commonly consist of stony, unconsolidated sedimentary deposits in which limestone, dolomite, chalk, flint and marl predominate, together with some fines (silt and clay). Sandstone and granite debris have also been reported to contribute to Reg formation. Less frequently, they form on sedimentary bedrock (Fig. 1.5).

The predominant landforms in which Reg soils occur are those of level or slightly undulating plains and plateaus, less frequently with a slightly sloping, hilly, physiography.

The vegetation of Reg soils is generally very poor. Plant growth is restricted to depressions and run-off channels.

The most conspicuous single morphological feature of Reg soils is the surface layer of gravel. Both size and composition of that gravel differ among the soils. The most common average size is between 2 and 5 cm diameter. Up to 10 cm diameter gravel is occasionally present but not common. As common components of that gravel occur flint, chalk, limestone and dolomite. The gravel layer is embedded into the soil to a depth of about 1 cm. The gravel cover is usually complete. Commonly, the gravel is coated with a brilliant, dark brown desert varnish.

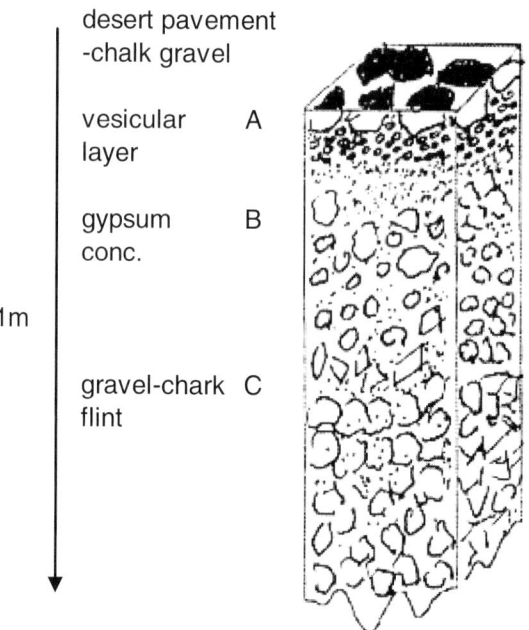

Figure 1.5. Schematic Reg soil pedon: the gravel is chalk, underneath the gravel cover is the porous vesicular A horizon. (%) (from Singer, 2007. Reprinted from The Soils of the Land of Israel, Copyright (2007), with permission from Springer)

Underneath the gravel cover, the A horizon is a sandy loam or silt loam of a light yellowish brown to very pale brown color. Some coarse sand is also nearly always present. The structure of that 2–3 cm thick horizon is porous, unstable and often slightly vesicular. The vesicles are particularly evident in the contact zone between the gravel and the underlying earth. The following horizon is somewhat heavier in texture, darker in color, and of a very loose structure. The gravel content increases sharply with depth. Frequently the flint gravel predominates in the surface and in the upper part of the profile, and calcareous gravel in the lower parts of the profile (Singer, 2007).

Local rock formations are the major source for the stones associated with Reg soils. Only stones of a relatively great resistance towards physical weathering, as for example flint, and to a lesser extent limestone, are common as stone cover. That implies a certain extent of sorting. *In situ* sorting is possible, but would require the weathering of a deep zone, that includes several different rock strata. Such an extensive weathering activity is not likely, nor is there any evidence for it. More acceptable is therefore the suggestion of Evenari et al. (1982) that the stones forming the desert pavement are not formed *in situ*, but rather had been transported and distributed from nearby sources. While other, less resistant rocks disintegrate and disappear, these more resistant ones persist and are gradually distributed in the environs of the source area by solifluction, creep, and flood water.

When, however, the bedrock itself is composed of a mixture of components with varying degrees of resistance towards weathering, such as for example a conglomerate containing flint, limestone and chalk cemented by sand, *in situ* formation of the stone cover is quite feasible. Such would also be the case with a bedrock in which non-resistant layers such as chalk, alternated frequently with resistant flint. With soil development, the stones or gravel, both in the desert pavement and beneath it, decrease in size by a process called "gravel shattering" induced by salt weathering (Amit et al., 1993).

It can be concluded that the Reg soil in its well-developed form indicates a state of erosive balance of the slopes under arid conditions. This balance is characterized by a high degree of stability. The undisturbed Reg cover seals the surface off against the influence of external forces and the influences of the erosional processes on the surfaces are reduced in this way to a minimum. Disturbance of this stability, as for example removal of the stone cover, does not create a new state, but leads to a relatively quick process of regaining its former state and the re-establishment of the balance. The relative proportion of the more resistant flint gravel increases with the development of the desert pavement and in some of them no other stones or gravel are found in this layer. The relative proportion of the more resistant flint gravel increases with the development of the desert pavement and in some of them no other stones or gravel are found in this layer. The relative proportion of limestone and other less resistant rocks increases with soil depth. The sizes of the pavement gravel and stones also decrease with development. The disintegration of larger into smaller gravel and stones may be related to salt crystal growth.

One of the characteristic features of Reg soils is the vesicular nature of the uppermost soil horizon. The size distribution of the vesicles is up to a few mm in diameter. Similar vesicular structures were also observed in lithosols and takyr-like alluvial soils and were always associated with the presence of stones or thin, hard crusts that sealed the soil surface. It forms mostly through accumulation of aeolian dust (McFadden et al., 1998).

Trapped air and expansion of heated air in the soil are the cause of vesicle formation. When the air is driven out by infiltrating rain or floodwater, and cannot escape downwards, it escapes through the upper surface of the soil. When the soil surface is neither covered by stone, nor sealed by crusts, the vesicles are of a temporary nature only. When, however, stones or hard crusts cover the soil surface, the air cannot escape freely, and stable vesicles are eventually formed. The vesicles remain protected by the stones or crusts and are stabilized by soil particles forming the walls of the vesicles. This stabilization is likely to be furthered by the precipitation of carbonate during the drying out of the soil solution, the $CaCO_3$ acting as a cementing agent. Algal crusts that form below the stone cover may contribute to vesicle formation. The formation of an incipient vesicular layer takes only a few years, but thicker layers are found only in much older soils.

Most of the desert pavement stones are covered with a brown-black and shiny crust. When the stones are composed of limestone, the dark crust contrasts strongly with the much lighter inside color exposed on fracture surfaces. The crust forms on various stones, both sedimentary and igneous and is also known under the name of "desert varnish", or "desert patina". The varnish is less common on non-resistant rocks such as a soft limestone. These, apparently, disintegrate before the crust has time to develop.

The dark color of the crust is due to iron and manganese oxides, and also some trace metals such as copper and cobalt. Dew, and to some degree rainwater, wet the rocks and partially penetrate, dissolving some of the rock components. Evaporation of the solutions by sun heat later leaves a precipitate that gives rise to the varnish. Also biological processes may play a part in varnish development, since lichens, fungi and blue green algae that are able to oxidize manganese and iron have been found living below the varnished crust. The varnish develops relatively fast, as it is missing only in barely developed Reg soils.

1.5.1.2 Dust Accretion – Aeolian Material Derived Soil

On a desert surface with little or no protection from vegetation, the effects of wind action may be important. Because of variations in atmospheric pressure, winds are often very strong, especially during the day. Some of these winds occur at more or less fixed periods in the year and get special names: *harmattan* (south of the Sahara), *sirocco* (North Africa and the Western Mediterranean), *chamsin* (the Middle East), etc.

Wind action affects soil formation in 3 ways: (1) *deflation*, (2) *abrasion and erosion*, and (3) *transport and accumulation* (Verheye, 2006).

Deflation is the process whereby soil particles are taken up by the wind and displaced at another location. The process is affected by wind speed, nature of the soil surface, and particle-size or aggregation status of the surface.

The result in terms of soil formation is a loss of soil. In desert areas covered by physically weathered shallow soils, deflation removes mainly the fine and medium-sized particles – clay and silt first, the somewhat coarser sand afterwards – and leaves behind a desert pavement, variously called *reg* (Sahara), *serir* (Libya) or *gibber plains* (Australia).

In semi-arid areas, deflation is less important, except in places where the sparse vegetation is removed by Man. The intensification of deflation processes in these parts of the world can therefore, to a large extent, be attributed to Man. It has been estimated that between 20 and 50% of the soils in Iraq are impoverished in fine elements by wind action. In North China wind deflation has tremendously increased between 1975 and 1985.

Abrasion and erosion. The dust-loaded wind has an erosive action and contributes to a physical disintegration of rock surfaces or to a polishing of the components of the desert pavement, giving them a characteristic patina (*desert varnish*) and shape (*ventifacts*).

Wind transport. Wind-blown components are carried away over a more or less important distance as a function of wind velocity and particle size of the material. Wind speeds up till 6.5 m/sec transport dust and fine sand with a diameter of less than 0.25 mm; sand grains up to 1 mm diameter are uplifted at wind speeds of 10 m/sec. At 20 m/sec also particles of 4–5 mm may be removed. Based on these physical laws, the transportation of coarse fragments, *in casu* the sand fraction, occurs over-relatively short distances from the deflation zones. These sand grains settle then in more or less continuous layers and either become progressively mixed with the underlying soil layers, or concentrate in dune formations.

Medium-coarse particles are transported over relatively important distances of the order of a few 100 meters to more than 1000 km. Hence, eolian deposits of silt-size material settle into a loess belt along desert fringes. This transported material may affect the chemical composition of the original material. The relatively high base saturation (and high exchangeable Ca^{2+} content) in the surface layers of soils south of the Sahara may be due to the addition of Ca-saturated aerosol brought along with the *harmattan*.

The finest particles composed of fine silt and clay are carried over much larger distances (Fig. 1.6). Using oxygen isotopes of aeolian quartz, it has been shown that significant quantities of dust with an average diameter of 1–10 microns are transported by jet streams from desert zones over the rest of the world. Eolian admixtures have been referred to explain the presence of SiO_2 in soils developed in non-quartz-containing volcanic rocks in Hawaii and Israel.

Figure 1.6. The transport of fine particles by duststorm (after Singer, 2007. Reprinted from The Soils of the Land of Israel, Copyright (2007), with permission from Springer)

Though the process of wind action and transport of material is clearly recognized, it is difficult to measure its impact in the accumulation zones. Such measurements are most successful in the immediate neighborhood of the deflation zones where the thickness and volume of sand or loess deposits can easily be calculated. In the northern periphery of the Negev, aeolian deposits range from a few cm to several meters, which corresponds to an average accumulation of 10 to 100 mm/millennium since the Lower Pleistocene.

Dan (1990) distinguishes four degrees in the influence of aeolian material on soil formation: 1. Soils in which accession of atmospheric dust has acted as a modifying agent. 2. Soils in which accretion of atmospheric dust has proceeded simultaneously with the process of soil development, and where it has significantly altered the nature of the soils. 3. Soils which have received a thin surface aeolian layer, which is thinner than the depth of the solum. 4. Soils formed from thick aeolian sediments. While the soils of southern Israel, and particularly the Negev, belong to one of the three latter categories, those in the central and northern parts of the country fall into the first category. Long distance transport of large dust clouds, up to 5 km high, distributes mainly the finer grained dust (10 μm) over distances exceeding 1000 km. Five to ten dust storms per year is a common occurrence in Israel. With a deposition rate of less than 2 μm per event (2.5 g m^{-2}), this kind of accretion is generally incorporated into the local soil and assimilated by an ongoing soil forming process. Medium distance transport is mainly downwind from the major wadis and river valleys for a distance of 50–200 km and includes particles of up to 50 μm in diameter in suspension. In such cases, both thickness of deposition and grain size decrease (exponentially) from the source. Where deposition rates exceeds 40 μm per year (50 g m^{-2}), over a long period of time, distinct loess deposits several meters thick are formed (over 10 m in the Negev) (Singer, 2007).

Analyses of airborne dust concentration measurements carried out in Israel showed that the average airborne dust concentration was 150.2 μg m^{-3}. The major mineral constituents of the airborne dust are quartz and calcites: the minor mineral constituents are gypsum and halite, and traces of plagioclase, kaolinite, and illite. Suspended dust over the Dead Sea was measured and analyzed during three dust storms in the spring of 2001. Suspended dust concentration varied from 300 μg m^{-3} in two moderate storms to >400 μg m^{-3} in a stronger storm. Particle size distribution had a mode at 2–3 μm, characteristic of long distance plume dusts from a single source. Most common minerals included quartz and kaolinite. Some feldspar, apatite and dolomite also were identified. The particle size distribution differs from that of sedimented dust at the same location and suggests a longer migration path. The clay mineral population suggests a relative enrichment in kaolinite relative to smectite clay minerals (Singer et al., 2004).

Rates of dust accretion and deposition are dependent on the amount of available dust and the trap efficiency of a particular site. Several types of dust-trapping terrains are widespread in deserts: (1) gravelly (Serir) surfaces that turn with time into Reg soils; (2) vegetated surfaces in the desert fringe that may turn into loessial terrains; (3) stabilized sand dunes; (4) playa surfaces. Even mosses have been shown to be capable of trapping dust (Danin and Ganor, 1991). Loessial terrains exhibited a high rate of dust accretion during the late Pleistocene – 0.07–0.15 mm y^{-1} on the interfluves and <0.5 mm y^{-1} along the flood plains. Gravelly surfaces usually trap about 0.1 mm y^{-1} of dust initially but the rates decrease to several μm y^{-1} due to plugging with dust and salts, and may ultimately remain constant as a gravel-free B horizon develops. The amounts of imported dust, from both local and distant sources, have changed during the Quaternary due to climatic fluctuations.

It seems that clay translocation (argilluviation) from the upper into lower horizons takes place only to a limited extent in arid zone soils. B_1 horizons are only rarely identified. That limited argilluviation may indeed have taken place in some profiles is also indicated by the clay coatings (argillans) that were observed in subsurface horizons of some soils. Clay migration might possibly have been facilitated by the strong dispersion associated with the alkalinization that occurs in the lower horizons of many aeolian material-derived soils. Granted clay migration had taken place, it must have proceeded at a very slow rate, since indications of the process can be revealed only in the older, more developed soils.

Mineral surface crusts – Some arid and semi-arid zone soils are known for their capacity of forming surface crusts upon wetting and drying. No important differences were found between the chemical properties of the crust and those of the underlying soil layers. The differences were solely of a physical nature: the bulk density of the crust is higher, total porosity lower: microporosity is often higher and also mechanical strength in the dry state is higher. Crusting was found to be a positive function of the wetting duration and moisture content of wetting.

Crust formation thus is attributed to the disaggregation of the uppermost soil layer, initiated by the mechanical impact of the rain drops, and the subsequent dispersion of the clay fraction facilitated by the high ESR of the soil and by the low electrolyte content of rain water (Shainberg, 1990). Upon drying, the dispersed clay is responsible for the formation of the hard crust and decreases in infiltration rate.

Biological surface crusts – The main role of vegetation and biological activity as a soil forming factor in arid/semi-arid areas is that their impact is comparatively small, though somewhat specific. Vegetation is composed mainly of short and ephemeral grasses which can survive after a long dormancy period and then germinate after a local rainstorm, and complete their growth cycle in a few weeks. These plants and the sclerophytes that survive aridity provide little organic material to the surface and do almost not play a role in soil weathering and horizon differentiation. In addition, there exist some salt-tolerant plants like *Atriplex* and a number of others that accumulate salts in their above-ground system and thus modify the salt profile.

Biological activity from insects, lizards, snakes and rodents is limited as well, and mainly concentrates in the deeper soil layers where moisture remains relatively high and temperature fluctuations are reduced.

The most important and most particular biological activity is that by microorganisms which concentrate by preference on surface stones and rocks, where they are called *lithophytes*. The number of these micro-biota in various deserts throughout the world is in the order of 10^3 to 10^6 per gram of soil, and in the Sahara alone French microbiologists have described more than 45 types of cyanophyceae, 70 chlorophyceae, 90 lichens and more than 300 diatoms. Hence, deserts cannot be considered an a-biotic environment.

The first colonizers of a bare rock are often bacteria and green-blue algae. They dissolve some of the more soluble components in the rock through an exchange reaction between the acids they secrete and the nutrients they absorb (K from feldspars; P from apatite). Some bacteria are even capable of taking up Si from aluminosilicate clays. Diatoms are known to absorb amorphous silica and to store it in their structures.

Algae, in particular green-blue algae, and lichens form the bulk of the micro-flora in desert environments. Many of them can stand long periods of drought, but can also rapidly rehydrate with a minimal amount of moisture, e.g. dew. They are less affected by temperature or light intensity for their photosynthesis. Algae form these dense populations on the surface of arid soils and influence in many ways the soil properties. They can absorb and concentrate N from the air and make it available to plants. They can largely modify the soil pH, and can disintegrate organo-mineral compounds, silicate clays and even silica (Verheye, 2006).

On the other hand, algae are also influenced in their development and activities by environmental conditions. Green-blue algae *Microleus* and *Nostoc* are predominant colonizers of a number of arid areas in the American South West, India and New South Wales (Australia), but are almost absent in many other parts of the world. *Azotobacter* is observed in many arid soils in Arizona, but disappears once the concentration of soluble Na in the soil exceeds 3000 ppm. Algae populations may also vary as a function of the season, as is the case in *takyr* soils where green-blue algae colonize the surface after the floodwaters have retreated, but where they are rapidly replaced by a reddish variant at the moment the profile starts to dry out. Some algae, whether or not in association with lichens, form a soil crust and protect the soil from further drying out and from wind and water erosion.

The water regime in the sandy dunal areas of deserts is highly dependent on a fragile cryptogamic crust (BSC), only several millimeters thick. This crust develops due to the presence of cyanobacteria which agglomerate the sand grains and trap aeolian dust particles. Not only does this semi-permeable crust increase runoff but the water which does infiltrate the soil is protected from excessive evaporation. The grain size distribution shows a concentration of silt and clay in the crust compared to the sand just beneath the crust. The microporosity shows that approximately 40% of the access pores can be blocked by the swelling of cyanobacteria that absorb water, which limits rainwater infiltration. The removal of a thin cyanobacterial-dominated crust from a loess-covered hillslope in the Northern Negev resulted in three to five-fold increases in sorptivity and steady-state infiltration under both ponding and tension. The removal of a depositional crust colonized by cyanobacteria from a loess floodplain in the Central Negev resulted in an increased infiltration under tension. The removal of the crusts in all three landscapes affects water flow, particularly the redistribution of runoff water, which is essential for the maintenance of desert soil surface patterning (Kidron et al., 2000). Studies revealed that the BSC in the Nizzana dune fields had hydrophobic properties which inhibit the infiltration rate and enhance runoff generation.

1.5.1.3 Salt Accumulation and Distribution – Saline Soils

All soils contain soluble salts, but their concentration is low. The salt content of most arid soils is, however, much higher. Salts in desert soils are usually derived from three main sources: (1) deposition of wind-blown salt spray or dust; (2) *in situ* weathering of salt-containing rocks or sediments, and (3) upward movement with the capillary flow from a shallow salty groundwater. Along the coastline, some salinization may occur through intrusion and flooding by seawater.

Saline soils vary considerably in their salt content, type of salt, structure and ease to be reclaimed. Dominant anions are chlorides, sulfates and carbonates, sometimes nitrates and bicarbonates. Sodium salts occur most frequently, but calcium and magnesium compounds are common as well, while mixtures of various salts and complex minerals are not exceptional. The non-salt solution contains mainly calcium salts (50–80%); magnesium (15–35%), potassium (2–5%) and sodium (1–5%) make up the remaining cations. In saline soils, however, the percentage of Ca^{2+} is lower, and the values of K^+, Mg^{2+} and Na^+ is higher.

Saline soils are often recognized in the field by the presence of a white surface crust, by damp oily-looking surfaces devoid of vegetation, stunted plant growth, and sometimes by tipburn and firing of leaf margins. Soil analysis rather than visual observations are nevertheless needed to properly assess salinity.

One of the prominent features of arid soils is the translocation of salts along the profile. Frequently, soluble salts accumulate at some depth or other. The salt accumulation appears to be associated with maximum depth of wetting of the soils and sometimes also with their texture. Thus, while in a loessial Serozem with a silty loam texture from the environs of Beer Sheva where rainfall is 210 mm y^{-1}, EC passed 1 Sm^{-1} after 60 cm soil depth, in a similar soil about 40 km south of Beer Sheva, where rainfall is only 91 mm annually, EC passed 4 Sm^{-1} after only 22 cm soil depth. On the other hand, a sandy loam soil of aeolian origin in the north-western Negev has an EC of only 0.4 Sm^{-1} throughout the profile up to a depth of 250 cm, although rainfall in the area is very similar to that in the Beer Sheva basin (Singer, 2007). A close relationship was found between the ESP and EC values and the depth of rainwater penetration in soils of the northern Negev with a rainfall of about 280 mm y^{-1} (Fig. 1.7).

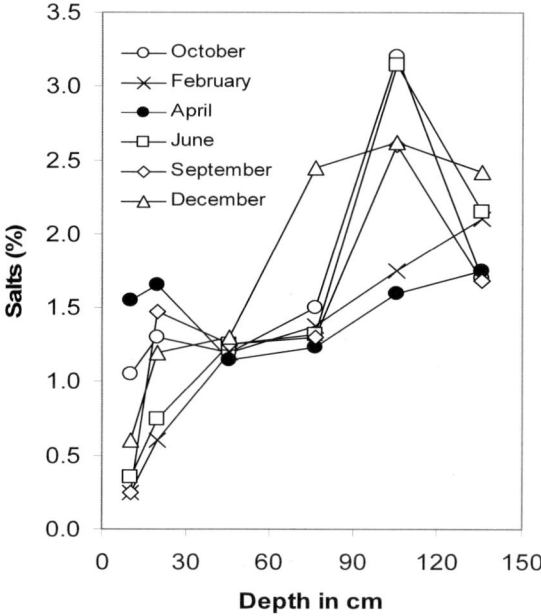

Figure 1.7. Distribution of salts in chloride-sulfate Solonchak during the various seasons of the year (after Ravikovitch and Bidner-Bar Hava, 1948) (from Singer, 2007, after Ravikovitch and Bidner-BarHava, 1948. Reprinted from The Soils of the Land of Israel, Copyright (2007), with permission from Springer)

The salt accumulations are usually accompanied by considerable increases in the exchangeable Na, and frequently also in exchangeable Mg. The origin of the salts is atmospheric deposition or saline groundwater. Rainwater was shown to carry significant amounts of soluble salts. Atmospheric dust too carries significant amounts of salt. While in the upper soil horizons chlorides are dominant, sulphates become prominent lower down. The sulfates, mainly in the form of gypsum, are commonly concentrated in well-defined horizons. Due to their lower solubility, the gypsum horizons are more stable and less likely to be redistributed by water within the profile than the more soluble salts.

Another prominent feature of arid/semi-arid soils is the translocation of carbonates. Invariably, carbonates, mainly calcite, are present throughout all the profile. Commonly, however, there is a distinct horizon of accumulation, frequently in the form of concretions. More than one and up to three such horizons of accumulation can be found in some soils. It appears that clay translocation from the upper into lower horizons does take place, though to a limited extent, in arid soils. Soils with B_2 horizons are not rare.

In summary, as with humid soils, the processes of soil genesis (i.e., gains, transfers, transformations, and losses) operate in arid soils, but the magnitude and direction of these processes are different (Monger et al., 2004). The shallow depth-of-wetting and incomplete leaching have a major impact on gains because authigenic minerals, such as calcite, silica, and gypsum, accumulate in the profile and give rise to the formation of calcic, petrocalcic, duripans, gypsic, and petrogypic horizons. Gains of dust are also important. Dust, containing clay for example, contributes to the formation of argillic horizons, and carbonate dust to the formation of calcic horizons. Gains of photosynthetic carbon in the form of soil organic matter are lower than in humid soils. But gains of photosynthetic carbon released as respired soil CO_2 that leads to HCO^-_3 and $CaCO_3$ formation are higher than in humid soils.

Transfers of material down the profiles of arid soils include illuvial clay, carbonate, and salts. Arid soils typically display a chromatographic pattern of an argillic horizon overlying a calcic horizon. If gypsum and soluble salts are also present, the profile can contain an argillic overlying a gypsic overlying a salic horizon. Transfers of materials also occur up the profiles of some arid and semi-arid soils. These include capillary rise of soluble material and particles moved upward by ants and termites. In some cases, desert pavements are formed by the upward movement of coarse fragments lifted by silts and fine sands that accumulate beneath them. Important transformations in arid soils include rock shattering and disintegration resulting from the crystallization of salts and thermal fluctuations.

1.5.2 Processes in Semi-Arid Areas

Three additional processes are prominent in semi-arid (and also subhumid) soils:

1. *Melanization*, responsible for the Mollisol soil order, 2. *Pedoturbation*, responsible for the Vertisol soil order, and 3. *Argilluviation* (mobilization and transport of clay to an accumulation horizon) responsible for the Alfisol order.

1.5.2.1 Melanization – Mollisols

Melanization is defined as a process of darkening of the soil by addition of organic matter and is probably the dominant process in Mollisol formation. Most of the soils classified as Mollisols had a grass vegetation at some time. It is generally agreed that the total amount of organic matter produced, its distribution in the soil, and relative rate of decomposition are important differences among different vegetative types and environments. The types of grasslands associated with Mollisols include desert grassland, short-grass prairie, mixed tall-grass prairie, and tall-grass prairie (Fenton, 1983).

Plant distribution is correlated with depth below the surface of the layer of carbonate accumulation. Where this depth is less than two feet, the plains type of vegetation predominates. Where greater than about 30 inches or where lacking entirely, the prairie type of grassland occurs. The important point here is the depth of soil periodically moistened by rainfall and the total moisture supply available. Short grass characterizes areas where each season all available soil moisture is consumed by plant growth. All available soil moisture is also consumed along the margins of the tall grass areas. Over the tall-grass area as a whole, however, moisture during the rainy period penetrates so deep into the soil that it is not all recovered and brought to the surface by plants. Consequently, the carbonates are carried down and away entirely with the drainage water. At the beginning of the growth period, the soil is moist to the layer of carbonate accumulation, the equivalent of four to six inches of rainfall in the tall grass and from two to four inches in the short grass.

Mollisols are soils in which it is thought there has been decomposition and accumulation of relatively large amounts of organic matter in the presence of calcium, producing calcium saturated or calcium-rich forms of humus. According to Smith, this requires decomposition *in the soil* and not on the soil. The Mollisols have high base saturation (greater than 50 percent as measured by the ammonium-acetate method) with abundant calcium.

The difference in the distribution and amount of organic matter in the soils developed under prairie vegetation and forest vegetation can be explained in part by its mode of addition to the soil. Under prairie vegetation, more than 50 percent of the biomass is added to the soil annually – almost all the above ground parts and at least 30 percent of the underground parts.

The residue from aerial plant parts partially decomposes on the surface and enriches the upper part of the A horizon through incorporation by soil fauna. Earthworms and many kinds of small insects are considered to be important agents in promoting the incorporation and breakdown of litter into the soil while the activities of burrowing rodents and other fauna are considered to extend the distribution of organic matter into the subsoil. Other factors that may tend to promote a greater content of organic matter in soils under grasses relate to processes favoring production of humic acid.

Products of this type seem to protect the humus from rapid incorporation into new biological processes. Additional factors that appear to be associated with the accumulation of organic matter in Mollisols are high exchange capacities, saturation with calcium, an abundance of mineral colloids and a high content of minerals of the smectite group (Fenton, 1983).

1.5.2.2 Pedoturbation – Vertisols

Pedoturbation is the constant internal turnover of soil material. Vertisols are churning heavy clay soils with a high proportion of swelling 2:1 clay minerals (FAO, 2001). These soils form deep wide cracks from the surface downward when they are desiccated, which happens at least once in each year.

The environmental conditions that lead to the formation of a *vertic* soil structure are also conducive to the formation of suitable parent materials:

1. Rainfall must be sufficient to enable weathering but not so high that leaching of bases occurs; 2. Dry periods must allow crystallization of clay minerals that form upon rock or sediment weathering; 3. Drainage must be impeded to the extent that leaching and loss of weathering products are curbed; 4. High temperatures, finally, promote weathering processes. Under such conditions smectite clays can be formed in the presence of silica and basic cations – especially Ca^{2+} and Mg^{2+} - if the soil-pH is above neutral

Smectite is the first secondary mineral to form upon rock weathering in the semi-arid to sub-humid tropics. Smectite clay retains most of the ions, notably Ca^{2+} and Mg^{2+}, released from weathering primary silicates. Iron, present as Fe^{2+} in primary minerals, is preserved in the smectite crystal lattice as Fe^{3+}. The smectites become unstable as weathering proceeds and basic cations and silica are removed by leaching. Fe^{3+}-compounds however remain in the soil, lending it a reddish color; aluminum is retained in kaolinite and Al-oxides. Leached soil components accumulate at poorly drained, lower terrain positions where they precipitate and form new smectitic clays that remain stable as long as the pH is above neutral. Additional circumstances for the dominance of clays are:

1. *fine clay* in which the proportion of smectites is greater than in coarse clay, is *transported laterally*, through surface and subsurface layers, and 2. *drainage and leaching of soluble compounds decrease* from high to low terrain positions. Internal drainage is impeded by the formation of smectites. It is increased when kaolinite forms: ferric iron, released from the smectite lattice, cements soil particles to stable structural peds and maintains a permanent system of pores in the soil (FAO, 2001).

The formation of characteristic structural aggregates (*"vertic structure"*) is an additional soil forming process in Vertisols. This typical structure may occur in most of the solum but has its strongest expression in the *"vertic horizon"*; the grade of development and the sizes of peds change only gradually with depth.

In a clay plain that is flooded at the end of the rainy season, most of the standing water evaporates eventually. When the saturated surface soil starts to dry out, shrinkage of the clayey topsoil is initially one-dimensional and the soil surface subsides without cracking. Upon further drying, the soil loses its plasticity and tension builds up until the tensile strength of the soil material is locally exceeded and the soil cracks. Cracks are formed in a pattern that becomes finer as desiccation proceeds. In most Vertisols, the surface soil turns into a *"surface mulch"* with a granular or crumb structure. Vertisols, which develop surface mulch, are called *"self-mulching"* (Fig. 1.8). Granules or crumbs of the mulch fall into cracks. Upon re-wetting, part of the space that the soil requires for its increased volume is occupied by mulch material. Continued water uptake generates pressures that result in shearing: the sliding of soil masses against each other.

Shearing occurs as soon as the *"shear stress"* that acts upon a given volume of soil exceeds its *"shear strength"*. The swelling pressure acts in all directions. Mass movement along oblique planes at an angle of 20 to 30 degrees with the horizontal plane resolves this pressure. The shear planes are known as *"slickensides"*, polished surfaces that are grooved in the direction of shear. Such ped surfaces are known as *"pressure faces"*. Intersecting shear planes define wedge-shaped angular blocky peds. Although the structure conforms to the definition of an angular blocky structure, the specific shape of the peds has prompted authors to coin special names such as "lentils", "wedge-shaped peds", "tilted wedges", "parellelepipeds" and "bicuneate peds".

Sliding of crumb surface soil into cracks and the resultant shearing have important consequences:

1. Subsurface soil is pushed upwards as surface soil falls into the cracks. In this way, surface soil and subsurface soil are mixed, a process known as "churning" or *"pedoturbation"*. Churning has long been considered an essential item in Vertisol formation. However, recent morphological studies and radiocarbon dating have shown that many Vertisols do not exhibit strong homogenization. In such Vertisols, shearing is not necessarily absent but it may be limited to up-and-down sliding of soil bodies along shear planes.

2. In churning Vertisols, coarse fragments such as quartz gravel and hard, rounded, carbonatic nodules are concentrated at the surface, leaving the solum virtually gravel free. The coarse fragments are pushed upwards

with the swelling soil, but most of the desiccation fissures that develop in the dry season are to narrow to let them fall back.

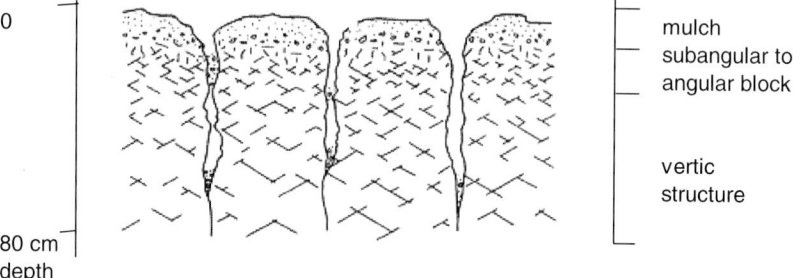

Figure 1.8. Cracks, surface mulch and soil structure in a Vertisol during the dry season (from FAO, 2001. Reprinted from World Soil Resources Reports 94, P. Driessen, J. Deckers, O. Spaargaren, F. Nachtergaele, eds., p80, Copyright (2001), with permission from FAO)

3. Aggregates of soft powdery lime indicate absence of churning, unless such aggregates are very small and form rapidly. Soft powdery lime is a substratum feature in Vertisols.

A typical self-mulching Vertisol has an uneven surface topography: the edges of crack-bounded soil prisms crumble, whereas the centers are pushed upward. The scale of this surface irregularity is that of the cracking pattern, usually a few decimeters. "Gilgai" represents micro-relief at a large scale, superimposed on this unevenness. Gilgai on level terrain consists of small mounds in a continuous pattern of small depressions, or depressions surrounded by a continuous network of narrow ridges.

1.5.2.3 Argilluviation – Alfisols

Argilluviation is the formation of an argillic illuviation horizon, as a result of translocation of clay from the surface soil to the depth of accumulation. It is the dominant process in the Alfisol order of soils. The process consists of three essential phases:

1. *mobilization* of clay in the surface soil; 2. *transport* of clay to the accumulation horizon; 3. *immobilization* of transported clay.

Normally, clay in soil is not present as individual particles but is clustered to aggregates that consist wholly of clay or of a mixture of clay and other mineral and/or organic soil material. Mass transport of soil material along cracks and pores, common in cracking soils in regions with alternating wet and dry periods, does not necessarily enrich the subsoil horizons with clay.

For an argillic horizon to form, the (coagulated) clay must disperse in the horizon of eluviation before it is transported to the depth of accumulation by percolating water.

Mobilization of clay. Mobilization of clay can take place if the thickness of the electric "double layer", i.e. the shell around individual clay particles that is influenced by the charged sides of the clay plates, becomes sufficiently wide. If the double layers increase in width, the bonds between negatively charged sides and positive charges at the edges of clay plates become weaker until individual clay particles are no longer held together in aggregates. The strength of aggregation is influenced by: the *ionic strength of the solution*, the *composition of the ions adsorbed at the exchange complex*, the *specific charge characteristics of the clay in the soil*.

At high electrolyte concentrations of the soil solution, the double layer is compressed so that clay remains flocculated. A decrease in ion concentration, e.g. as a result of dilution by percolating rain water, can result in dispersion of clay and collapse of aggregates. If the exchange complex is dominated by polyvalent ions, the double layer may remain narrow even at low electrolyte concentrations and consequently aggregates remain intact (FAO, 2001).

Soil-pH may influence both the concentrations of ions in the soil solution and the charge characteristics of the clay. Dispersion of clays is thus, to some extent, a pH-dependent process. At soil-$pH_{(H2O,1:1)}$ values below 5, the aluminum concentration of the soil solution is normally sufficiently high to keep clay flocculated (Al^{3+} is preferentially adsorbed over divalent and monovalent ions in the soil solution). Between pH 5.5 and 7.0, the content of exchangeable aluminum is "low". If concentrations of divalent ions are low, clay can disperse. At still higher pH values, divalent bases will normally keep the clay flocculated unless there is a strong dominance of Na^+-ions in the soil solution.

Certain organic compounds, especially polyphenols, stimulate mobilization of clay by neutralizing positive charges at the edges of clay minerals. As iron-saturated organic complexes are insoluble, this process might be of little importance in Fe-rich Luvisols (particularly common in the subtropics).

Transport of clay through the soil body. Transport of peptized clay particles requires downward percolation of water through wide (>20 μm) pores and voids. Clay translocation is particularly prominent in soils that shrink and crack in the dry season but become wet during occasional downpours.

Smectite clays that disperse more easily than non-swelling clays are a common constituent of Alfisols.

Precipitation and accumulation of clay. Precipitation of clay particles takes place at some depth in the soil as a result of *1. flocculation* of clay particles, or 2. (mechanical) *filtration of clay* in suspension by fine capillary pores.

Flocculation can be initiated by an increase in the electrolyte concentration of the soil solution or by an increase of the content of divalent cations (e.g. in a $CaCO_3$-rich subsurface horizon).

Filtration occurs where a clay suspension percolates through relatively dry soil; it forces the clay plates against the faces of peds or against the walls of (bio)pores where skins of strongly "oriented" clay ("cutans") are formed. With time, the cutans may wholly or partly disappear through homogenization of the soil by soil fauna, or the cutans may be destroyed mechanically in soils with a high content of swelling clays. This explains why there is often less oriented clay in the argic subsurface horizon than one would expect on the basis of a budget analysis of the clay profile. There could also be more illuviated clay than expected viz. if (part of) the eluviated surface soil is lost through erosion.

1.6 CLASSIFICATION OF ARID AND SEMI-ARID SOILS

The main criterion for the classification of arid and semi-arid soils is soil dryness, or the aridic (torric) moisture regime, which is defined as soils too dry for agricultural crops unless irrigated. Further taxonomic subdivisions are based on diagnostic horizons. In contrast to the notion that arid/semi-arid soils are poorly developed, many soils are strongly developed with a variety of diagnostic subsurface horizons. These horizons include the argillic, nitric, salic, gypsic, petrogypsic, calcic, and petrocalcic horizons, and the duripan. Diagnostic surface horizons include the ochric epipedon with minor occurrences of the mollic and anthropic epidons.

Arid soils that have diagnostic subsurface horizons are generally classified as Aridisols. These include many of the older soils on piedmont slopes, basin floors, mountain uplands. Various types of Aridisols occur on the landscape because of lateral changes in particle size, truncation of diagnostic horizons, degradation of diagnostic horizons, moisture heterogeneity across the landscape, and age differences, which can range from Holocene to Pliocene within small geographical areas.

Arid soils that lack diagnostic subsurface horizons are generally classified as Entisols, which fall into the azonal concept of Sibirtsev. These include many of the younger soils on floodplains, dunes, and erosional surfaces. In Soil Taxonomy, floodplain soils are mainly classified as

Fluvents or, more specifically, Torrifluvents. Arid soils associated with dunes are commonly Torripsamments and those associated with erosional surfaces are commonly Torriorthents.

Table 1.3. Global extent of arid and semi-arid soils (km^2) based on the Soil Taxonomy system) (after Monger et al., 2004)

Soil order	Suborder	Africa	Asia	Australia Oceania	Europe	South America	Central America	North America	Global
				Arid soils					
Aridisols									15,798,100
	Cryids	653	417,587	-	103	165,909	-	499,351	1,083,601
	Salids	95,249	595,400	919	130	59,756	-	17,698	769,152
	Durids	-	-	-	3,454	-	-	-	-
	Gypsids	347,458	326,136	-	3454	-	-	-	677,050
	Argids	450,884	1,635,130	1,710,271	1,218	520,175	1,111	1,093,291	5,412,080
	Calcids	1,818,639	2,204,270	613,547	1,013	98,680	-	202,235	4,938,384
	Cambids	842,681	1,125,259	346,253	5,573	422,429	2,259	173,378	2,917,832
Vertisols	Torrerts	196,570	65,210	602,630	-	6,233	1,754	42,435	914,832
Oxisols	Torroxes	9,346	-	4,247	-	16,550	-	-	30,143
Andisols	Torrands	834	-	-	-	95	-	150	1,078
Entisols	In aridic (torric) moisture regimes								12,600,308
Total arid soils									29,344,460
				Semi-arid soils					
Mollisols									
	Ustolls	4,371	1,587,526	21,689	303,635	409,796	-	1,744,438	4,071,455
	Xerolls	72,125	403,161	55,502	236,656	769	-	165,132	933,346
Vertisols									
	Usterts	722,630	600,544	14,863	29,001	38,107	18,247	114,387	1,763,776
	Xererts	11,692	47,348	20,258	17,105	1,050	-	969	98,423
Oxisols	Ustoxes	1,718,701	18,457	39,249	-	1,338,805	1,295	360	3,116,866
Andisols									
	Ustands	11,235	12,308	-	1,975	11,414	11,004	7,750	55,686
	Xeralfs	80,071	209,377	270,520	118,833	25,340	-	176,034	880,174
Ultisols									
	Ustults	1,649,310	824,575	101,301	-	1,091,366	58,903	131,048	3,856,502
	Xerults	933	2,672	-	-	22	-	15,958	19,586
Inceptisols									
	Ustepts	1,675,118	832,200	257,463	377,338	611,443	83,255	331,262	4,168,080
	Xerepts	163,793	178,312	264	310,697	8,342	-	9,246	670,654
Entisols	In ustic moisture regimes								4,620,113
Entisols	In xeric moisture regimes								840,021
Total semi-arid soils									30,802,179
Total ice-free land area									130,268,185

Aridisols (14,942,000 km^2) and Entisols (12,682,000 km^2) are the dominant soil types in arid regions, although other soil types include Vertisols (889,000 km^2) and very minor amounts of Mollisols, Andisols, Oxisols and Spodosols (Table 1.3). Arid soils grade into semi-arid soils across three climatic transects: laterally into wetter regions, upslope into wetter climates at higher elevations, or downslope into run-in areas with wetter microclimates. Taxonomically, changes in soil types from dry region aridic to wetter region ustic or xeric moisture regimes are expressed at the Suborder and Great Group level (Fig. 1.9). Linked to this climatic transition is a progressive change in vegetation – desert shrublands give way to grasslands that in turn give way to woodlands.

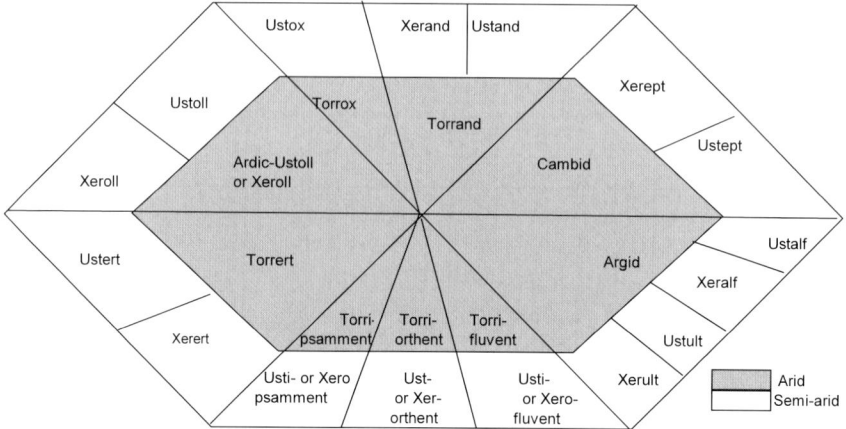

Figure 1.9. Illustration of soil taxonomy suborders and great groups that have aridic moisture regimes (shaded) and their moisture counterparts that have ustic and xeric moisture regimes (after Monger, 2002. Reprinted from Encyclopedia of Soil Science, H. Tan, ed., Arid Soils, Monger, p. 86, Copyright (2002), with permission from Marcel Dekker)

Also across this transition, soils have progressively deeper carbonate horizons. In the Chihuahuan Desert, for instance, carbonate zones are 50 cm deep at 230 mm of annual rainfall and 100 cm deep at 320 mm of annual rainfall. Likewise, gypsum zones progressively deepen from about 50 cm depth at 150 mm of annual rainfall to about 100 cm depth at 250 mm annual rainfall. Accompanying an increase in rainfall is an increase in soil organic matter. Although the amount of organic matter depends on the clay content, organic matter ordinarily increases from less than 0.5% in A-horizons of arid shrubland soils to 2–5% in semi-arid and sub-humid grassland soils (Monger, 2002). Thus, the major soil orders in the semi-arid areas include Mollisols (5,000,000 km^2), Alfisols (6,600,000 km^2), Inceptisols (4,800,000 km^2) and smaller areas of Entisols, Ultisols, Oxisols and Vertisols (Table 1.3).

Acknowledgement

Use excerpts from World Soil Resource Reports 94, 2001 (Lectures notes on the major soils of the world) (p. 185–220) – pages 59, 60, 61, 188, 189, 223, 224, and 225. P. Driessen, J. Deckers, O. Spaargaren, F. Nachtergaele, eds. Rome, Copyright (2001). With perimission from FAO.

Section II
Biogeochemistry of Trace Elements in Arid Environments

By

Fengxiang X. Han

Institute for Clean Energy Technology and
Department of Plant and Soil Sciences
Mississippi State University
205 Research Blvd
Starkville, MS 39759
U.S.A.

Chapter 2

TRACE ELEMENT DISTRIBUTION IN ARID ZONE SOILS

Trace elements refer to a number of elements that occur in natural systems in small concentrations (Page, 1974). Trace elements include chemical elements used by organisms in small quantities that are essential to their physiology as well as those elements with no known physiological function, which are toxic to living organisms at high concentrations (Bradford et al., 1996). Trace elements are present in less than 0.1% average abundance in the earth's crust (Mitchell, 1964; Adriano, 2001). According to the concentrations of elements in the earth's crust, elements can be divided into major (10,000–1,000,000 mg/kg), intermediate (1,000–10,000 mg/kg), minor (100–1,000 mg/kg), and trace elements (<100 mg/kg) (Fig. 2.1). Trace elements that are of most concern in arid and semi-arid soils throughout the world include Cd, Cu, Co, Cr, Ni, Pb, Hg, Mn, As, Se, Mo and B. This chapter addresses natural sources, contents and distribution of trace elements in arid and semi-arid soils. Factors affecting the content and distribution of trace elements, including climate, parent materials, soil processes/properties, agricultural management and industrial activities will be also discussed.

2.1 SOURCES OF TRACE ELEMENTS IN SOILS

The primary sources of trace elements in soils are the parent materials from which soils are derived. These parent materials constitute the reserve for trace elements. Concentrations of trace elements in soils are directly dependent upon their abundance in the earth crust. In general, concentrations of most trace elements in global soils are from one third to three times those in the earth's crust. The logarithm ratios of their concentrations in the global soils over the earth's crust are in the range ± 0.5 (Fig. 2.2).

This is especially important for trace elements in soils in undeveloped arid and semi-arid zones. In agricultural ecosystems, two more major routes for inputs of trace elements are aerial and terrestrial (land) pathways (Adriano, 1986). The aerial pathways include aerosols, particulate matter, and re-suspended and airborne dusts, which are important for arid and semi-arid soils. The land pathways include the application of fertilizers and pesticides, solid wastes, and other anthropogenic and industrial activities. Anthropogenic pollution of trace elements will be discussed in Chapters 8 and 9.

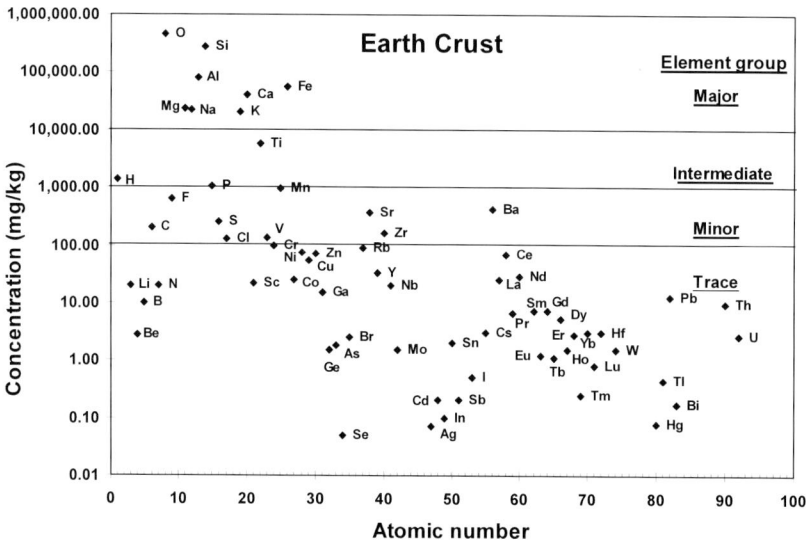

Figure 2.1. Distribution of chemical elements in the earth's crust (mg/kg)

Trace elements can be transported out of arid soils through plant biomass removal and erosion/leaching. Plant biomass removal is through production of food, feedstuff, and fiber plants. Wind erosion is a serious loss of surface soils and trace element outputs from arid and semi-arid soils as discussed in the first chapter.

2.2 TRACE ELEMENTS IN SOILS AND ROCKS: A GLOBAL VIEW

The pedosphere is defined as the loose surface of the earth and the interface among the lithosphere, biosphere, atmosphere and hydrosphere (Merritts et al., 1997; Han et al., 2002a). Banin and Navrot (1975) found a similar pattern of the distribution of elements as indicated by ionic potentials for all major groups of organisms from bacteria to fungi to plants to land animals and for the ocean (Figs. 2.3–2.5). Ionic potential (IP) is defined as the positive charge of an ion divided by its radius. Ionic potential has been used as a quantitative criterion for explaining trends in the distribution of elements, especially minor and trace elements in sedimentary rocks (Banin and Navrot, 1975). A general pattern is observed for all living organisms and oceans: (1) For elements of low ionic potential values (IP < 3), the log of the enrichment factor (EF, the ratio of the concentration of the element in an organism or ocean to its concentration in the earth's crust) is in the range of –1 to +1, indicating small enrichment or small depletion relative to the crust, (2) For intermediate IP values (3 < IP < 10), the log of EF is –3 to –4, indicating large depletion in living organisms, and (3) For large IP values (IP > 10), the log of EF increases as the ionic potential increases and varies from –4 to +4 (Banin and Navrot, 1975). Seawater and bacteria are completely depleted of all trace elements, while plants, animals, and fungi are depleted of most of trace elements relative to the crust. Thus the basic similarity of elemental composition and distribution pattern of all groups of living organisms and oceans indicates that the pattern (i.e., the distribution of elements) was determined at the initial steps of the development of life (Banin and Navrot, 1975).

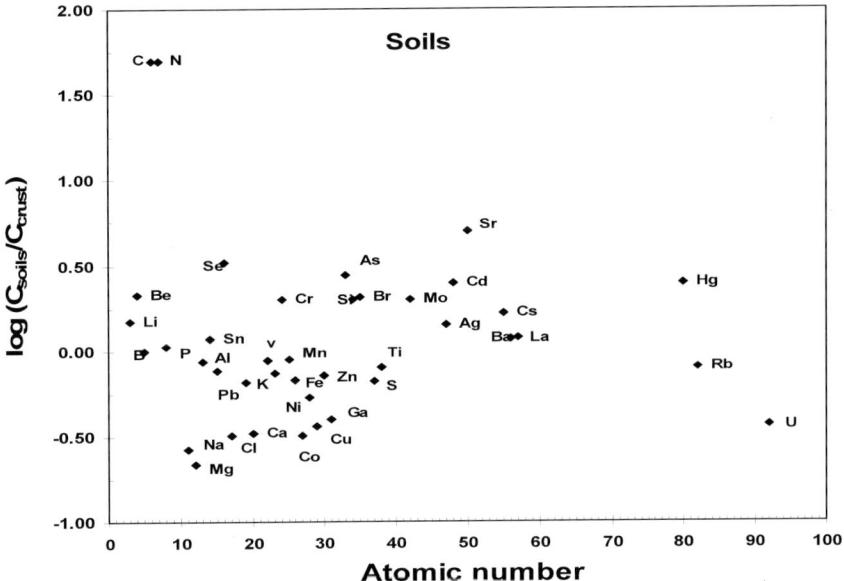

Figure 2.2. Elemental enrichment factors in soils, plotted on a log scale against the ionic potential of the elements

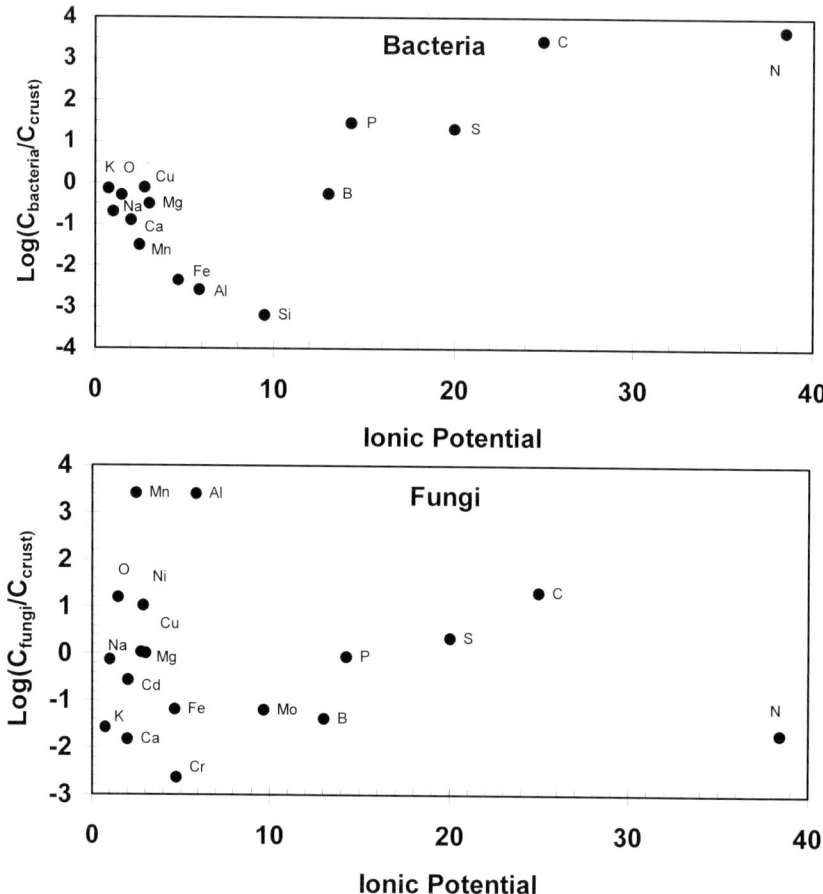

Figure 2.3. Elemental enrichment factors in baterial and fungi, plotted on a log scale against the ionic potential of the elements (after Banin and Navrot, 1975. Reprinted from Science, 189, Banin A. and Navrot J., Origin of Life: Clues from relations between chemical compositions of living organisms and natural environments, pp 550–551, Copyright (1975), with permission from AAAS)

The range and average of trace elements in the earth's crust, rocks and world soils are presented in Table 2.1. The average Cr concentration in the lithosphere is around 200 mg/kg. The basic, metamorphic and sedimentary rocks contain 100–300 mg/kg Cr. But some ultra basic volcanic rocks (dunite and peridotite) and serpentite have the highest concentration of Cr, in the range of 2000–3000 mg/kg. Acid igneous rocks (granite) contain 20–40 mg/kg (Aubert and Pinta, 1977). The total Cr concentrations in soils throughout the world vary drastically from trace quantities to 3000–4000 mg/kg, with the average concentration measuring 40–300 mg/kg (Aubert and Pinta, 1977; Adriano, 2001; Ure and Berrow, 1982). Arid and semi-arid soils usually contain higher Cr contents than other soils.

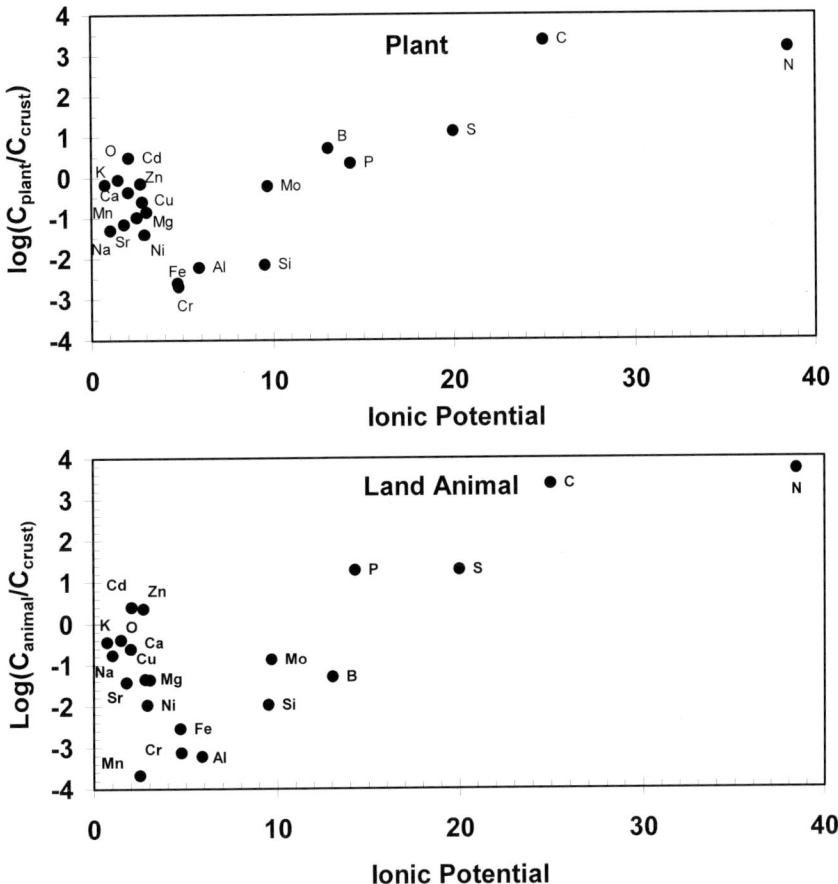

Figure 2.4. Elemental enrichment factors in plants and land animals, plotted on a log scale against the ionic potential of the elements (after Banin and Navrot, 1975. Reprinted from Science, 189, Banin A. and Navrot J., Origin of Life: Clues from relations between chemical compositions of living organisms and natural environments, pp 550–551, Copyright (1975), with permission from AAAS)

Cobalt is the 30th element in the earth's crust with an average of 20–30 mg/kg (Bowen, 1979; Vinogradov, 1959). Ultrabasic rocks and their metamorphized serpentines contain 100–200 mg/kg Co, followed by basic igneous rocks (30–45 mg/kg), while acid rocks have the lowest Co concentration (5–10 mg/kg). Metamorphic rocks and sedimentary rocks contain 20–30 mg/kg. Sands and limestone contain very low Co (0.1–5 mg/kg). The total Co concentration in the world's soils varies from trace quantities to 300 mg/kg with an average value of 10–15 mg/kg (Aubert and Pinta, 1977). Arid and semi-arid soils have higher average Co concentration than those of temperate and boreal regions as well as the humid tropic zones.

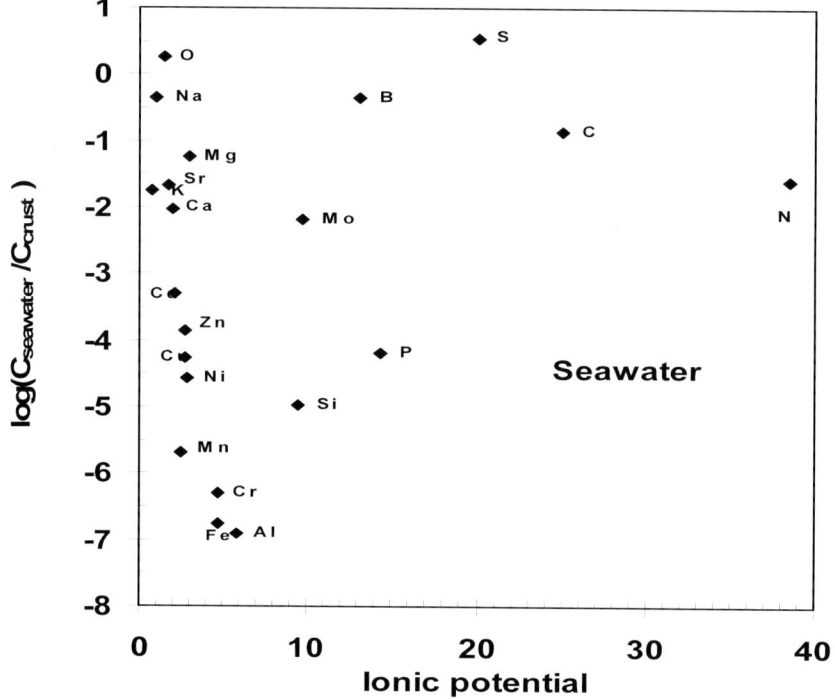

Figure 2.5. Elemental enrichment factors in seawater, related to the ionic potential of the elements (after Banin and Navrot, 1975. Reprinted from Science, 189, Banin A. and Navrot J., Origin of Life: Clues from relations between chemical compositions of living organisms and natural environments, pp 550–551, Copyright (1975), with permission from AAAS)

Copper, the 26th element in the earth's crust, has 100 mg/kg in the lithosphere (Vinogradov, 1959). Basic igneous rocks (basalt and dolerite) contain 100–200 mg/kg Cu, higher than acid igneous rocks (granite) (10–20 mg/kg). Metamorphic and sedimentary rocks contain 30–40 mg/kg Cu, while sandstone, sands and limestone have the lowest total Cu (3–15 mg/kg) (Aubert and Pinta, 1977). The total Cu in the world's soils ranges from trace to 200–250 mg/kg and averages 15–40 mg/kg (Aubert and Pinta, 1977), while in arable soils, the total Cu concentration varies from 1 to 50 mg/kg

Table 2.1. Concentrations of trace elements in the earth's crust, rocks, and world soils[a]

		Cd	Cr	Pb	Hg	Cu	Zn
				(mg/kg)			
Lithosphere[b]			(200)	(16)		100	(20-70)
Rocks	Igneous rocks	0.001-0.60 (0.082)	2-3400	2-30 (15)	0.005-0.25	10-100	5-1070
	Metamorphic rocks	0.005-0.87 (0.06)	100-300[b]	15-20[b]		30-40[b]	
	Sedimentary rocks	0.05-500 (3.42)	<1-1000	15-20[b]	<0.005-3.25	0.6-67	<1-1500
Soil	World soils	0.01-2.0 (0.35) (0.62)[d]	10-150 (40)	10-100 (15-25[b]) (33.7)[d]	0.02-0.25 (0.07)	2-250 (30) (59.8)[d]	1-900 (90) 10-300[c], 50-100[b], 29.2[d]
	Global arid and semi-arid soils						Trace-900

		Ni	Co[b]	Hg[b]	As[b]	Se	B[b]	Mo
Lithosphere[b]		(100)		(0.05)		(0.8[b])	(50)	(1-2[b])
Rocks	Igneous rocks	2-3600 (75)	20-30	0.005-0.25	1-3.4	(0.05)	1-10	0.9-7
	Metamorphic rocks		5-200		15-3.0			
	Sedimentary rocks	0.1-30	20-30	<3.0	0.3-500			
Soil	World nonpolluted soils	5-500 (20) (25.8)[d]	Trace-300 (10-15)	0.01-0.06	5-10	<0.1-2 (0.4[b])	2-100 (20-50[b])	0.2-5
	Global arid and semi-arid soils	5-300[b] (50[b])						(2-5)[b]

[a] Data from Adriano, 2001. Data in the parenthesis are averages.
[b] Aubert and Pinta, 1977
[c] Swaine, 1955
[d] Ure and Berrow, 1982.

(Adriano, 2001). Arid and semi-arid soils contain, in general, from average to high Cu in the range of 15–200 mg/kg. The concentration of Cu in the arid and semi-arid soils of the U.S. (22.2 mg/kg) is higher than the average Cu in all U.S. soils (18 mg/kg) (Holmgren et al., 1993).

The average Pb concentration in the earth's crust is 16 mg/kg (Vinogradov, 1959; Swaine, 1955). Igneous rocks contain 2–30 mg/kg, with an average of 8 mg/kg for basic igneous rocks and 20 mg/kg for acid igneous rocks. Metamorphic and sedimentary rocks contain 15–20 mg/kg. The average total Pb concentration in the world's soils ranges from 15 to 25 mg/kg (Aubert and Pinta, 1977). Lead concentrations in arid and semi-arid soils vary widely, depending on the nature of the parent materials and the soil processes present. The Pb concentrations in the arid and semi-arid soils of the U.S. (10.4 mg/kg) are similar to the average Pb concentration in U.S. soils (10.6 mg/kg) (Holmgren et al., 1993).

Nickel is the 23rd element in the earth's crust. The average Ni content in the lithosphere is about 100 mg/kg (Vinogradov, 1959). Total Ni contents are the highest in ultrabasic rocks (dunite and peridotite) and metamorphic rocks (serpentine), containing 500–3000 mg/kg. Total Ni concentrations in igneous rocks and shales are generally higher than those in sedimentary rocks. Sandstone and limestone contain low total Ni content. The average Ni in soils throughout the world was reported to be from 20 to 25 mg/kg (Adriano, 2001; Berrow and Reaves, 1984). The average Ni concentrations in the United States, Canada, Italy, England, and Wales are in the range of 20–28 mg/kg (Adriano, 2001). Arid and semi-arid soils generally contain higher quantities of Ni than soils of temperate and boreal regions. Nickel concentrations in the world's arid and semi-arid regions are in the range 5–300 mg/kg with an average of 50 mg/kg (Aubert and Pinta, 1977). The Ni concentration in the arid and semi-arid soils of the U.S. (24.7 mg/kg) is higher than the average Ni in all the U.S. soils combined (16.5 mg/kg) (Holmgren et al., 1993).

Zinc is the 24th most abundant element in the earth's crust. The Zn concentration in the lithosphere is 50–70 mg/kg (Vinogradoc, 1959; Adriano, 2001). Basic igneous rocks contain higher Zn (70–130 mg/kg) than metamorphic and sedimentary rocks (80 mg/kg). Carbonate and limestones contain low Zn (16–20 mg/kg) (Aubert and Pinta, 1977). The total Zn concentration in the soils of the world ranges from 10 to 300 mg/kg (Swaine, 1955), with average concentrations from 50 to 100 mg/kg (Aubert and Pinta, 1977). Arid and semi-arid soils vary from trace levels (subdesert soils) to 900 mg/kg (saline alkali soils) (Aubert and Pinta, 1977). The average Zn concentration in the arid and semi-arid soils of the U.S. (62.9 mg/kg) is higher than the average quantity of Zn measured in all U.S. soils (42.9 mg/kg) (Holmgren et al., 1993).

Manganese concentrations of the earth's crust are about 900–1000 mg/kg (Vinogradov, 1959). Basic igneous rocks (basalt) contain the highest Mn content (1000–2000 mg/kg). Manganese minerals are widely distributed. Oxides, silicates, and carbonates are the most common. The most important Mn mineral is MnO_2 (pyrolusite). Manganese concentrations in most soils are in the range of 200–3000 mg/kg with an average of 850–1000 mg/kg (Swaine, 1955; Bowen, 1979; Adriano, 2001). In the arid and semi-arid soils, Mn varies from trace amounts (chernozems and chestnut soils) to 10,000 mg/kg (saline alkali soils of Chad) (Aubert and Pinta, 1977).

Molybdenum, the 53rd element in the earth's crust, varies from 1 to 2 mg/kg in the lithosphere (Vinogradov, 1959). Sedimentary and metamorphic rocks contain higher Mo concentrations than igneous rocks. Concentrations of Mo in acid igneous rock (granite) are higher than basic (basalt) rocks, while ultrabasic rocks (peridotite and dunite) contain lower Mo. Total Mo content depends upon the nature of the parent materials, the soil processes involved, and the soil properties, especially organic matter. In general, Mo concentrations in soils are higher than those of the parent rocks. The normal range of soil Mo is 0.5–5 mg/kg (Adriano, 2001). In general, arid and semi-arid soils contain higher total Mo levels than soils of temperate and boreal regions, which average in the range of 2–5 mg/kg (Aubert and Pinta, 1977).

Mercury in the lithosphere is around 0.05 mg/kg (Bowen, 1979). Mercury content in igneous rock is in the range of 0.005–0.25 mg/kg. Most sedimentary rocks (< 0.30 mg/kg Hg) contain higher Hg than igneous rocks (Adriano, 2001). Shales with high organic matter content contain Hg as high as 3.0 mg/kg. Average Hg concentrations in the world's soils have been reported to be in the range of 0.01–0.06 mg/kg (Bowen, 1979; Brooks, 1977). In the U.S., the mean Hg concentration is 0.112 mg/kg, but soils in the western states (mostly arid and semi-arid soils) contain 0.083 mg/kg (as a mean Hg).

Cadmium in the lithosphere is 0.098 mg/kg (Taylor and McLennan, 1985), but Adriano (2001) pointed out that the most often-quoted average concentrations of Cd in the earth's crust are 0.15–0.20 mg/kg. Sedimentary rocks contain higher Cd content than igneous rocks and metamorphic rocks. The average Cd concentrations in the world's soils are 0.30–0.62 mg/kg (Bowen, 1979; Ure and Berrow, 1982; Lee and Keeney, 1975). Cadmium and Zn coexist in the lithosphere and soils. In the U.S. soils, the average Cd concentration (as a geometric mean) is 0.175 mg/kg (Holmgren et al., 1993). In arid and semi-arid U.S. soils, total Cd concentrations (0.27 mg/kg) are higher than the national average.

The average selenium content of the earth's crust is about 0.8 mg/kg (Aubert and Pinta, 1977). Carbonated rocks contain higher Se (0.8 mg/kg) than metamorphic and sedimentary rocks (0.6 mg/kg) and igneous rocks (0.5 mg/kg) (Aubert and Pinta, 1977). Most of the world's soils contain 0.03–2 mg/kg Se (Ure and Berrow, 1982) with a mean of 0.4 mg/kg. In the U.S., the total Se concentration ranges from trace amounts to 82 mg/kg (Adriano, 2001).

Arsenic is the 52nd element in the earth's crust, averaging from 1.0–3.4 mg/kg (Taylor and McLennan, 1995; Wedepohl, 1995; Lide, 1996; Bowen, 1979; Bockris, 1977). The average As content in igneous rocks is between 1.5–3.0 mg/kg, but As in sedimentary rocks varies from 0.3 to 500 mg/kg. In general, sedimentary rocks contain higher As than igneous rocks. Shale contains very high levels of As. The average total As in the world's soils is estimated to be between 5 and 10 mg/kg (Koljonen, 1992; Bowen, 1979; Neff, 1997; Allard, 1995; Baur and Onishi, 1969).

Boron is the 37th element in the earth's crust with an average of 50 mg/kg (Vinogradov, 1959). Sedimentary rocks contain higher B concentrations than igneous rocks. Basic igneous rocks contain 1–5 mg/kg B, while acid igneous rocks have 3–10 mg/kg. Some types of shale have up to 100 mg/kg (Adriano, 2001). The total B in the soils varies from 1 to 270 mg/kg, with the average ranging from 20 to 50 mg/kg. Arid and semi-arid soils generally have average to high total boron concentrations (Aubert and Pinta, 1977).

2.3 DISTRIBUTION OF TRACE ELEMENTS IN ARID AND SEMI-ARID SOILS

2.3.1 Africa

In the semi-arid soils of the Nakuru District, Kenya, which are derived from igneous and volcanic materials, the average Cd and Mo concentrations are in the narrow range of 0.3–0.5 mg/kg, while Cu varies from 2.2–4.0 mg/kg (Simyu et al., 2005). Lead averages from 4.3–7.1 mg/kg and Zn from 15.7–62.3 mg/kg. Corresponding to soils, concentrations of trace elements in water samples from the regions are < 1 µg/L for Cd and 2.0–6.0 mg/L for Cu. Zinc, Pb and Mn in these water samples are in the ranges of 4–10, 3–21 and 0.7–39 µg/L, respectively.

In arid soils from the United Arab Emirates with 5–75% $CaCO_3$, Cd, Cr and Cu range from 0–0.2 mg/kg, 15.1–21.2 and 2.6–300 mg/kg, respectively. The averages and standard deviations of theses trace elements are Cd, 0.043 ± 0.07; Cr, 19.1 ± 1.9; and Cu, 53.3 ± 92.3 mg/kg. Nickel, Pb and Zn are in the ranges of 3.2–19, 8.8–21 and 1.3–48.5 mg/kg and average 4.9, 4.3 and 12 mg/kg, respectively. Copper has a large variation due to possible contamination. The variations of the other trace elements are small, indicating there is no anthropogenic pollution in these areas (Howari, 2004).

Vertisols in Madagascar formed on basalt and marl contain very high Cr (200–540 mg/kg), while alluvial soils have 190 mg/kg Cr (Aubert and Pinta, 1977). In vertisols derived on alluvions, Pb is in the range of 20–45 mg/kg, while the average Cr in saline soils of the region is from 1–22 mg/kg. Vertic soils on marl and basalt contain 700–2400 mg/kg Mn. In soils derived on clayey sands and alluvions, Zn content is 105 mg/kg. Soils formed on alluvions contain 115 mg/kg Ni.

In the vertisols of Chad, which are from sandy-clay sediments, total Cr and Mn concentrations are found to be in the range of 180–300 and 60–180 mg/kg, respectively (Aubert and Pinta, 1977). Total Co, Pb, Cu and Ni are in the range of 10–95, 20–45, 30–90 and 30–90 mg/kg, respectively, while saline soils of the regions contain higher Mn (700 mg/kg). Saline alkali soils have as high as 10000 mg/kg Mn and 90 mg/kg Zn.

Vertisols formed from amphibolite in the Central African Republic contain 300–1000 mg/kg Cr, 11–200 mg/kg Cu, trace amounts to 6 mg/kg Mo and 60–300 mg/kg Ni (Aubert and Pinta, 1977). In chernozem soils of the flood plains of the Amur region, the average Cr is 400 mg/kg. High Co concentrations (100–300 mg/kg) have been found in soils. Manganese concentrations in lithomorphic vertisols are in the range of 3000–5000 mg/kg. Soils on clayey sediments, solonetses and saline alkali soils contain 50–75, 40–100 and 10–50 mg/kg Ni, respectively.

2.3.2 Asia

2.3.2.1 India (Aubert and Pinta, 1977)

In the vertisols of India, total Mn is in the range of 650–2950 mg/kg, while the exchangeable Mn is 90–730 mg/kg, accounting for 7.5–30% of the total Mn. The exchangeable Mn in vertic, saline and alluvial soils is in the range of 16–17.7 mg/kg (0.1–19.5% of total Mn). In saline and alkaline soils of Rajasthan of India, Mn varies from 4.6–20.5 mg/kg (1.9–6.4% of total Mn). However, exchangeable Mn is much higher in alluvil soils (61–258 mg/kg, 16.6–37.5% of the total Mn). Total Cu in vertisols derived from sands and limestone of the Gujarat region is in the range of 80–155 mg/kg, while in vertisols on marls and basalts, total Ni is 30–90 mg/kg. EDTA-extractable Cu in alluvial and vertic soils varies from 0.45–12.3

mg/kg. In soils on sandstone and igneous rocks of Gujarat, the total Zn is 60–70 mg/kg. Vertisols on basaltic rocks and coastal alluvions contain 25–50 mg/kg B. In calcareous alluvial soils with sand and clayey texture, total B is in the range of 15–50 mg/kg. In vertisols on sediments derived from basalts and from lava and limestone in the Gurajat region, total Mo contents vary from 1.2–4 mg/kg.

2.3.2.2 China (Liu, 1996)

The distribution of selected trace elements in arid and semi-arid soils of China is summarized in Table 2.2. Total Mn in soils of China varies from 10–5532 mg/kg, with some soils very rich in Mn (9478 mg/kg). The average Mn in China's soils is 710 mg/kg. In general, arid and semi-arid soils have lower Mn contents than in humid and temperate soils. Total Mn in arid and semi-arid soils ranges from 59–1500 mg/kg with an estimated average of 677 mg/kg. Total Mn in soils developed on loess varies from 116–1065 mg/kg with an average of 537 mg/kg. Soils in the North China Plain contain 210–1219 mg/kg. Arid soils in the North West region have 59–1500 mg/kg Mn with an average of 564 mg/kg. In soils of the Tibet region, total Mn is in the range of 126–2183 mg/kg and averages 423 mg/kg. In calcareous paddy soils, total Mn varies from 262–675 mg/kg and the average Mn content is 464 mg/kg.

Table 2.2. Concentrations of selected trace elements in representative arid and semi-arid soils of North China[a] (from Liu, 1996)

Soil	Mn	Zn	Cu	Ni	Mo	B	Se
				Range, mg/kg			
Blown sand soils	116-500	30.3-92.2	10-22		0.21-0.50	17-71	
Brown calc soils	143-973	24-215	6-48	9.6-37.6			0.006-0.612
Chao soils	441-776	8.7-184	9-36	16.2-23.8	0.3-1.4	14-141	0.014-9.135
Chernozems	730-1200	58-153	16-53	5.5-33.7	2.0-4.2	49-64	0.018-0.598
Chestnut soils	220-944	34-150	0-145	4.2-95.8	0.47-1.30	12-75	0.014-0.352
Cinnamon soils	210-1010	58-128	11-36	18.1-25.5	0.2-3.0	41-78	0.024-0.540
Heilu soils	455-675	50-101	14-27	8.4-41.3	0.42-0.72	22-96	0.019-0.466
Loessal soils	660-1170	40-75	15-42		0.4-1.1	44-128	0.054-0.185
Shajiang black soils	243-1219	21-194					
Sierozems	180-940	34-118	14-31	12.9-81.8	0.48-1.45	48-105	0.053-0.328
Solonchaks			5-48	0.1-50.0		23-150	0.012-0.366
Solonetz	59-1035	16.5-123		4.3-42.8			0.143-0.313
Tier soils	320-1065	54-99	17-37		0.4-1.1	37-117	0.052-0.418

[a]Data from Liu, 1996. The data cover North China Plain, Loess Plateau, and Xinjiang Province.

Soils of China have total Zn of < 3–790 mg/kg with an average of 100 mg/kg. Arid and semi-arid soils contain Zn in the range of 8.7–215 mg/kg. The average Zn in arid and semi-arid soils is 73.7 mg/kg. In soils derived on loess of North China, total Zn ranges from 34–101 mg/kg with an average of 67 mg/kg. Soils in the North West region contain 9–216 mg/kg and average 75 mg/kg total Zn. Total Zn in calcareous paddy soils is generally lower than acid and neutral paddy soils, varying from 44–135 mg/kg with an average of 82 mg/kg.

Total Cu in soils of China varies from 3–500 mg/kg with an average of 22 mg/kg. Arid and semi-arid soils contain an average of 20.8 mg/kg with the range of 0–145 mg/kg. Total Cu in soils derived on loess varies from 10–48 mg/kg (22.5 mg/kg as an average). In the North China Plain, soils have an average total Cu of 20–23 mg/kg and range from 10–85 mg/kg. Arid soils in the North West region contain average Cu contents of 25–28 mg/kg, depending upon soil types and parent materials, and some arid soils may have up to 145 mg/kg. Total Cu contents in Tibetan soils are from 3 - 42 mg/kg with an average of 13 - 21 mg/kg.

Total Ni contents in soils of China are in the range of 7.7–71 mg/kg with an average of 23.4 mg/kg. But some soils contain as much as 627 mg/kg total Ni. Arid and semi-arid soils have 0.1–85.8 mg/kg with an average of 31.4 total Ni.

Soils of China contain an average of 12.7 mg/kg total Co. Arid and semi-arid soils contain higher total Co than soils in humid and temperate regions.

Total B is in the range of 18–88 mg/kg in soils of China. Arid and semi-arid soils contain higher B than humid and temperate soils. Total B in arid and semi-arid regions varies from 12–150 mg/kg with an average of 62.5 mg/kg, while soils in tropic and subtropic humid zones contain an average of 34 mg/kg of total B. In soils derived on loess of North China, total B is 13–150 mg/kg with an average of 60.2 mg/kg. Soils in the Tibet Plateau contain very high B in the range of 11–500 mg/kg with an average of 154 mg/kg total B, while water soluble B is in the range of 0.18–2.54 mg/kg. Saline and alkali saline soils in arid and semi-arid regions contain high B, as high as 1050 mg/kg with 195 mg/kg of water soluble B. Total B in calcareous paddy soils varies from 12–61 mg/kg and averages 45 mg/kg of total B and 0.72 mg/kg of water soluble B.

Total Mo in soils of China varies from 0.1–6 mg/kg with an average of 1.7 mg/kg. Arid and semi-arid soils have 0.2–4.2 mg/kg total Mo. Average Mo contents in arid and semi-arid regions are 1.58 mg/kg. In general, lower Mo contents are found in soils of arid and semi-arid regions than the humid and tropic/subtropic soils. In soils derived on loess, total Mo is from 0.21–1.45 mg/kg with an average of 0.62 mg/kg. Soils in the North

China Plain contain 0.1–4.5 mg/kg, and average total Mo contents are in the range of 0.48–2.4 mg/kg. In soils of the North West region, total Mo is 0.4–3.1 mg/kg with an average of 0.985 mg/kg. Soils in Tibetan areas contain 0.1–5.5 mg/kg and have average total Mo between 0.9–1.1 mg/kg. Calcareous paddy soils vary in total Mo concentrations from 0.26–1.21 mg/kg with an average of 0.57 mg/kg.

Soils of China contain 0.006–9.135 mg/kg total Se with an average of 0.296 mg/kg. Most soils have 0.012–3.37 mg/kg total Se. Arid and semi-arid soils contain an average of 0.171 mg/kg total Se. Soils in the North West region contain 0.014–0.60 mg/kg with an average of 0.16 mg/kg total Se. The average Se in soils of the North China Plain is 0.138 mg/kg.

Soils of China contain a total Cd in the range of 0.017–0.333 mg/kg, As in the range of 2.5–33.5 mg/kg, Hg in the range of 0.006–0.272 mg/kg, Pb in the range of 10–56.1 mg/kg, and Cr in the range of 14.3–150.2 mg/kg (95% samples). The average Cd, As, Hg, Pb and Cr are 0.097, 11.1, 0.065, 26 and 61 mg/kg, respectively. In arid and semi-arid soils, total Cd, As, Hg, Pb and Cr are strongly affected by parent materials due to weak weathering and soil formation processes. Arid and semi-arid soils derived on loess in North West regions contain 0.103 ± 1.24 mg/kg (average ± standard deviation) total Cd, 11.8 ± 1.2 mg/kg total As, 22.6 ± 1.2 mg/kg total Pb and 66.70 ± 1.15 mg/kg total Cr. The variation of total Cd is larger, while As, Pb and Cr contents in arid soils are in the narrow ranges with less variation.

2.3.3 Australia (Aubert and Pinta, 1977)

In vertisols of Queensland and Tasmania derived on basalt, dolerite and alluvium, total Co is in the range of 7–70 mg/kg, but in soils derived on granodiorite, total Co is 3.5 mg/kg. In soils of the Adelaide and Southeast regions, Co varies from 1–30 mg/kg.

In vertisols on diorite, basalt, alluvions, and dolerite, total Mn is in the range of 1250–2750 mg/kg, while brown isohumic soils on calcareous sandstone and clayey sediments contain 550–1670 mg/kg Mn. Soils in the Adelaide and Southeast regions contain 140–1400 mg/kg total Mn. In solods on granitic rocks and alluvions, total Mn is 60–990 mg/kg.

In Queensland, soils on clayey sediments have 3–12 mg/kg total Mo, and mediterranean red soils on granodiorite and basalt contain 2–8.5 mg/kg Mo. In vertisols on dolerite in Tasmania, total Mo is 2.5–3.5 mg/kg.

In vertisols on diorite and basalt of Queensland, total Zn is 120 mg/kg, while Mediterranean red soils average 11–86 mg/kg of total Zn. Soils on calcareous sandstone and on clayey sediments contain 45–100 mg/kg of total Zn.

In vertisols of Tasmania, the average total Ni is 30 mg/kg. Saline soils on limestone contain 1–20 mg/kg of total Ni, and Ni in red Mediterranean soils varies from 5–54 mg/kg

2.3.4 Europe and the Former Soviet Union (Aubert and Pinta, 1977)

2.3.4.1 Bulgaria

In chernozems formed on serpentinite diluvium, Co content is in the range of 10–30 mg/kg, while in chestnut and chestnut vertic soils, Co concentrations vary from 3–15 and 15–45 mg/kg, respectively. Soils on basalt, andesite and gabbro contain 15–68 mg/kg total Cu. Total Mn in chernozems is in the range of 520–850 mg/kg. Chestnut soils have 42–106 mg/kg Zn content. Total Zn in saline alkali soils is in the range of 40–60 mg/kg Zn. Bioavailable Zn (ammonium acetate-extractable Zn) in chernozems, chestnut soils and saline alkali soils of the steppe zones varies from trace amounts to 3.8 mg/kg (1–8.3% of total Zn). In chernozems of Northern Bulgaria, total B is in the range of 25–53 mg/kg. Boron increases in saline soils and saline alkali soils.

2.3.4.2 The Former Soviet Union

Soils in Uzbekistan have an average Co of 17 mg/kg. Total Cu in humic soils is in the range of 18–22 mg/kg, and total Zn is 83 mg/kg. In subdesert soils formed on loessial clayey loams and saline alkali soils, total Mo is 2.6–4.5 mg/kg, and Mo in some soils is as high as 8 mg/kg. Total Zn in subdesert soils varies from 60–112 mg/kg. Sodium acetate-extractable Zn in these subdesert soils is 2.1–3.2 mg/kg, accounting for 2.3–5.1% of total Zn. In saline alkali soils formed on loess and marine clays, total B content is 160 mg/kg. The exchangeable Mn in arid and semi-arid soils is 7–50 mg/kg, accounting for 0.7–7.8% of total Mn. Saline soils in the Ustyurt region contain 42–80 mg/kg of Pb.

Total Mo in soils of Russia varies from 1–20 mg/kg. In the Stavropol region, total Zn is in the range of 14.4–71 mg/kg. In the Rostov regions, total Zn is 680 mg/kg. Manganese content varies from 600–1000 mg/kg. Saline soils have low concentrations of total Co. In saline soils, total Cu varies from 60–64 mg/kg and in the very saline alkali soils, Cu is 164 mg/kg.

In the Ural-Sakmara basin, total Cu is 60–70 mg/kg in chernozems on secondary sediments. Chestnut soils on weathered basic rocks in the Or-Kumak basin contain higher Cu (88–96 mg/kg). The average Pb concentrations are 11–25 mg/kg in chernozems derived from serpentinite and secondary/tertiary sediments of the Ural-Sakmara basin. Soils in the Ural-Sakmara basin on serpentinite contain 133 mg/kg total Ni.

In soils of Azerbaidzhan, Co varies from 10–80 mg/kg. Some chernozems contain 100 mg/kg total Co. Total Mo is 0.5–3.3 mg/kg in the chernozems of southern Poland.

2.3.4.3 Romania, Spain, and Others

In chernozems derived on loess of Cluj and Dobrudja regions of Romania, total Cu is in the range of 25–45 mg/kg. Chernozems formed on loess contain 73 mg/kg of total Zn and 17–30 mg/kg of total Ni.

In vertisols on quaternary alluvions, total Mn is 500 mg/kg, while saline soils on the same parent material have 1000 mg/kg of Mn. Mediterranean red soils on calcareous rocks and quaternary alluvions contain 340–390 mg/kg of Mn. In the Delta of the Guadalquiver River of southwestern Spain, Cu, Pb, Cd and Zn are in the range of the following: Cu, 1.62–46.8 mg/kg; Pb, 3.15–44.3 mg/kg; Cd, 0.23–7.10 mg/kg and Zn 5.13–669 mg/kg (Ramos et al., 1994).

In cherozems of Serbia, soil boron is in the range of 25–40 mg/kg. The average B content is 30 mg/kg. Boron content increases with salts in arid soils. Saline salts and saline alkali soils contain 40–65 mg/kg total B.

2.3.5 North America

In the arid and semi-arid soils of the U.S., total Cd, Zn, Cu, Ni and Pb concentrations are 0.27, 62.9, 22.2, 24.7 and 10.4 mg/kg, respectively (Table 2.3) (Holmgren, et al., 1993). Concentrations of these elements in arid and semi-arid soils are higher than the national averages of mineral soils (0.155, 41.1, 15.5, 17.1 and 10.4 mg/kg for Cd, Zn, Cu, Ni and Pb, respectively). Lead concentration in arid and semi-arid soils is close to the national average. At the regional levels of arid and semi-arid areas, soils in the Rocky Mountains contain the highest Cd (0.302 mg/kg), Zn (105 mg/kg), and Pb (13.2 mg/kg) concentration. Soils in the subtropical regions of California have the highest Cu (43.4 mg/kg) and Ni (64.4 mg/kg) concentration. In general, concentrations of these elements in soils of the Rocky Mountains and California's subtropical regions are higher than in the

Western Great Plains and the Western range and in irrigated regions, while the Southwest Plateau has the lowest metal concentrations. In the Southwest Plateau, soils have 0.143, 38.1, 10, 12.5 and 7.0 mg/kg Cd, Zn, Cu, Ni and Pb, respectively. Moreover, soils in the Western Great Plains have higher concentrations of these elements, except for Pb, than those in the Western range and irrigated regions. Soils in the Western Great Plains contain 0.271 mg/kg Cd, 54.3 mg/kg Zn, 16.3 mg/kg Cu, 17.2 mg/kg Ni and 11.8 mg/kg Pb. Soils in the Western range and irrigated regions have 0.291 mg/kg Cd, 73.8 mg/kg Zn, 26.8 mg/kg Cu, 25.2 mg/kg Ni and 9.6 mg/kg Pb. In soils of the western U.S., total Se is 0.23 mg/kg with an observed range of <0.1 to 4.3 mg/kg (Shacklette and Boerngen, 1984).

Table 2.3. Geometric means for selected trace elements in arid and semi-arid soils of the U.S

States	Cd	Zn	Cu	Ni	Pb
			mg/kg		
Arizona	0.233	70.5	38.1	27.9	13.3
California	0.243	82.7	37.3	50.5	9.7
Colorado	0.309	76.1	18	14.4	12.8
Idaho	0.338	64.3	20.9	24.4	10.4
Nevada	0.316	58.7	17.8	25.9	8.6
New Mexica	0.200	46.5	15.4	16.2	10.5
Texas	0.123	30.4	9.5	12.5	7.4
Montata	0.367	74	20.6	25.8	10.5
Average	0.266	62.9	22.2	24.7	10.4
U.S.	0.155	41.1	15.5	17.1	10.4

[a]Data from Holmgren et al., 1993.

In soils of Arizona, the geometric means of Cd, Zn, Cu, Ni and Pb concentrations in surface soils are 0.233, 70.5, 38.1, 27.9 and 13.3 mg/kg, respectively (Holmgren, et al., 1993). The average concentrations in surface soils of Colorado are 0.309 mg/kg for Cd, 76.1 mg/kg for Zn, 18 mg/kg for Cu, 14.4 mg/kg for Ni and 12.8 mg/kg for Pb (Holmgren, et al., 1993), and surface soils in Idaho have 0.338 mg/kg Cd, 64.3 mg/kg Zn, 20.9 mg/kg Cu, 24.4 mg/kg Ni and 10.4 mg/kg Pb. Soils in New Mexico contain 0.200 mg/kg Cd, 46.5 mg/kg Zn, 15.4 mg/kg Cu, 16.2 mg/kg Ni and 10.5 mg/kg Pb. The average concentrations of these trace metals are 0.316, 58.7, 17.8, 25.9 and 8.6 mg/kg in soils of Nevada and 0.123, 30.4, 9.5, 12.5 and 7.4 in soils of Texas for Cd, Zn, Cu, Ni and Pb, respectively (Holmgren, et al., 1993). In Montana, soils have 0.367 mg/kg of Cd, 74 mg/kg of Zn, 20.6 mg/kg of Cu, 25.8 mg/kg of Ni and 10.5 mg/kg of Pb (Holmgren, et al., 1993).

In Californian soils, Cd concentrations in uncontaminated soils range from 0.05–1.70 mg/kg with an average of 0.36 ± 0.31 mg/kg (Bradford et al., 1996). Lead concentrations are in the range of 12.4–97.1 mg/kg with an average of 23.9 ± 13.8 mg/kg. Nickel content is from 9–509 mg/kg with an average of 57 ± 80 mg/kg. Concentrations of Zn average 149 ± 32 mg/kg with the range of 133–236 mg/kg. Chromium content is in the range from 23–1579 mg/kg and averages 122 ± 223 mg/kg. Cobalt concentrations range from 2.7–46.9 mg/kg with an average of 14.9 ± 9.2 mg/kg. Soils contain 9.1–96.4 Cu and average 28.7 ± 19.3 mg/kg. Mercury concentrations average 0.26 ± 0.21 mg/kg (in the range of 0.05–0.90 mg/kg). Manganese content is in the range of 253–1684 mg/kg and averages 646 ± 285 mg/kg. Arsenic concentrations range from 0.6–11 mg/kg with an average of 3.5 ± 2.5 mg/kg. Selenium concentration is from 0.015–0.43 mg/kg with an average of 0.058 ± 0.084 mg/kg. Molybdenum concentration averages 1.3 ± 1.5 mg/kg in the range of 0.1–9.6 mg/kg. In general, the ranges of trace elements in Californian soils are similar to those determined at the global levels (Bradford et al., 1996).

Among these trace elements in Californian soils, Zn and, to some extent, Mn, Pb, Co, Cu and As are evenly distributed in soils with little variation. However, Ni, Cr, Mo and Se are mostly unevenly distributed in soils with a large variation in their concentrations. Mercury and Cd are between these two types of distributions and variations, as shown below. The variation of concentrations of trace elements in the soils increases in the order: Zn (21 %) < Mn (44 %) < Pb (58 %) <Co (62 %), Cu (67 %) < As (71 %) < Hg (80 %), Cd (88 %) < Mo (113 %) < Ni (141 %), Se (147 %) < Cr (183 %) (Bradford et al., 1996).

Strong correlations occur between concentrations of trace elements in Californian soils. Nickel concentrations in soils are strongly correlated with Cr ($r = 0.95$); Cu contents are also significantly correlated with Co ($r = 0.81$). Strong correlations between Ni and Cr and between Cu and Co are observed as well (Marrett et al., 1992). This strong correlation between trace elements indicates that these elements associate in parent materials and suggests similar physical-chemical processes governing soil formation (Bradford et al., 1996).

Northern Californian soils have higher concentrations of Cr, Co, Cu and Ni than southern California since there are volcanic ultramafic rocks in northern California. Ultramafic rocks are mostly serpentine, a magnesium silicate with associated high amounts of Ni and Cr. Soils formed on serpentine parent materials contain high to extremely high Ni and Cr concentrations in soils. Soil parent material is a factor mostly controlling trace element concentrations in soils.

2.3.6 The Middle East

Banin and his colleagues (Banin et al., 1997a) have studied the distribution of trace elements in 45 representative Israeli soils with 0.5–68% of CaCO$_3$ and 7.0–8.25 pH (Table 2.4). The total Cu concentration ranges from 3.21 mg/kg in a loessial soil to 62.05 mg/kg in a desert stony soil. In general, alluvial soils, colluvial-alluvial soils, and Terra-Rossa soils contain higher Cu than soils in Mediterranean brown and rendzina. The EDTA-extractable Cu content varies from 2–7.5 mg/kg in rendzinas, alluvial soils, and saline soils.

Table 2.4. Averages and ranges of concentrations of selected trace elements in 45 arid and semi-arid soils of Israel[a]

	Cd	Ni	Cr	Cu	Zn	Mn[b]
			mg/kg			
Average	0.36	31.0	41.0	18.2	56.1	367
Standard Deviation	0.33	18.5	32.2	11.7	28.7	260
Maximum	2.13	98.7	182	73.9	144	810
Minmum	0.07	3.93	7.51	4.41	7.94	53

[a]From Banin et al., 1997a (unpublished).
[b]From Han, 1998

Cadmium concentrations in Israeli soils are in the range of 0.07–2.13 and average 0.37 ± 0.34 mg/kg, while cadmium content in Terra-Rossa and Desert Stony soils is higher than that in Rendzina.

The total Cr concentration in the soils of Israel varies from 7.51–181.8 mg/kg with an average of 41.0 ± 32.2 mg/kg. Terra-Rossa, Mediterranean brown and alluvial soils have higher Cr concentrations than soils in rendzina, brown-red sandy and loessial soils.

Soil Ni is in the range of 3.93–98.7 mg/kg and averages 31.0 ± 18.5 mg/kg. Colluvial-alluvial soils, alluvial soils, Terra-Rossa and Mediterranean brown soils have higher Ni than brown-red sandy and loessial soils.

The total Zn concentration in 45 Israeli soils varies from 7.94–144.2 mg/kg with an average of 56.1 ± 28.7 mg/kg. Terra-Rossa soils contain higher Zn than rendzina soils. Total Zn concentrations in Mediterranean red soils vary from 200–215 mg/kg, while brown isohumic soils on calcareous sandstone contain 48 mg/kg total Zn. However, soils on alluvians from aeolian deposits have 82–90 mg/kg, and saline alkali soils contain 100–200 mg/kg of total Zn. EDTA-extractable Zn varies from 1.9–13 mg/kg, representing 1.7–9.6% of the total Zn in Mediterranean red soils, reddish-brown isohumic soils and rendinas soils (Aubert and Pinta, 1977).

The total Mn concentration in Israeli soils varies from 52.6–810 mg/kg and averages 367 ± 259.8 mg/kg (Han, 1998). The average total Co content is 9.45 ± 4.88 mg/ kg, varying from 3.35 mg/kg in a very sandy soil to 15.9 mg/kg in alluvial soil samples (Han et al., 2002b). In soils on alluvions and derived from aeolian deposits, the Co concentration is 8 mg/kg, and soils from hard limestone contain 10 mg/kg Co. The B concentration in alluvial soils is in the range of 25–85 mg/kg, while brown isohumic soils formed on alluvions contain 25–40 mg/kg B. Mediterranean red soils on limestone are rich in B (190 mg/kg). In soils on calcareous sandstone, alluvions, and aeolian deposits, total Mo content ranges from 4.6–6 mg/kg (Aubert and Pinta, 1977).

In soils from Syria with 19.2–43% of $CaCO_3$, DTPA-extractable Zn, Cu and Mn are in the range of 0.4–0.90, 0.22–2.0 and 2.3–12.0 mg/kg, respectively (Ryan et al., 1996). In the surface soils of agricultural, industrial and urban regions of Isfahan, central Iran, the average total Cd is 1.74 mg/kg. Urban soils contain higher Cd content (2.07 mg/kg) than agricultural soils (1.79 mg/kg) and uncultivated soils (1.61 mg/kg) (Amini et al., 2005).

2.4 FACTORS AFFECTING CONTENTS OF TRACE ELEMENTS IN ARID AND SEMI-ARID SOILS

2.4.1 Climate

In general, soils in arid and semi-arid zones contain higher concentrations of Cr, B, Co, Cu, Se and Mo than temperate and tropic regions. In China, total Mo concentrations in arid and semi-arid regions (2–5 mg/kg) are higher than those in soils of temperate and cold temperate regions (Liu, 1996). The total B concentration decreases from South to North and from East to West, which is consistent with changes of climate and aridity. Soils in the North and West parts of China contain higher total B than soils in the South and East regions (Liu, 1996). Therefore, most B deficiency occurs in the acidic soils of South China.

Moreover, salts in arid soils increase concentrations of boron. In Israeli soils, total B content in saline alluvial soil is in the range of 150–170 mg/kg, four to six times higher than that in alluvial soils (25–40 mg/kg) (Aubert and Pinta, 1977). The soluble boron contents tend to increase with decreases in rainfall. Bioavailable Mo represents an average 2–20% of total Mo. The percentages of bioavailable Mo are higher in arid and semi-arid zones than in humid zones. Bioavailable Mo is 35.8% of the total Mo in soils of Israel and increases to 50–60% in saline alkali soils in India (Aubert and Pinta, 1977). Total Se concentrations in the western United States are higher than the rest of the regions of the U.S. Selenium, which often accumulates in ponds and reservoirs of arid and semi-arid regions, will be discussed in detail in Chapter 8.

2.4.2 Parent Materials

The total contents of most trace elements in soils primarily depend upon their concentrations in the parent materials as well as the soil formation processes imposed. Parent materials strongly control concentrations of trace elements (Cd, Hg, As, Pb, Cr, and other elements) in arid and semi-arid soils since weathering and leaching processes in soils and the soil profile are very weak. Parent material type primarily determines concentrations of trace elements in soils. In North and Northeast China, the soils derived from basalt, volcanic materials and shale contain higher Zn than soils from granite, quartz, loess, and sand materials (Liu, 1996). Soils derived from limestone are rich in Zn. Soils on basalt are higher in Ni and Co content than soils on granite. Selenium concentrations in soils on sedimentary rocks, including shale and limestone, are much higher than soils on other acidic igneous rocks, including granite, quartz, and loess materials. In the U.S., soils derived from shale and granite in the western regions contain higher Mo than soils in the eastern regions formed on marine and glacial materials. Serpentine soils contain higher Cr, Ni and Co content than other soils.

2.4.3 Soil Processes and Properties

Soil texture, clay and organic matter contents are the most important soil properties, affecting the total amounts of trace elements in soils. In general, soils with clayey texture contain higher Cr, Co, B, Mn and Mo than light-textured soils. Soils high in organic matter content have higher Co and Zn concentrations. Zinc contents in soils of North China increase with clay content. There is a significant correlation between total Zn contents and clay particles (< 0.01 mm) (Liu, 1996). Total Se increases with clay content. In soil profiles, trace elements (Cd, As, Hg, Cr, Mo and Pb) are accumulated in B horizons as clay and humic substances accumulate. Both total and bioavailable Mn are accumulated in the humiferous upper horizons. In 39 Israeli arid and semi-arid soils, total Cu significantly increased with soil pH (R: 0.50, N: 39), as depicted in Fig. 2.6.

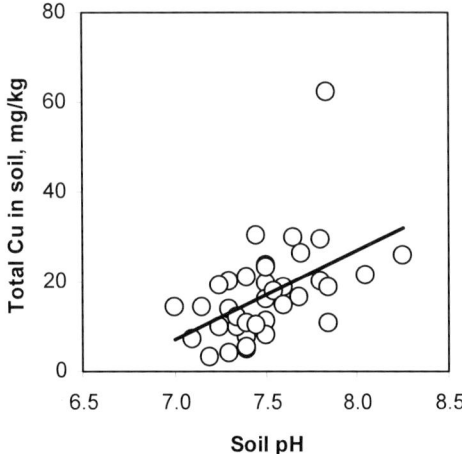

Figure 2.6. Effects of soil pH on total Cu concentrations in Israeli arid and semi-arid soils (R: 0.50, N: 39)

2.4.4 Agricultural Management and Industrial Activities

Agricultural management and industrial activities strongly affect the distribution of trace elements in soils. Long-term applications of fertilizers, herbicides, and pesticides containing metal and metalloids increase Cd, As and Pb accumulation in agricultural soils. It has been shown that long-term application of phosphate fertilizer increased Cd contents in agricultural soils, as most phosphate fertilizers contain trace amounts of Cd (Mortvedt, 1996). Most earlier herbicides and pesticides contained As and Pb. Arsenic and Pb have been found to accumulate in orchard soils and cotton fields after long-term application of these chemicals. Application of sewage sludge and irrigation with reclaimed sewage increases the concentrations of most trace elements in arid and semi-arid soils. Mining and industrial activities as well as traffic increase trace elements, such as Cd, Pb, Cu, Zn and As in soils, and this will be discussed in detail in Chapter 8.

Chapter 3

SOLUTION CHEMISTRY OF TRACE ELEMENTS IN ARID ZONE SOILS

Soil solution is the aqueous phase of soil. It is in the pore space of soils and includes soil water and soluble constituents, such as dissolved inorganic ions and dissolved organic solutes. Soil solution accommodates and nourishes many surface and solution reactions and soil processes, such as soil formation and decomposition of organic matter. Soil solution provides the source and a channel for movement and transport of nutrients and trace elements and regulates their bioavailability in soils to plants. Trace element uptake by organisms and transport in natural systems typically occurs through the solution phase (Traina and Laperche, 1999).

Soil solution analysis can be used for prediction of plant response (bioavailability) to nutrients and trace elements in the soil. An understanding of soil solution composition can also be helpful in estimating the speciation and forms of trace elements and nutrients which may be transported into surface and ground water (Wolt, 1994). Trace elements in soil solution equilibrate with various solid phase components. Therefore, soil solution is used to predict binding of nutrients and trace elements to solid phase components. Soil solution in either a conceptual or operational context is determined by the fraction of soil water to which soil solution composition is related (Wolt, 1994). Therefore, soil solution composition and chemistry must be interpreted with a soil solution displacement method. The details on these discussions can be found in the literature (Wolt, 1994). As stated in Chapter 1, arid and semi-arid zones have less than a 0.5 aridity index (defined as a ratio of precipitation over potential evapotranspiration). Surface soils in these areas contain very low water content, resulting in a challenge when studying solution chemistry of trace elements in soils. This chapter will discuss soil solution composition, concentrations, and speciation of trace elements in arid and semi-arid soils. The solubility of trace elements, governing factors, as well as major physicochemical and biological processes in arid and semi-arid soils will also be discussed.

3.1 SOIL SOLUTION COMPOSITION OF ARID ZONES

Soil solution composition reflects the intensity and distribution of trace elements in the soil aqueous phase and represents the integration of multiple physical, chemical, and biological processes occurring concurrently within the soil (Wolt, 1994). However, composition reflects the soil moisture content, sample handling, and the displacement technique. Major cations and anions in soil solution include Ca^{2+}, Mg^{2+}, K^+, Na^+, NO_3^-, Cl^-, SO_4^{2-}, HCO_3^-, CO_3^{2-}, and $H_2PO_4^-$, of which Ca^{2+} and Mg^{2+} are the dominant soil solution cations and HCO_3^- and CO_3^{2-} are the major anionic species (Hirsh and Banin, 1990). In a typical arid soil with pH 8.0 (California, U.S.), Ca and Mg concentrations in soil saturation extracts range from $1.0–2.0 \times 10^{-3}$ M. Potassium is 0.2×10^{-3} M, Na is in the range of $8.5–10.7 \times 10^{-3}$ M, while Cl, NO_3, SO_4 and HCO_3 range from $0.9–7.1 \times 10^{-3}$ M (Villarroel, et al., 1993). Concentrations of these major elements decrease in the following order: Na > Ca > Mg > K and Cl > SO_4 > NO_3 > HCO_3. In salt soils, Na is the major cation, while Cl is the major anion. Chloride in soil solution ranges from a few milligrams per liter to several hundred milligrams per liter. The major ion composition of dispersed soil solution and saturation extracts for selected soils from California are listed in Table 3.1.

In addition, dissolved organic carbon (DOC) is also an important soil solution solute affecting speciation and bioavailability of many trace elements in soil solution. Many trace elements and heavy metals complex with dissolved organic carbon. This is especially important in arid and semi-arid environments since high soil pH increases the solubility of organic molecules and accordingly increases concentrations of dissolved organic carbon in soil solution.

Soil solution composition is affected by soil moisture, seasonal changes, temperature, and agricultural practices, including crop systems, fertilization, land application of sewage sludge and irrigation. Fertilization with a specific nutrient source may increase nutrients in the soil solution with less effect on the concentrations of other components. Levels of phosphate, potassium and nitrate or ammonia ions in soil solution of agricultural soils frequently fluctuate with fertilization and other agronomic practices. Long-term no-tillage management decreases concentrations of soil solution compositions due to build-up of other materials, such as organic matter. Soils with less moisture contain high concentrations of soil solution components. High temperatures in the summer seasons may increase soil solution components due to high weathering rates, but high biological activities usually deplete soil solution concentrations.

Table 3.1. The major ion composition of dispersed soil solutions and saturation extracts from selected soils in California, U.S.

Soil	Taxonomic Descriptor	pH	EC ds m^{-1}	Ca	Mg	K	Na	Cl	NO$_3$	HCO$_3$	SO$_4$
						mmol L^{-1}					
			Saturation extracts, California USA (Villarroel et al., 1993)								
Ramona	Typic Haploxeralf[a]	8.0	1.5	2.0	1.0	0.2	10.7	6.3	2.0	1.6	3.2
	Typic Haploxeralf[b]	7.0	1.4	2.2	1.2	0.2	8.5	7.1	1.4	0.9	2.5
			Pressure Plate Extraction at 10 kPa Moisture, California USA (Eaton et al., 1960)								
Hanford	Typic Xerorthents	7.5	4.96	21.1	1.8	0.8	3.0	4.6	32	3.5	5.9
Escondodo	Typic Xerochrepts	7.4	0.7	1.0	0.5	0.2	2.8	2.9	1.2	1.0	0.7
Tujunga	Typic Xeropsamments	7.0	8.9	41.6	8.3	2.6	7.4	5.6	65	4.2	18.9
Greenfield	Typic Haploxeralfs	7.7	1.96	6.4	1.89	1.1	1.2	1.0	13.2	1.9	1.4
Yolo	Typic Xerorthents	6.9	2.74	8.7	1.95	3.7	1.0	1.9	16.9	2.3	3.0
Corona	Pachic Agrixerolls	8.1	3.36	5.85	1.85	7.3	9.2	3.2	12.4	6.1	3.8
Chino	Aquic Haploxerolls	7.5	1.68	5.55	1.1	0.3	2.3	2.3	8.8	3.8	1.4
Traver	Natric Haploxeralfs	7.4	143.3	118.8	105	5.6	1777	2205	25	32	39.4

[a]Control soil without sludge application.
[b]Sludge-treated soil.

3.2 CONCENTRATIONS OF TRACE ELEMENTS IN SOIL SOLUTION OF ARID ZONES

Very limited data have been available on concentrations of trace elements in soil solution of arid and semi-arid zones. Several factors contribute to this problem of limited data. First, most trace elements have a low natural abundance in uncontaminated soils. Second, trace elements have limited solubility in soils. Third, high soil pH and high carbonates in the arid regions further lower solubility of most trace elements, such as Cu, Cr(III), Ni, Pb, Cd, Zn, Co, Mn and many other cations. Fourth, most trace elements have high degrees of complexation and chelation with various inorganic and organic ligands in soils. Therefore, the concentrations of most trace elements in soil solution of arid and semi-arid zones are at parts per billion (ppb) levels and frequently below the detection limits of most current instruments (e.g., atomic absorption spectroscopy (AAS) and inductively coupled plasma atomic emission spectroscopy (ICP-AES)).

Among various trace elements in noncontaminated arid soils, Cd concentrations in soil solution are usually lower than Cr and Ni, while Cu and Zn have higher solution concentrations due to their natural abundance in soils. The Zn activities in 10 arid soils from Colorado are in the range of $10^{-7.9}$ to $10^{-10.9}$ M and are inversely related with pH ($r = 0.94$) (Ma and Lindsay, 1990). The concentrations of trace elements in the 20 Colorado soils with pH 6.6–7.9 are as follows: Cu solution ranges from 0.055–0.62 µM/L, with an average of 0.17 µM/L, while Zn concentrations are less than 0.03 µM/L (Hodgson et al., 1966). Total Cu in soil solution increases with the amount of dissolved organic matter. Jeffery and Uren (1983) reported that total Cu linearly increased with DOC in limed soils due to the fact that Cu is mostly present in complexed form, which is strongly related to dissolved organic matter. Sanders (1983) reported that Mn, Zn and Co concentrations in soil solution of limed soils decreased as the pH increased.

In soil solution below the sludge-soil layers of California soils, Cd is in the range of 0.003–0.005, Cu is in the range of 1.1–5.35, Ni is 0.17–1.70 and Zn is 0.92–2.29 µM/L (Emmerich et al., 1982) (Table 3.2). Bradford et al. (1975) reported that various trace elements in saturated extracts from 68 soil samples representing 30 soil series in California were in the order of: B (282 µM/L, as an average) > Mo (7.6 µM/L) > Zn (1.07 µM/L) > Co (1.01 µM/L) > Cu (0.63 µM/L) > Ni (0.34 µM/L) > Pb (0.24 µM/L) > Cr (0.19 µM/L) > Cd (<0.09 µM/L). In saturated extracts, Cd was in the range of < 0.09, Co < 0.17–2.38, Cu < 0.16–3.15, Cr < 0.19–0.33, Ni < 0.17–1.53, Pb < 0.15–1.45 and Zn < 0.15–6.11 µM/L.

Solution Chemistry

Table 3.2. Concentrations of trace elements in soil solutions of the California soils that received sludge applications.

Sample		Taxonomic Classification	Cd	Cu	Ni	Zn
				mM/L (dry soil basis)		
Holland	Sludge-soil layer	Ultic Haploxeralf	0.011-0.014	2.83-3.15	1.53-2.22	0.76-1.22
	5-10 cm below sludge-soil layer		0.004	1.1	0.17-0.34	0.92
Romona	Sludge-soil layer	Typic Hapoloxeralf	0.042-0.038	21.2-23.6	7.50-7.84	3.06-5.51
	5-10 cm below sludge-soil layer		0.003-0.004	1.42-1.89	1.02-1.70	1.38-1.53
Helendale	Sludge-soil layer	Typic Haplargid	0.042-0.10	17.8-25.0	5.79-10.6	3.52-6.27
	5-10 cm below sludge-soil layer		0	4.40-5.35	0.85-1.36	1.38-2.29

[a]From Emmerich et al., 1982.

In four soils from South Australia with pH 7.59–8.99, Zn and Cu concentrations in soil solution range from 0.009–0.218 and 0.058–4.425 µM/kg, respectively (Fotovat et al., 1997). The total quantities of Zn and Cu extracted increase with the solution to soil ratio and show considerable variation among the soils. Zinc concentration decreases with increasing pH, but for Cu no such trend is obvious. The Cd concentration in soil solution from South Australia increases ($p < 0.05$) more consistently with increasing concentrations of SO_4 (compared to NO_3) in soil solution (McLaughlin et al., 1998b). In another eight agricultural soils (pH 7.15–8.52) from eastern and southern Australia, soluble Zn and Cu in soil pore water are 14.7–804 and 61–370 µM/L, respectively (Nolan et al., 2003).

In noncontaminated Israeli arid and semi-arid soils with pH 7.4–8.4 and 0.25–12% $CaCO_3$, Cd, Pb and Cu in soil solution (saturated paste extracts) range from 0.142–0.218, 0.801–1.230 and 0.69–3.6 µM/kg, respectively (Hardiman et al., 1984a, b). Hirsh and Banin (1990) reported slightly higher Cd concentrations in three native Israeli arid soils in the range of 0.014–0.032 µM/L. In two other Israeli noncontaminated soils with 0.5–23% $CaCO_3$, Cd solution concentrations range from 0.007–0.013 µM/L, Cr concentration is 0.12 µM/L, while Ni, Cu and Zn concentrations are in the ranges of 0.34–0.51, 0.52–1.6 and 0.46–1.72 µM/L, respectively (Table 3.3) (Han, 1998).

Badawy et al. (2002) reported that in near neutral and alkaline soils, representative alluvial, desertic and calcareous soils of Egypt, lead activity ranged from $10^{-6.73}$ to $10^{-4.83}$ M, and was negatively correlated with soil and soil solution pH. It could be predicted in soil solution from the equation: log (Pb^{2+}) = 9.9–2pH. In German noncontaminated soil with 2.3% $CaCO_3$ and soil pH 8.5, Zn, Cu and Cd are 1.87, 0.66, and 0.20 µM/L in soil solution, respectively (Helal et al., 1996).

Sewage sludge application, however, significantly increases soil solution concentrations of trace elements. In a sewage sludge-amended arid soil with pH 7.2–7.7 from California, Cd solution concentrations range from 0.07–80 µM/L and linearly increase with increases in total Cd inputs as sewage sludge (Mahler et al., 1980). Emmerich et al. (1982) reported that the application of sewage sludge increased total concentrations of Cd, Cu, Ni and Zn in soil solution of arid soils (Table 3.2). Cadmium in soil solution of sewage sludge amended soils is in the range of 0.011–0.10, Cu is in the range of 2.83–25, Ni is 1.53–10.6 and Zn is 0.76–6.27 µM/L (Table 3.2). The free metal ions in sewage sludge amended soil solution are 1.4–21.6 $\times 10^{-9}$ M for Cd, 2.0–11.9 $\times 10^{-8}$ M for Cu, 1.8–24.1 $\times 10^{-7}$ M for Ni and 2.5–17.4 $\times 10^{-7}$ M for Zn. Total soluble Cu, Zn and Ni concentrations in soil solution from Washington, U.S. of an arid soil amended with sludge are in the ranges of $10^{-4.12}$ to $10^{-5.06}$, $10^{-5.37}$ to $10^{-6.77}$ and $10^{-4.71}$ to $10^{-5.17}$ M, respectively (Dudley et al., 1987).

Similarly, long-term irrigation with sewage and reclaimed sewage also increases trace element concentrations in soil solution of arid and semi-arid zones. In Israeli arid soils with pH 7.4–8.4 and 0.25–12% $CaCO_3$, long-term irrigation (25 years) with reclaimed sewage effluents increases trace element concentrations in soil solution. Cadmium, Pb and Cu in soil solution (saturated paste extracts) range from 1.5–2.4, 2.0–3.1 and 11.4–23.4 µM/kg (ppb), respectively (Hardiman et al., 1984a, b). Both Cd and Cu concentrations in soil solution of soils with long-term irrigation are 10 times higher than those in control soils, and Pb is 2–3 times higher than those in control soils. Candelaria and Chang (1997) reported that Cd concentrations in soil solution of California soils amended with Cd nitrate were higher than those with sewage sludge.

Concentrations of four trace elements (Cu, Zn, Ni and Cr(III)) in saturated paste extracts of two contaminated Israeli soils are presented in Table 3.3. Clearly, metal concentrations in soil solution are dependent upon metal species, loading levels, soil properties, and the reaction time. In general, metal concentrations in metal salt-spiked soil solution are, initially, higher than those in the non-amended soil solution. They increased linearly with the loading levels in both soils. The solution concentrations of trace metals in the amended soils sharply decreased with time. In the non-amended soils, Cr, Cu, Ni and Zn concentrations are in the range of 10^{-7} to 10^{-8} M. In the metal salt-spiked soils, the initial solution concentrations are in the range of 10^{-5} to 10^{-6} M. After 30 days of incubation, metal solution concentrations decreased to 10^{-6} to 10^{-7} M. Soil solution concentrations of heavy metals continued to decrease during the rest of the incubation period (Table 3.3). At similar metal loading levels, concentrations of most heavy metals in the loessial soil were lower than those in the sandy soil. This indicates that strong immobilization of added soluble metals occurs in the loessial soil due to its richness in many active components. In general, after one day the concentrations of four heavy metals in soil solution were in the order: Cr > Ni > Cu > Zn, and after one month, Ni > Cu > Zn and Cr. This implies that Cr is more strongly immobilized by solid-phase in the two soils during the first month. Hirsh and Banin (1990) reported that even 30–45 minutes was sufficient to reduce the Cd concentration to its native value in soil solution of three Israeli arid soils receiving 0.1 mg/kg Cd. They found that Cd concentrations in soil solution were between 0.012–0.029 µM. The data obtained in Israeli arid soils contaminated by metal salts are similar to those found in the equilibrium solution of California soils (Bradford et al., 1975). Although Zn in soil solution in non-amended soils and in metal salt-spiked soils was less than in Californian soil (Bradford et al., 1975) after 6–18 days, it was in the range reported by Barber (1984).

Table 3.3. Concentrations of trace elements in soil solutions extracted by saturated paste from two metal salt-spiked Israeli soils incubated at the saturated paste regime

Metal	Time Days	Sandy Treatment ($\mu M/L$)			Time Days	Loessial Treatment ($\mu M/L$)		
		CK	0.5T[a]	3T		CK	0.5T	3T
Cr	1	0.12	4.61	20.3		NA[b]		
	18	0.16	0.08	0.76				
	30	0.06	0.04	0.43				
	746	0.09	0.009	0.03				
Cu	1	1.6	2.5	2.0	1	0.52	1.3	3.4
	18	1.3	0.72	0.68	6	0.25	0.55	3.2
	30	0.24	0.35	0.62	30	0.001	0.14	0.91
	746	0.013	0.06	0.07	816	NA	0.01	0.03
Ni	1	0.51	6.3	10.1	1	0.34	1.0	7.1
	18	0.33	1.4	2.5	6	0.58	1.3	4.7
	30	0.36	2.1	1.7	30	0.12	0.95	4.6
	746	0.25	0.54	0.57	816	0.00	0.22	0.3
Zn	1	1.72	1.23	3.80	1	0.46	0.16	0.76
	18	0.009	0.03	0.11	6	0.16	0.10	0.15
	30	0.11	0.36	0.05	30	0.06	0.19	0.34
	746	NA[1]	NA	NA	816	NA	NA	103

[a]T is the concentration of trace elements in native soils.
[b]NA: not available

In eight contaminated soils with pH 7.05–8.20 from eastern and southern Australia, soluble Zn, Cu and Cd concentrations in soil pore water are 0.19–422, 36–4840, and 7.2–4260 μM, respectively (Nolan et al., 2003).

3.3 SOLUTION SPECIATION OF TRACE ELEMENTS IN ARID ZONES

Since carbonate and high pH are unique characteristics of arid and semi-arid soils, we will first examine solution speciation and the equilibrium reactions of the CO_2-H_2O system. We will then examine the solution speciation of Ca and Mg, Zn, Cu, Ni, Cd, Pb, Cr(III) and Cr(IV), Hg, and Se.

Solution Chemistry

3.3.1 The CO_2-H_2O System

Carbon dioxide-water systems play an important role in controlling the pH of alkaline and calcareous soils as well as adjusting solubility of most trace elements and their compounds. Carbon dioxide dissolves in water to form dissolved CO_2 and dissociated carbonic acid, $H_2CO_3^0$:

$$CO_2(g) + H_2O \rightleftharpoons H_2CO_3^0 \qquad \log K = -1.46 \qquad (3.1)$$
$$H_2CO_3^0 = 10^{-1.46} \, CO_2(g) \qquad (3.2)$$

The partial pressure of CO_2 in the soil air controls the concentration of both dissolved CO_2 and undissociated carbonic acid. At 0.003 atm of CO_2 (g) as a reference level for soils, $[H_2CO_3^0]$ is about 1.04×10^{-4} M (Lindsay, 1979). At a normal atmospheric level of 0.0003 atm CO_2 (g), $[H_2CO_3^0]$ is approximately 1.04×10^{-5} M. In most soils, CO_2 (g) is higher than in the atmosphere. CO_2 is released from soil and plant root respiration. In flooded soils, CO_2 (g) partial pressure increases to 0.01–0.3 atm, about 1000-fold higher than normal upland soils due to strong microbiological activity (Lindsay, 1979).

The dissolved CO_2(g) dissociates to give bicarbonate and carbonate:

$$H_2CO_3 \rightleftharpoons H^+ + HCO_3^- \qquad \log K = -6.36 \qquad (3.3)$$
$$HCO_3^- \rightleftharpoons H^+ + CO_3^{2-} \qquad \log K = -10.33 \qquad (3.4)$$

At pH 6.36, the molar ratio of $(HCO_3^-)/(H_2CO_3^0)$ is unity. At pH 10.33, the molar ratio of CO_3^{2-}/HCO_3^- is unity. For each pH unit increase, the two ratios increase 10-fold.

Combining equations 3.1, 3.3 and 3.4, the following two expressions are given:

$$CO_2(g) + H_2O \rightleftharpoons H^+ + HCO_3^- \qquad \log k = -7.82 \qquad (3.5)$$

Thus
$$\text{Log } HCO_3^- = -7.82 + pH + \log CO_2(g) \qquad (3.6)$$

$$CO_2(g) + H_2O \rightleftharpoons 2H^+ + CO_3^{2-} \qquad \log k = -18.15 \qquad (3.7)$$

Thus
$$\text{Log } CO_3^{2-} = -18.15 + 2pH + \log CO_2(g) \qquad (3.8)$$

Higher CO_2 concentrations in soil solution result in higher dissolved H_2CO_3, HCO_3^- and CO_3^{2-}. For each pH unit increase, $[HCO_3^-]$ and $[CO_3^{2-}]$ increase 10-fold and 100-fold, respectively. For the ranges of soil pH in arid and semi-arid zones, the bicarbonate HCO_3^- form is the major solution speciation. Only when pH is beyond 10.33 does the carbonate ion $[CO_3^{2-}]$ become the dominant solution species.

3.3.2 Solution Speciation of Ca and Mg

Calcium in soil solution can occur as free Ca ion and various inorganic and organic complexes, such as Ca^{2+}, $CaCl^+$, $CaCl_2^0$, $CaHCO_3^+$, $CaCO_3^0$, $CaNO_3^+$, $Ca(NO_3)_2^0$, $CaOH^+$, $Ca(OH)_2^0$, $CaH_2PO_4^+$, $CaHPO_4^0$, $CaPO_4^-$, $CaSO_4^0$ and Ca-DOC. The complexing reactions and equilibrium constants are listed in Table 3.4. In the presence of calcite, a major calcium carbonate in arid and semi-arid soils, the activity of $CaCO_3^0$ (neutral ion pair) is constant.

$$CaCO_3 \text{ (calcite)} \rightleftharpoons CaCO_3^0 \qquad \log K \ -5.27 \qquad (3.9)$$

In arid and semi-arid soils with a pH range of 6–9, free calcium is the major Ca speciation form in soil solution. When pH > 9.2–9.5, $CaPO_4^-$ becomes a major calcium species in soil solution of neutral and calcareous soils, especially when the activity of $H_2PO_4^-$ is > 10^{-5} M (Lindsay, 1979), such as after phosphate fertilizers are used. Lindsay (1979) further pointed out that $CaSO_4^0$ contributes significantly to the total calcium in solution when SO_4^{2-} is > 10^{-4} M.

Magnesium can occur in forms of Mg^{2+}, $MgSO_4^0$, $MgHPO_4^0$, $MgHCl_2^+$, $MgOH^+$, $MgCO_3^0$, $Mg(NO_3)_2^0$ and $MgCl_2^0$ in soil solution (Table 3.4). Magnesium is mainly present as Mg^{2+} in soil solution in soils with pH < 9. In arid alkaline soils with pH > 9 and in submerged soils which result in high CO_2, Mg in soil solution is predominant as $MgCO_3^0$. However, $MgSO_4$ and $MgHPO_4$ ion pairs contribute significantly to total Mg in soil solution when SO_4 increases above 10^{-4} M (Lindsay, 1979). As pH and $H_2PO_4^-$ activity increase, the $MgHPO_4^0$ ion pair becomes important in soil solution. In arid soils with dolomite and calcite, the $MgCO_3^0$ ion pair is independent of pH and CO_2 (Lindsay, 1979). All other complexes such as $Mg(NO_3)_2^0$, $MgCl_2^0$, $Mg(OH)^+$ and $Mg(OH)_2^0$ are insignificant in normal soil solution.

3.3.3 Solution Speciation of Zn

Zinc in soil solution can occur in the forms of Zn^{2+}, $Zn(OH)^+$, $Zn(OH)_2^0$, $Zn(OH)_3^-$, $Zn(OH)_4^{2-}$, $ZnCl^+$, $ZnCl_2^0$, $ZnCl_3^-$, $ZnCl_4^{2-}$, $ZnH_2PO_4^+$,

$ZnHPO_4^0$, $ZnNO_3^+$, $Zn(NO_3)_2^0$, $ZnSO_4^0$ and Zn-DOC. The possible reactions and equilibrium constants of Zn species in soil solution are listed in Table 3.5. In soil solution with pH < 7.7, free Zn^{2+} is the predominant inorganic Zn species. In 10 Colorado arid soils with pH 6.70–8.14 and 0.4–11.6% $CaCO_3$, the Zn activities (Zn^{2+}) are highly pH-dependent (Ma and Lindsay, 1990). The Zn activities decrease linearly with soil pH ($R^2 = 0.94$) (Fig. 3.1) (Ma and Lindsay, 1990). With an increase in pH, $Zn(OH)^+$ becomes a predominant inorganic species in soil solution (Table 3.6). When pH is > 9.11, a neutral species $Zn(OH)_2^0$ is the major inorganic species of soil solution in arid and semi-arid zones. Other zinc complexes with chloride, phosphate, nitrate and sulfate are not significant species in arid soil solution compared to the prevalence of zinc hydrolysis species. However, $ZnHPO_4^0$ becomes important when $H_2PO_4^- = 10^{-4}$ M; then the activity of both $ZnHPO_4^0$ and $Zn(OH)^+$ are equal (Lindsay, 1979).

Table 3.4. Solution complexes of calcium and magnesium at $25°C$[a]

Solution Complexes	log K°
$Ca^{2+} + Cl^- \rightleftharpoons CaCl^+$	-1
$Ca^{2+} + 2Cl^- \rightleftharpoons CaCl_2^0$	0
$Ca^{2+} + CO_2(g) + H_2O \rightleftharpoons CaHCO_3^+ + H^+$	-6.7
$Ca^{2+} + CO_2(g) + H_2O \rightleftharpoons CaCO_3^0 + 2H^+$	-15.01
$Ca^{2+} + NO_3^- \rightleftharpoons CaNO_3^+$	-4.8
$Ca^{2+} + 2NO_3^- \rightleftharpoons Ca(NO_3)_2^0$	-4.5
$Ca^{2+} + H_2O \rightleftharpoons CaOH^+ + H^+$	-12.7
$Ca^{2+} + 2H_2O \rightleftharpoons Ca(OH)_2^0 + 2H^+$	-27.99
$Ca^{2+} + H_2PO_4^- \rightleftharpoons CaH_2PO_4^+$	1.4
$Ca^{2+} + H_2PO_4^- \rightleftharpoons CaHPO_4^0 + H^+$	-4.46
$Ca^{2+} + H_2PO_4^- \rightleftharpoons CaPO_4^- + 2H^+$	-13.09
$Ca^{2+} + SO_4^- \rightleftharpoons CaSO_4^0$	2.31
$Mg^{2+} + 2Cl^- \rightleftharpoons MgCl_2^0$	-0.03
$Mg^{2+} + CO_2(g) + H_2O \rightleftharpoons MgHCO_3^+ + H^+$	-6.76
$Mg^{2+} + CO_2(g) + H_2O \rightleftharpoons MgCO_3^0 + 2H^+$	-14.92
$Mg^{2+} + 2NO_3^- \rightleftharpoons Mg(NO_3)_2^0$	-0.01
$Mg^{2+} + H_2O \rightleftharpoons MgOH^+ + H^+$	-11.45
$Mg^{2+} + 2H_2O \rightleftharpoons Mg(OH)_2^0 + 2H^+$	-27.99
$Mg^{2+} + H_2PO_4^- \rightleftharpoons MgHPO_4^0 + H^+$	-4.29
$Mg^{2+} + SO_4^- \rightleftharpoons MgSO_4^0$	2.23

[a]From Lindsay W.L., 1979.

Table 3.5. Solution complexes of selected trace elements at 25°C[a]

Element	Equlibrium Reaction	log K⁰
Zn	$Zn^{2+} + H_2O \rightleftharpoons ZnOH^+ + H^+$	-7.69
	$Zn^{2+} + 2H_2O \rightleftharpoons Zn(OH)_2^0 + 2H^+$	-16.80
	$Zn^{2+} + 3H_2O \rightleftharpoons Zn(OH)_3^- + 3H^+$	-27.68
	$Zn^{2+} + 4H_2O \rightleftharpoons Zn(OH)_4^{2-} + 4H^+$	-38.29
	$Zn^{2+} + Cl^- \rightleftharpoons ZnCl^+$	0.43
	$Zn^{2+} + 2Cl^- \rightleftharpoons ZnCl_2^0$	0.00
	$Zn^{2+} + 3Cl^- \rightleftharpoons ZnCl_3^-$	0.50
	$Zn^{2+} + 4Cl^- \rightleftharpoons ZnCl_4^{2-}$	0.20
	$Zn^{2+} + H_2PO_4^- \rightleftharpoons ZnH_2PO_4^+$	1.60
	$Zn^{2+} + H_2PO_4^- \rightleftharpoons ZnHPO_4^0 + H^+$	-3.90
	$Zn^{2+} + NO_3^- \rightleftharpoons ZnNO_3^+$	0.40
	$Zn^{2+} + 2NO_3^- \rightleftharpoons Zn(NO_3)_2^0$	-0.30
	$Zn^{2+} + SO_4^{2-} \rightleftharpoons ZnSO_4^0$	2.33
Cu	$Cu^{2+} + H_2O \rightleftharpoons CuOH^+ + H^+$	-7.70
	$Cu^{2+} + 2H_2O \rightleftharpoons Cu(OH)_2^0 + 2H^+$	-13.78
	$Cu^{2+} + 3H_2O \rightleftharpoons Cu(OH)_3^- + 3H^+$	-26.75
	$Cu^{2+} + 4H_2O \rightleftharpoons Cu(OH)_4^{2-} + 4H^+$	-39.59
	$2Cu^{2+} + 2H_2O \rightleftharpoons Cu_2(OH)_2^{2+} + 2H^+$	-10.68
	$Cu^{2+} + Cl^- \rightleftharpoons CuCl^+$	0.40
	$Cu^{2+} + 2Cl^- \rightleftharpoons CuCl_2^0$	-0.12
	$Cu^{2+} + 3Cl^- \rightleftharpoons CuCl_3^-$	-1.57
	$Cu^{2+} + CO_{2(g)} + H_2O \rightleftharpoons CuHCO_3^+ + H^+$	-5.73
	$Cu^{2+} + CO_{2(g)} + H_2O \rightleftharpoons CuCO_3^0 + 2H^+$	-11.43
	$Cu^{2+} + 2CO_{2(g)} + 2H_2O \rightleftharpoons Cu(CO_3)_2^{2-} + 4H^+$	-26.48
	$Cu^{2+} + H_2PO_4^- \rightleftharpoons CuH_2PO_4^+$	1.59
	$Cu^{2+} + H_2PO_4^- \rightleftharpoons CuHPO_4^0 + H^+$	-4.00
	$Cu^{2+} + 2H^+ + P_2O_7^{4-} \rightleftharpoons CuH_2P_2O_7^0$	18.67
	$Cu^{2+} + H^+ + P_2O_7^{4-} \rightleftharpoons CuHP_2O_7^-$	14.78
	$Cu^{2+} + P_2O_7^{4-} \rightleftharpoons CuP_2O_7^{2-}$	6.64
	$2Cu^{2+} + P_2O_7^{4-} \rightleftharpoons Cu_2P_2O_7^0$	-0.03
	$Cu^{2+} + NO_3^- \rightleftharpoons CuNO_3^+$	0.50
	$Cu^{2+} + 2NO_3^- \rightleftharpoons Cu(NO_3)_2^0$	-0.40
	$Cu^{2+} + SO_4^{2-} \rightleftharpoons CuSO_4^0$	2.36
Hg	$Hg^{2+} + H_2O \rightleftharpoons HgOH^+ + H^+$	-3.40
	$Hg^{2+} + 2H_2O \rightleftharpoons Hg(OH)_2^0 + 2H^+$	-6.19
	$Hg^{2+} + 3H_2O \rightleftharpoons Hg(OH)_3^- + 3H^+$	-21.10
	$Hg^{2+} + Cl^- + H_2O \rightleftharpoons HgClOH^0 + H^+$	3.23
	$Hg^{2+} + Cl^- \rightleftharpoons HgCl^+$	6.76
	$Hg^{2+} + 2Cl^- \rightleftharpoons HgCl_2^0$	13.16
	$Hg^{2+} + 3Cl^- \rightleftharpoons HgCl_3^-$	14.05
	$Hg^{2+} + 4Cl^- \rightleftharpoons HgCl_4^{2-}$	14.47
	$Hg^{2+} + NO_3^- \rightleftharpoons HgNO_3^+$	0.33
	$Hg^{2+} + 2NO_3^- \rightleftharpoons Hg(NO_3)_2^0$	-1.36
	$Hg^{2+} + SO_4^{2-} \rightleftharpoons HgSO_4^0$	1.41
	$Hg^{2+} + 2H^+ + 2S^{2-} \rightleftharpoons Hg(HS)_2^0$	63.58
	$Hg^{2+} + 2S^{2-} \rightleftharpoons HgS_2^{2-}$	51.00
	$Hg^{2+} + NH_4^+ \rightleftharpoons HgNH_3^{2+} + H^+$	-0.48
	$Hg^{2+} + 2NH_4^+ \rightleftharpoons Hg(NH_3)_2^{2+} + 2H^+$	-1.06
	$Hg^{2+} + 3NH_4^+ \rightleftharpoons Hg(NH_3)_3^{2+} + 3H^+$	-9.34
	$Hg^{2+} + 4NH_4^+ \rightleftharpoons Hg(NH_3)_4^{2+} + 4H^+$	-17.83

[a]From Lindsay W.L., 1979

Table 3.5. Continued

Element	Equlibrium Reaction	$\log K^0$
Cd	$Cd^{2+} + H_2O \rightleftharpoons CdOH^+ + H^+$	-10.10
	$Cd^{2+} + 2H_2O \rightleftharpoons Cd(OH)_2^0 + 2H^+$	-20.30
	$Cd^{2+} + 3H_2O \rightleftharpoons Cd(OH)_3^- + 3H^+$	-33.01
	$Cd^{2+} + 4H_2O \rightleftharpoons Cd(OH)_4^{2-} + 4H^+$	-47.29
	$Cd^{2+} + 5H_2O \rightleftharpoons Cd(OH)_5^{3-} + 5H^+$	-61.93
	$Cd^{2+} + 6H_2O \rightleftharpoons Cd(OH)_6^{4-} + 6H^+$	-76.81
	$2Cd^{2+} + H_2O \rightleftharpoons Cd_2OH^{3+} + H^+$	-6.40
	$4Cd^{2+} + 4H_2O \rightleftharpoons Cd_4(OH)_4^{4+} + 4H^+$	-27.92
	$Cd^{2+} + Cl^- \rightleftharpoons CdCl^-$	1.98
	$Cd^{2+} + 2Cl^- \rightleftharpoons CdCl_2^0$	2.60
	$Cd^{2+} + 3Cl^- \rightleftharpoons CdCl_3^-$	2.40
	$Cd^{2+} + 4Cl^- \rightleftharpoons CdCl_4^{2-}$	2.50
	$Cd^{2+} + P_2O_7^{4-} \rightleftharpoons CdP_2O_7^{2-}$	8.70
	$Cd^{2+} + H_2PO_4^- \rightleftharpoons CdHPO_4^0 + H^+$	-4.00
	$Cd^{2+} + NO_3^- \rightleftharpoons CdNO_3^+$	0.31
	$Cd^{2+} + 2NO_3^- \rightleftharpoons Cd(NO_3)_2^0$	0.00
	$Cd^{2+} + SO_4^{2-} \rightleftharpoons CdSO_4^0$	2.45
	$Cd^{2+} + CO_{2(g)} + H_2O \rightleftharpoons CdHCO_3^+ + H^+$	-5.73
	$Cd^{2+} + CO_{2(g)} + H_2O \rightleftharpoons CdCO_3^0 + 2H^+$	-14.06
	$Cd^{2+} + NH_4^+ \rightleftharpoons CdNH_3^{2+} + H^+$	-6.73
	$Cd^{2+} + 2NH_4^+ \rightleftharpoons Cd(NH_3)_2^{2+} + 2H^+$	-14.00
	$Cd^{2+} + 3NH_4^+ \rightleftharpoons Cd(NH_3)_3^{2+} + 3H^+$	-21.95
	$Cd^{2+} + 4NH_4^+ \rightleftharpoons Cd(NH_3)_4^{2+} + 4H^+$	-30.39
Pb	$Pb^{2+} + H_2O \rightleftharpoons PbOH^+ + H^+$	-7.70
	$Pb^{2+} + 2H_2O \rightleftharpoons Pb(OH)_2^0 + 2H^+$	-17.75
	$Pb^{2+} + 3H_2O \rightleftharpoons Pb(OH)_3^- + 3H^+$	-28.09
	$Pb^{2+} + 4H_2O \rightleftharpoons Pb(OH)_4^{2-} + 4H^+$	-39.49
	$2Pb^{2+} + H_2O \rightleftharpoons Pb_2OH^{3+} + H^+$	-6.40
	$3Pb^{2+} + 4H_2O \rightleftharpoons Pb_3(OH)_4^{2+} + 4H^+$	-23.89
	$4Pb^{2+} + 4H_2O \rightleftharpoons Pb_4(OH)_4^{4+} + 4H^+$	-20.89
	$6Pb^{2+} + 8H_2O \rightleftharpoons Pb_6(OH)_8^{4+} + 8H^+$	-43.58
	$Pb^{2+} + Cl^- \rightleftharpoons PbCl^-$	1.60
	$Pb^{2+} + 2Cl^- \rightleftharpoons PbCl_2^0$	1.78
	$Pb^{2+} + 3Cl^- \rightleftharpoons PbCl_3^-$	1.68
	$Pb^{2+} + 4Cl^- \rightleftharpoons PbCl_4^{2-}$	1.38
	$Pb^{2+} + F^- \rightleftharpoons PbF^-$	1.49
	$Pb^{2+} + 2F^- \rightleftharpoons PbF_2^0$	2.27
	$Pb^{2+} + 3F^- \rightleftharpoons PbF_3^-$	3.42
	$Pb^{2+} + 4F^- \rightleftharpoons PbF_4^{2-}$	3.10
	$Pb^{2+} + H_2PO_4^- \rightleftharpoons PbH_2PO_4^+$	1.50
	$Pb^{2+} + H_2PO_4^- \rightleftharpoons PbHPO_4^0 + H^+$	-4.10
	$Pb^{2+} + P_2O_4^{4+} \rightleftharpoons PbP_2O_7^{2-}$	11.30
	$Pb^{2+} + NO_3^- \rightleftharpoons PbNO_3^+$	1.17
	$Pb^{2+} + 2NO_3^- \rightleftharpoons Pb(NO_3)_2^0$	1.40
	$Pb^{2+} + SO_4^{2-} \rightleftharpoons PbSO_4^0$	2.62
	$Pb^{2+} + 2SO_4^{2-} \rightleftharpoons Pb(SO_4)_2^{2-}$	3.47

[a]From Lindsay W.L., 1979

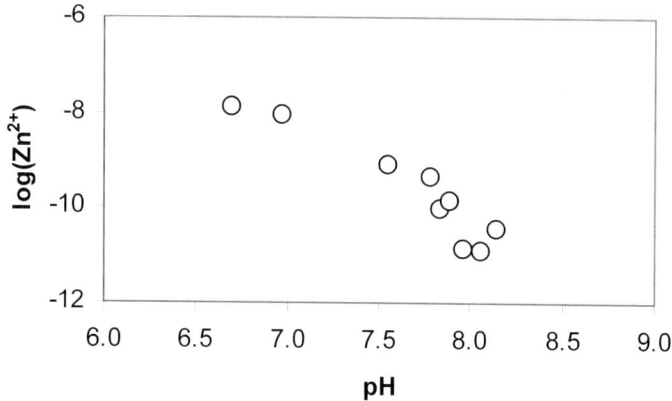

Figure 3.1. Changes of Zn activity in 10 Colorado soils with pH (data extracted from Ma and Lindsay, 1990)

Table 3.6. The first hydrolysis constants of selected elements in soil chemistry[a]

Elements	Ionic radius	pK 1
Co^{2+}	0.73	10.2
Ni^{2+}	0.77	9.9
Cu^{2+}	0.81	8.0
Zn^{2+}	0.83	9.0
Cd^{2+}	1.03	10.1
Hg^{2+}	1.10	3.7
Pb^{2+}	1.26	8.4
Cr^{3+}	0.70	4.0
Ca^{2+}	1.08	12.7
Mg^{2+}	0.80	11.4
K^+	1.46	14.5
Na^+	1.10	14.1

[a]From Hayes and Traina, 1998

Concentrations of free Zn in soil solution are strongly dependent upon soil pH. Zinc solubility expressed as free Zn increases with the concentration of H^+ (Norvel and Lindsay, 1969; Dang et al., 1994; Tiller, 1983; Jeffery and Uren, 1983). Norvel and Lindsay (1969) proposed that the solubility of soil Zn could be described by the equation:

$$(Zn^{2+}) = 10^6 (H^+)^2 \qquad (3.10)$$

Ma and Lindsay (1990) reported the similar relationships between Zn^{2+} and soil pH in 10 Colorado arid soils: $(Zn^{2+}) = 10^{5.7\pm0.38} (H^+)^2$ (Fig. 3.1).

However, Dang et al. (1994) reported that the relationships between soil solution pH and activity of Zn^{2+} of Australian soils with pH 7.45–8.98 and slight carbonate content (0.08–2.07%) were as follows:

$$Zn^{2+} = 10^{-3.52} (H^+)^{0.56} \text{ for unfertilized soils} \qquad (3.11)$$
$$Zn^{2+} = 10^{-5.47} (H^+)^{0.32} \text{ for fertilized soils} \qquad (3.12)$$

Tiller (1983, Eq 3.13) and Jeffery and Uren (1983, Eq 3.14) reported similar relationships between Zn^{2+} and soil pH.

$$Zn^{2+} = 10^{-2.8} (H^+)^{0.75} \qquad (3.13)$$
$$Zn^{2+} = 10^{-3.6} (H^+)^{0.75} \qquad (3.14)$$

In addition to inorganic zinc complexes, Zn can form moderate complexes with dissolved organic molecules including soil humic substances. Vulkan et al. (2002) reported that 82.8% of Zn in the water extracts of the sludge-amended arid sandy soil from Israel was found in low molecular weight complexes as zwitter ions (amphoteric species). These complexes have both negatively and positively charged functions, which are adsorbed on the cation and anion exchangers. But the relative content of positively charged Zn in soil solution increases as incubation of sewage-soil continues.

In 14 unfertilized vertisols from southern Queensland with soil pH from 7.45–8.98 and 0.08–2.07% $CaCO_3$, 47.8% (an average) of Zn in soil solution is in Zn^{2+}, followed by $ZnHCO_3^+$ (22.8%) and $ZnCO_3^0$ (9.4%) (Dang et al., 1994). However, Zn-fulvate complex is only 6.9%. The other inorganic complexes such as $ZnSO_4^0$, $Zn(OH)^+$ and $Zn(OH)_2^0$ range from 2.4–5%, while $ZnNO_3^+$ and $ZnCl^-$ only account for an insignificant portion (< 0.1%). Overall complexation of total soluble Zn by organic and inorganic ligands constitutes 40–50% of total soluble Zn in fertilized and unfertilized soil solution. In both agricultural soils and contaminated soils with pH 7.05–8.52 from eastern and southern Australia, free Zn^{2+} in soil pore water measured by the Donnan dialysis membrane technique is from 3–76% (Nolan et al., 2003). Compared to measured values, the GEOCHEM model underestimates distribution of free Zn^{2+} in soil pore water. Hodgson et al. (1966) reported 28–99% of soil solution Zn in soluble complexes of calcareous soils.

Soil pH and dissolved organic carbon strongly affect complexation of Zn in soil solution. Complexed Zn decreases with soil pH. McBride and

Blasiak (1979) reported that minimal amounts of soil solution Zn were complexed in acid soil solution (pH < 6.5). On the other hand, complexation of Zn increases with DOC in soil solution. Dang et al. (1994) found that the ratio of complexed to free Zn was positively related to soluble organic carbon in soil solution. They further reported an increase in the complexation of Zn with fulvic acid with an increase in soil solution pH. Organic complexed Zn increases with soluble organic matter. Mullins and Sommers (1986) pointed out that organo-Zn complexes increased with the application of sewage sludge. Sadig (1991) reported that Zn in arid soil solution with pH 6–8 from Saudi Arabia was mainly in forms of Zn^{2+}, $ZnSO_4^0$, $Zn(OH)^+$ and $ZnCl^+$; however, Zn^{2+}, $Zn(OH)^+$ and $Zn(OH)_2^0$ became important species for soils with pH above 8. Emmerich et al. (1982) reported that in soil solution of arid soils amended with sewage sludge, free Zn^{2+} ion accounted for 60–70% of the total Zn in soil solution. Villarroel et al. (1993) reported in a California sludge-treated soil, Zn was predominately present in the organic complexed form (77–90% of total Zn in soil solution), followed by the free Zn^{2+} ion (6–18%), with a very low presence of the rest of the inorganic complexes. Zinc activities and speciation are not significantly affected by P and N treatments. The major representative species of trace elements in arid soil solution are summarized in Table 3.7.

Table 3.7. Summary of major representative species of selected trace elements in arid soil solutions

Elements	Major Species in Soil Solution
Zn	$Zn(OH)^+$, $Zn(OH)_2^0$, Zn^{2+}, $ZnHCO_3^+$, $ZnCO_3^0$, $ZnSO_4^0$, $ZnCl^+$, Zn-DOC
Ni	$NiHCO_3^+$, $NiCO_3^0$, Ni^{2+}, DOC-Ni, $NiSO4^0$
Cu	Cu^{2+}, $Cu(OH)_2$, $CuCO_3^0$, $CuHCO_3^+$, $CuHPO_4^0$, Cu-DOC
Cd	Cd^{2+}, $Cd(OH)^+$, $Cd(OH)_2^0$, $CdHCO_3^+$, $CdCO_3^0$, $CdSO_4^0$, $CdCl^+$,
Hg	$Hg(OH)_2^0$, Hg-DOC
Pb	$PbOH^+$, Pb^{2+}, $Pb(OH)_2^0$, $Pb(OH)_3^-$, Pb-DOC, $PbSO_4^0$, $PbHCO_3^+$
Co	$Co(OH)_2^0$
Cr(III)	$Cr(OH)_4^-$
Cr(VI)	CrO_4^{2-}
Se(IV)	SeO_3^{2-}
Se(VI)	SeO_4^{2-}

3.3.4 Solution Speciation of Cu

In soil solution, Cu can occur in forms of hydrolyzed Cu species ($Cu(OH)^+$, $Cu(OH)_2^0$, $Cu(OH)_3^-$, $Cu(OH)_4^{2-}$, $Cu_2(OH)_2^{2+}$), copper chloride complexes ($CuCl^+$, $CuCl_2^0$, $CuCl_3^-$), copper carbonate ($CuHCO_3^+$, $CuCO_3^0$, $Cu(CO_3)_2^{2-}$), nitrate ($CuNO_3^+$, $Cu(NO_3)_2^0$), phosphate ($CuH_2PO_4^+$, $CuHPO_4^0$), sulfate species ($CuSO_4^0$) and organic complexes. The equilibrium reactions and constants of inorganic copper species are listed in Table 3.5. In soil solution with pH < 6.9, the predominant inorganic Cu ion species is Cu^{2+}, while $Cu(OH)_2^0$ is the major solution inorganic Cu species above this pH (Lindsay, 1979). $CuCO_3^0$ is also important in arid soil solution. At $10^{-2.35}$ atm of CO_2, the activity of $CuCO_3^0$ equals that of $Cu(OH)_2^0$. The activity of $CuCO_3^0$ increases with changes of CO_2 partial pressure in soil solution (Lindsay, 1979). The other possible copper species in soil solution of arid soils are $CuHCO_3^+$, $Cu(CO_3)_2^{2-}$ and $CuHPO_4^0$, which are normally low. However, they increase to more than 1% under specific conditions such as high phosphate, high pH and high CO_2 (Lindsay, 1979). In eastern and southern Australia, free Cu^{2+} in soil pore water measured by the Donnan dialysis membrane technique was usually less than 26% and 3% in agricultural soils and contaminated soils with pH 7.05–8.52, respectively (Nolan et al., 2003). Both measured values and the GEOCHEM model provide a similar estimation of distribution of free Cu^{2+} in soil pore water.

In contrast, organic complexes of Cu are the most important Cu species in soil solution in arid soils. Vulkan et al. (2002) reported that Cu in water extracts of the sludge-amended arid sandy soil from Israel was found almost exclusively in low molecular weight (below 1000 Da) complexes. These complexes are negatively charged and consist primarily of about two Cu ions attached to dissolved organic carbon species containing five or six carbon atoms. The researchers (Vulkan et al., 2002) suggested that amino acids and small peptides or polycarboxylic acids such as citric acid may be important complexing agents of Cu. Higher molecular weight (around 2500 Da) dissolved organic carbon is present in soil solution but with little complexation. Mingelgrin and Biggar (1986) reported that Cu in the saturation extract of sewage sludge was mainly in a complexed form. In most of the soil solutions, more than 98% of the soluble Cu in soils from Quebec and New York is bound to organic ligands (Sauve et al., 1997). In soil solution amended with sewage sludge, Cu and Ni are initially associated to the greatest degree with the small molecule-size fraction of organic compounds containing high amide content (Dudley et al., 1987). Emmerich et al. (1982) reported that in soil solution of arid soils amended with sewage sludge, Cu was almost exclusively in organically complexed forms.

3.3.5 Solution Speciation of Ni

Emmerich et al. (1982) reported that in soil solution of arid soils amended with sewage sludge, free Ni^{2+} ion accounted for 60–70% of the total Ni in soil solution. Sposito and Page (1984) indicated $NiHCO_3^+$ and $NiCO_3^0$ were also important Ni speciations in alkaline and calcareous soils.

3.3.6 Solution Speciation of Cd

In arid soil solution, Cd is predominately found in free Cd^{2+}. However, when pH > 7.5, $Cd(OH)^+$ and $Cd(OH)_2^0$ account for only a small percentage of the speciation (Lindsay, 1979). $CdHCO_3^+$ is significant near pH 8.0.

$$CdCO_3 \text{ (octavite)} + H^+ \rightleftharpoons CdHCO_3^+ \qquad \log K = 0.43 \qquad (3.15)$$

The activity of $CdHCO_3^+$ and $CdCO_3^0$ becomes important when pH is nearly 8, and the former species is approximately that of Cd^{2+} when pH is nearly 8 (Lindsay, 1979). The $CdHCO_3^-$ and $CdCO_3^0$ increase with increasing $CO_2(g)$. At the normal soil solution, halide complexes are not important. All other complexes such as $CdNO_3^+$, $Cd(NO_3)_2^0$ and $CdHPO_4^0$ are not important in normal arid soils as well. But when $H_2PO_4^-$ is >10^{-5} M, the $CdHPO_4^0$ ion pair becomes a significant contributor to soil solution species (Lindsay, 1979). Possible Cd speciation in soil solution is listed in Table 3.5.

In calcareous soils, Cd is mainly found in Cd^{2+} and $CdHCO_3^+$ species in soil solution (Fig. 3.2). Hirsh and Banin (1990) reported that in saturated pastes of three arid soils from Israel with pH 7.5–8.5 and treated with $Cd(NO_3)_2$, free Cd^{2+} and $CdHCO_3^+$ dominated the soil solution, accounting for 35 and 45% of Cd in soil solution, respectively. Other inorganic Cd complexes in the soilsolution are less important, decreasing in the order: $CdSO_4^0$ > $CdCl^+$ > $Cd(OH)^+$. The percentage of Cd^{2+} decreases with an increase in pH and drastically reduces after a pH of 7.5, while $CaCO_3^0$ increases with pH after pH 8, and $CdHCO_3^+$ peaks around pH 8. $CdSO_4^0$ and $CdCl^+$ are stable from pH 4–7 and decrease after pH 7.0 (Fig. 3.2). The computational mode for Cd speciation is in agreement with the values with Cd^{2+} selective electrode.

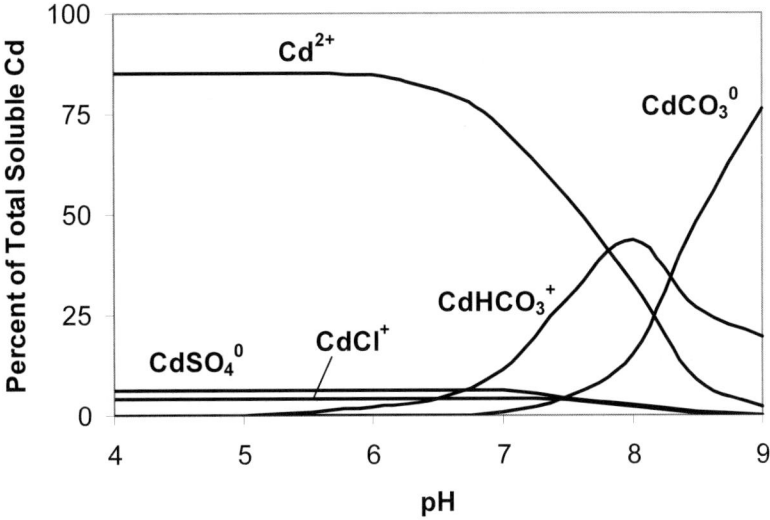

Figure 3.2. Changes of Cd speciation in soil solutions of a typical Israeli calcareous soil with pH 4–9 (after Hirsh and Banin, 1990, with permission from Soil Sci. Soc. Am)

Cadmium in a soil solution from South Australia is predominately present in Cd^{2+} (96%) (Weggler-Beaton et al., 2000). In contaminated soils with pH 7.05–8.20 from eastern and southern Australia, free Cd^{2+} in soil pore water measured by the Donnan dialysis membrane technique is from 20–86% (Nolan et al., 2003). Both measured values and the GEOCHEM model provide the similar prediction of distribution of free Cd^{2+} in soil pore water. Emmerich et al. (1982) reported that in soil solution of arid soils amended with sewage sludge, free Cd^{2+} ion accounted for 50–60% of the total Cd in soil solution. In a sewage sludge-amended arid soil with pH 7.2–7.7 from California, free Cd ion accounts for 65–68% of the total Cd in soil solution with total Cd in the range of 0.07–80.4 µM, followed by SO_4-complexed Cd (12–21%), CO_3-complexed Cd (2–7.1%) and fulvate-complexed Cd (2.5–7.2%) (Mahler et al., 1980). In both sewage sludge-amended and Cd nitrate-amended California soils, the solution speciation of Cd is similar and dominated by the free Cd^{2+} (51%), followed by chloride and sulfate complexes. The concentrations of all species increase with Cd treatment in both metal salt- (Fig. 3.3) and sludge-treated soils (Candelaria and Chang, 1997).

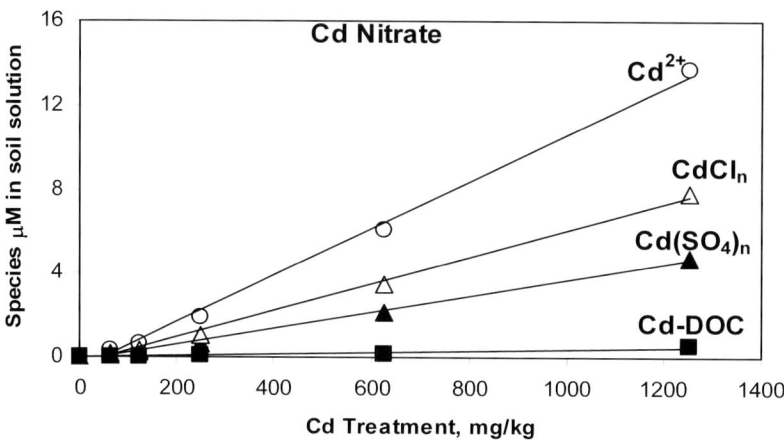

Figure 3.3. Changes of Cd speciation in soil solutions of Cd nitrate-treated Domino soil from California (data extracted from Candelaria and Chang, 1997)

Organic complexed Cd is not important in arid soil solution. Hirsh and Banin (1990) observed 5–10% of Cd bound to organic ligands in Israeli arid soil solution. Emmerich et al. (1982) found that organic-Cd complexes constituted 1–4% of Cd in California arid soil solution. However, Villarroel et al. (1993) reported that in a California sludge-treated soil, Cd was mainly present in both free ion and organic complex forms (each accounted for 32–40% and 30–45% of total Cd in soil solution, respectively), followed by the chloride complexes (8–20%), SO_4-complex (3–10%), and PO_4-Cd complex (1.5–7.7%). The nitrate Cd complexes were the lowest. Cadmium activities and speciation is not significantly affected by P and N treatments.

Solution speciation, calculated with the GEOCHEM-PC, shows significant increases of Cd^{2+} activities with increasing salt rates (McLaughlin et al., 1998b). Both Cd chloride complexes ($CdCl^+$ and $CdCl_2^0$) increase with an increase of NaCl salt in solution (Weggler-Beaton et al., 2000). At NaCl treatment of 1600 mg/L, $CdCl^+$ in soil solution accounts for 68–70%, followed by $CdCl_2^0$ (19–23%), while Cd^{2+} is 8–10%. High salinity in soils increases the chloride-complexes concentration of trace elements (such as $CdCl^+$ or $CdCl^{2+}$) in soil solution, which increases and correlates best with the Cd uptake of plant species (Weggler-Beaton et al., 2000). Complexation of Cd^{2+} by Cl increases soluble Cd in culture and soil solution, but the calculated activity of Cd^{2+} is not significantly affected by increasing concentrations of Cl in solution (Smolders and McLaughlin, 1996). Increasing concentrations of NaCl in the soil solution significantly increase the Cd concentration in solution from 64–400 nmol L^{-1}. However, increasing concentrations of $NaNO_3$ in solution have no effect on Cd concentrations in solution. Speciation calculations predict that the solution concentration of the free metal ion Cd^{2+} is not significantly affected by the NaCl or $NaNO_3$ treatments.

3.3.7 Solution Speciation of Pb

In soils with pH > 8, $PbOH^+$ significantly contributes to total inorganic Pb in solution, followed by Pb^{2+} (Lindsay, 1979). When soil pH is above 9, $Pb(OH)_2^0$ and $Pb(OH)_3^-$ become significant. Other inorganic complexes (halide, sulfate and nitrate) become important only when concentrations of these inorganic ligands increase to 10^{-3} to 10^{-2} M. Possible Pb speciation in soil solution is listed in Table 3.5. In general, the halide complexes of Pb^{2+} are not highly significant in soils (Lindsay, 1979). At halide activities > 10^{-4} M, halide Pb complexes begin to contribute significantly to total Pb in solution. At the range of 10^{-2} to $10^{-1.5}$ M Cl, the activity of the $PbCl^+$ complex is similar to that of free Pb^{2+}. The $PbSO_4^0$ and $Pb(SO_4)_2^{2-}$ complexes become important at $10^{-2.65}$ M and >10^{-3} SO_4^{2-} M, respectively (Lindsay, 1979). The $Pb(NO_3)^+$ and $Pb(NO_3)_2^0$ complexes become slightly significant as NO_3^- increase above 10^{-3} M and above 10^{-2} M, respectively. The $PbH_2PO_4^+$ complex is not significant in the phosphate range generally found in soils. Harter (1983) reported that in soil solution, $Pb(OH)^+$ predominated at pH around 8, and above 8, $Pb(OH)_2^0$ became important as well. In eastern and southern Australia, free Pb^{2+} in soil pore water measured by the Donnan dialysis membrane technique is from 4–54% in contaminated soils with pH of 7.05–8.20 (Nolan et al., 2003). Both measured values and the GEOCHEM model provide a similar estimation of distribution of free Pb^{2+} in soil pore water.

3.3.8 Solution Speciation of Cr(III) and Cr(VI)

The main aqueous species of Cr(III) are Cr^{3+}, $Cr(OH)_2^+$, $Cr(OH)_3^0$ and $Cr(OH)_4^-$, and the major aqueous species of Cr(VI) are $HCrO_4^-$, CrO_4^{2-} and $Cr_2O_7^{2-}$ (Bartlett and Kimble, 1976; Palmer and Puls, 1994). In arid and semi-arid soils, $Cr(OH)_4^-$ and $CrO_4^=$ dominate.

3.3.9 Solution Speciation of Hg

Mercury in soil solution can occur in the forms of Hg^{2+}, $HgCl^+$, $HgCl_2^0$, $HgCl_3^-$, $HgCl_4^{2-}$, $HgClOH$, $Hg(OH)^+$, $Hg(OH)_2^0$ and Hg-DOC. Mercury speciation in soil solution is strongly dependent upon soil pH and Cl^- concentration. Possible Hg speciation found in soil solution are listed in Table 3.5. In the pH range of arid and semi-arid soil solution, $Hg(OH)_2^0$ is the major species (Adriano, 2001), while at low pH and high Cl concentrations (especially in saline and saline sodic soils), Hg is mainly present as Hg^{2+}, $HgCl^+$ and neutral $HgCl_2^0$. Yin et al. (1996) systematically studied distribution of mercury species as a function of pH and Cl^- concentration. They reported that when soil pH was above pH 5 and Cl concentration was 10^{-5}–10^{-6} M, $Hg(OH)_2^0$ was the predominant species. $HgCl_2^0$ increases with Cl^- concentration from 10^{-6} to 10^{-2} M. However, further increases in Cl^- concentration higher than 10^{-2} M decrease $HgCl_2^0$ while it increases $HgCl_3^-$ and $HgCl_4^{2-}$. Also, Cl^- concentration higher than 10^{-2} increases HgClOH, which increases with pH (Yin et al., 1996; Han et al., 2005).

3.3.10 Speciation of Se

Selenium can exist as selenide (Se^{2-}), elemental Se (Se^0), selenite (SeO_3^{2-}) and selenate (SeO_4^{2-}). The speciation of selenium is controlled by the pH and redox conditions of soils. Selenide (Se^{2-}, HSe^-, H_2Se) and elemental Se occur in acidic, reducing and organic-rich environments (McNeal and Balistrieri, 1989). Selenides, Se-sulfides and elemental Se are insoluble and are not bioavailable. SeO_3^{2-} ($HSeO_3^-$, H_2SeO_3) and SeO_4^{2-} ($HSeO_4^-$, H_2SeO_4) are the dominant forms in most soils and aquatic systems. In soil solution of arid regions with high redox (pe + pH > 14.5), selenate (SeO_4^{2-}) is the major species (Elrashidi et al., 1989). There are no selenate mineral precipitates. However, in soil solution of humid regions with moderate redox (pe + pH: 7.5–14.5), $HSeO_3^-$ is predominant at low pH, and SeO_3^{2-} is the major species at high pH (Elrashidi et al., 1989). In soils with low redox (pe + pH < 7.5), HSe^- is the most important Se species in soil solution. Selenium adsorption may control Se concentrations in soil solution in arid and humid regions.

Selenate is the most mobile form of Se. Selenium becomes biologically unavailable by reduction to elemental Se or by formation of metal selenides or Se-sulfides. Inorganic Se compounds can be converted to volatile organic Se such as dimethyl selenide or dimethyl diselenide by plants, microorganisms and animals (McNeal and Balistrieri, 1989). Selenite can be slowly oxidized to SeO_4^{2-}, but at neutral and alkaline pH, the oxidation of selenite to selenate is much faster.

In addition to soil solution, speciation of trace elements in water of the Nahr-Ibrahim river valley of Lebanon was studied with the AQUACHEM model. The results indicate that a high percentage of Pb and Zn is present as carbonate species, but in low percentages in free hydrated ion species. Cadmium exhibits as a high percentage of a free hydrated Cd^{2+}.

3.4 FACTORS AFFECTING TRACE ELEMENT SPECIATION IN SOIL SOLUTION

3.4.1 Soil pH and Eh

Soil pH is the most important factor controlling solution speciation of trace elements in soil solution. The hydrolysis process of trace elements is an essential reaction in aqueous solution (Table 3.6). As a function of pH, trace metals undergo a series of protonation reactions to form metal hydroxide complexes. For a divalent metal cation, $Me(OH)^+$, $Me(OH)_2^0$ and $Me(OH)_3^-$ are the most common species in arid soil solution with high pH. Increasing pH increases the proportion of metal hydroxide ions. Table 3.6 lists the first hydrolysis reaction constant (K1). Metals with lower pK1 may form the metal hydroxide species $(Me(OH)^+)$ at lower pH. pK serves as an indicator for examining the tendency to form metal hydroxide ions.

Based on Table 3.6, Cu and Pb may form $Cu(OH)^+$ and $Pb(OH)^+$ at pH 8.0 and 8.4, respectively, while Zn may form $Zn(OH)^+$ at pH 9.0. However, Cd can form $Cd(OH)^+$ at pH 10.1. This indicates that at normal pH ranges of arid and semi-arid soils, Cu and Pb, and to some extent, Zn may be present in their hydroxide complex ions, while Cd is less likely to be in the hydroxide form unless at higher soil pH (Table 3.5). Sauve et al. (1997) reported that free Cu activity in soil solution of Quebec and New York decreased with higher pH. Fotovat et al. (1997) reported that $Zn(OH)_2^0$ and $ZnCO_3^0$ contributed to a considerable proportion of the total solution Zn (about 16–21%) in soils with high pH (7.56–8.99) from South Australia. Ma and Lindsay (1990) reported that the Zn activities measured in 10 Colorado arid soils can be expressed by the relationship with $\log K^\circ = 5.7 \pm 0.38$ (Fig. 3.1):

$$\text{Soil Zn} + 2H^+ \rightleftharpoons Zn^{2+} \tag{3.16}$$

where K^o is the equilibrium constant expressed in terms of activity. In addition, high pH increases the solubility of soil organic matter including humic substances, which strongly affect forms of trace elements in solution as well. Soil redox potential can be expressed as the negative logarithm of the electron activity pe. With increase in the pe, the predominant form is those ion species with higher oxidation states.

3.4.2 Dissolved Organic Matter

Dissolved organic carbon (DOC) in soil solution was found in the range of 12–266 mg/kg in soils of South Australia with pH 5.25–8.81 and 0–1.9% $CaCO_3$ (Fotovat et al., 1997). DOC concentrations increase as the amount of water to soil ratios increase. Dissolved organic matter may form soluble trace element organic complexes in soil solution. In soil solution, two categories of reactive natural organic chemicals include biochemical components of living organisms and the products of microbial decomposition of organic matter. Most functional groups in dissolved organic matter are dominated by carboxyl, carbonyl, amino, imidazole, phenolic OH, alcoholic OH and sulfhydryl groups. With an increase in soil pH, the solubility of soil organic matter increases. In addition, an increase in concentrations of soluble dissolved organic matter results in an increase in the metal organic complexes. Dudley et al. (1987) found that Cu and Ni were highly complexed by dissolved organic matter while Zn was less complexed compared to Cu and Ni in an arid soil solution amended with sewage sludge. They observed that soluble Cu increased despite the increasing pH during incubation of sewage sludge with soil, apparently due to organic complex formation, while soluble Zn declined during the incubation.

3.4.3 Total Inorganic Ligands

Trace elements can form ligand trace element complexes. Metal ligand complexes increase with increasing in concentrations of ligands. The dominant inorganic ligands are nitrate, sulfate, chloride, carbonate, biocarbonate and hydroxyl ion. Other significant inorganic ligands with low concentrations in soil solution include boric acid, silicic acid, arsenate, selenate, molybdate, fluoride and orthophosphate. These ligands are in the micromollar to nanomollar range in soil solution. Major complexation reactions for trace elements in soils are summarized in Table 3.5. Yin et al. (1996) reported that $HgCl_2^0$ was the major Hg species when Cl concentration was in the range of 10^{-6} to 10^{-2} M. Concentrations of Hg^{2+}, $Hg(OH)_2^0$ and $Hg(OH)^+$ decrease with an increase in Cl concentrations. $HgCl^+$, $Hg(OH)Cl$, $HgCl_3^-$ and $HgCl_4^{2-}$ increase with Cl concentration after Cl concentration is $> 10^{-2}$ M.

Hirsh and Banin (1990) reported that an increase in Cl concentration decreased Cd sorption due to formation of the $CdCl^+$ ion pair. Enhanced sorption in the presence of HCO_3^- was observed due to the formation of the $CdHCO_3^+$ ion pair. Mahler et al. (1980) found that in calcareous soils treated with sewage sludge spiked with $CdSO_4$, Cd complexes were mainly Cd sulfate and carbonate complexes, and the formation of Cd–Cl complexes increased in both soils as total Cd increased. In sludge-amended California soils with pH 7–8, both Cd and Zn organic complexes and Cd phosphate complexes increase with P levels, but free Cd^{2+} and Zn^{2+} decrease with P levels (Fig. 3.4) (Villarroel et al., 1993).

3.4.4 Other Factors

Trace element speciation in soil solution is affected by total metal concentrations in soils. Free Cu^{2+} activity increases with total Cu content in soils from Quebec and New York (Sauve et al., 1997). Total free Cu activity in soils could be predicted from total Cu content and soil pH:

$$pCu^{2+} = -1.7 \log(\text{Total Cu}) + 1.40 \text{ pH} + 3.42. \tag{3.17}$$

The calculated free Cu activity in soil solution is in agreement with the prediction with the solubility line soil-Cu by Norvell and Lindsay (1969) for uncontaminated soils with pH between 6 and 8.

Salt is a worldwide problem in arid and semi-arid soils. Salts, especially high concentrations of NaCl, increase the chloride-complexes of trace elements, thus increasing bioavailability and mobility of trace elements in arid soils. In the soil solution from South Australia, the Cd concentration in soil solution of the control soil is 96% in free ionic form, but when NaCl concentration is 1,600 mg/L, the solution Cd is mainly present in $CdCl^+$ (68–70%), followed by $CdCl_2^0$ (19–23%), while free Cd^{2+} activity is 8–10% (Weggler-Beaton et al., 2000). Bingham et al. (1983) reported that free Cd^{2+} decreased from 88.6% at 0.6 meq Cl/L of $NaCl/CaCl_2$ to 41% at 165 meq/L of $NaCl/CaCl_2$, while Cl complexes increased from 2–3% to 55–57% accordingly in soils from California (Fig. 3.5).

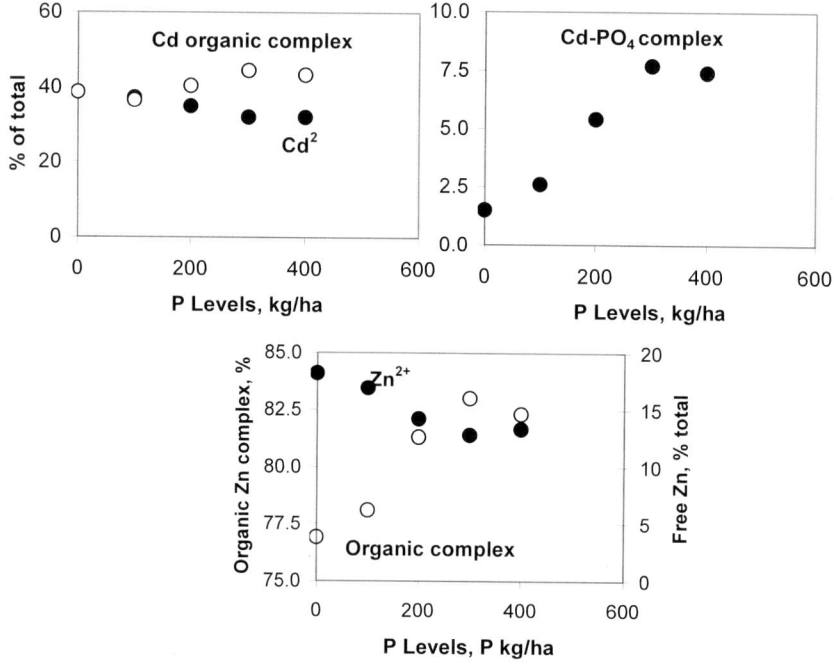

Figure 3.4. Effects of phosphate levels on Cd and Zn solution speciation in California soils that received sludge application (data extracted from Villarroel et al., 1993)

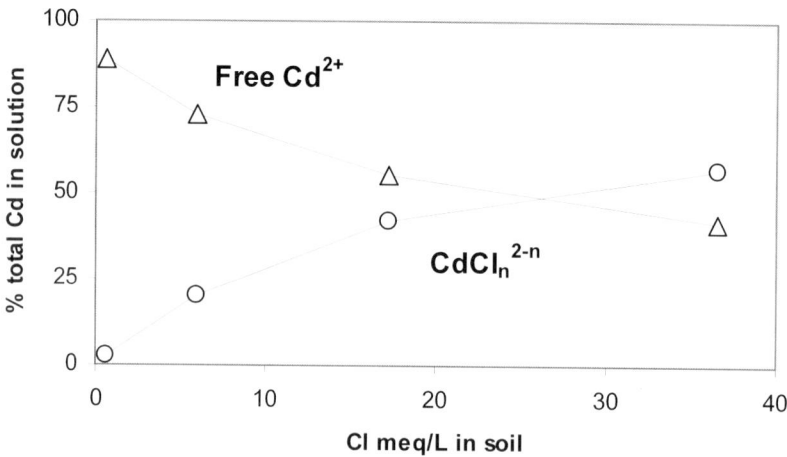

Figure 3.5. Effects of Cl concentrations (NaCl or $CaCl_2$) on Cd solution speciation in a California soil (data extracted from Bingham et al., 1983)

Soil solution to soil ratios also strongly affect distribution of some trace elements such as Zn speciation in arid and semi-arid soils. Fotovat et al. (1997) reported that the proportion of free hydrated Zn^{2+} to total Zn ranged from 20–65% at field capacity soil water content and decreased with increases in solution to soil ratios, while the proportion of Zn complexed with organic ligands increased dramatically in soils. However, solution to soil ratios do not strongly affect the distribution of Cu speciation in soil solution since Cu primarily occurs as organic complexes in these soil solutions.

3.5 MAJOR PHYSICO-CHEMICAL AND BIOLOGICAL PROCESSES GOVERNING SOLUBILITY OF TRACE ELEMENTS IN ARID AND SEMI-ARID SOILS

Concentrations of trace elements in soil solution may be controlled by the solubility of certain solid phases via dissolution/(co-)precipitation or by other physicochemical and biological processes such as adsorption-desorption, complexation, and redox reactions.

3.5.1 Dissolution/Precipitation/Co-precipitation

Dissolution and (co)precipitation are the most important processes controlling the solubility of most trace elements in arid and semi-arid soils. The aqueous concentrations of trace elements in soils, sediment, and aquatic environments frequently are controlled by the dissolution and precipitation of discrete mineral phases (Traina and Laperche, 1999). Precipitation refers to a selective accumulation of at least two or more constituent ions into an organized solid matrix. It includes homogeneous precipitation and heterogeneous precipitation. In soil systems, due to an abundance of various reactive constituents, heterogeneous nucleation is the major precipitation process controlling solubility of trace elements in arid and semi-arid soil. When soil solution is in supersaturation regarding a specific solid phase, homogeneous precipitation takes place. At a standard state, the solubility of a solid phase is referenced to an equilibrium constant. When soil solutions are over supersaturated with respect to a solid phase (the ion activity product, IAP, of the soil solution is > Ksp (the IAP at equilibrium) of a specific mineral), the soil solution favors precipitation. On the other hand, when soil solutions are undersaturated they promote dissolution of that specific solid phase mineral. When IAP = Ksp, the soil solution is saturated. Possible common soil solution solubility controls for various trace elements in aerobic and anaerobic soils are presented in Table 3.8.

When a foreign surface is present and constituent ions are adsorbed, heterogeneous precipitation occurs. The first step for precipitation is nuclei formation in either soil solution or on a foreign surface. Rapid kinetics of precipitation is expected for low interfacial energy solids. The second step is crystallite formation where nuclei coalesce into large particles, lowering the supersaturation of solution. The next step is crystal formation with continued surface area formation. In soil systems, precipitation may form amorphous precursors. This precursor phase may have a chemical composition different from that of a crystalline phase. A more stable solid phase may eventually nucleate through heterogeneous precipitation. Precipitation and dissolution have been reviewed in great detail by Robarge (1999).

Table 3.8. Possible soil solution solubility controls for selected trace elements[a]

Element	Aerobicc soils	Anaerobic soils
Cd	$Cd(OH)_2$, $CdCO_3$	Cd, CdS
Co	$Co(OH)_2$, $CoCO_3$, $CoSO_4$	$Co(OH)_2$, $CoCO_3$, $CoSO_4$
Cr	$Cr(OH)_3$	$Cr(OH)_3$
Cu	CuO, $CuCO_3$, $Cu(OH)_2CO_3$	Cu, CuS, Cu_2S
Hg	$HgCl_2$, HgO, $Cd(OH)_2$	Hg(I), HgS
Ni	NiO, $NiCO_3$, $Ni(OH)_2$	Ni, NiS
Pb	PbO, $PbCO_3$, $Pb_3(CO_3)(OH)_2$	Pb, PbS
Zn	ZnO, $ZnCO_3$, $Zn(OH)_2$, $ZnSO_4$	ZnS, Zn

[a]From Hayes and Traina, 1998.

Most primary and secondary minerals found in soil systems are barely soluble in the soil solution. The amount of mass from the bulk phase to hydrated ions in soil solution is negligible compared to the total mass of the solid phase. In arid and semi-arid soils, concentrations of most trace metals in soil solution may be controlled by their carbonates and to some extent by their hydroxides. Other than carbonates, trace elements in arid and semi-arid soils may also occur as sulfate, phosphate or siliceous compounds, or as a minor component adsorbed on the surface of various solid phase components. The solubility of carbonates, sulfates and other common minerals of trace elements in arid and semi-arid soils will be discussed in Chapter 5. Badawy et al. (2002) reported that in near neutral and alkaline soils representative of alluvial, desertic and calcareous soils of Egypt, the measured Pb^{2+} activities were undersaturated with regard to the solubility of

PbSiO$_3$ in equilibrium with SiO$_2$ (soil). However, they were supersaturated with regard to the solubilities of the Pb carbonate minerals PbCO$_3$ (cerussite) and Pb$_3$(CO$_3$)$_2$(OH)$_2$ in equilibrium with atmospheric CO$_2$ and hydroxide Pb(OH)$_2$. They also were supersaturated with regard to the solubilities of the Pb phosphate minerals Pb$_3$(PO$_4$)$_2$, Pb$_5$(PO$_4$)$_3$OH and Pb$_4$O(PO$_4$)$_2$ in equilibrium with tri-calcium phosphate and CaCO$_3$. The activity of Pb^{2+} was not regulated by any mineral of known solubility in soils, but possibly by a mixture of Pb carbonate and phosphate minerals.

3.5.1.1 Carbonate Dissolution/Precipitation (The Ca/MgCO$_3$-CO$_2$-H$_2$O System)

Calcium and magnesium are the major cations (co-)precipitating trace elements as carbonate. Trace elements are also precipitated as sulfate or phosphate. Solubility and reactions of carbonates, sulfates and phosphates of selected major and trace elements are in Table 3.9.

$$\text{Soil Ca} \rightleftharpoons \text{Ca}^{2+} \quad \log K = -2.50 \quad (3.18)$$
$$\text{CaCO}_3 \rightleftharpoons \text{Ca}^{2+} + \text{CO}_3^{2-} \quad \log K = -8.35 \quad (3.19)$$
$$\text{MgCO}_3 \text{ (magnesite)} + 2\text{H}^+ \rightleftharpoons \text{Mg}^{2+} + \text{CO}_2 + \text{H}_2\text{O}$$
$$\log K = 10.69 \quad (3.20)$$
$$\text{CaCO}_3 \text{ (calcite)} + 2\text{H}^+ \rightleftharpoons \text{Ca}^{2+} + \text{CO}_2 \text{(g)} + \text{H}_2\text{O}$$
$$\log K = 9.74 \quad (3.21)$$
$$\text{CaMg(CO}_3)_2 \text{ (dolomite)} + 4\text{H}^+ \rightleftharpoons \text{Ca}^{2+} + \text{Mg}^{2+} + 2\text{CO}_2 + 2\text{H}_2\text{O}$$
$$\log K = 18.46 \quad (3.22)$$

In arid and semi-arid soils, calcite, dolomite, leonhardite (Ca$_2$Al$_4$Si$_8$O$_{24}$·7H$_2$O) and lawsonite (CaAl$_2$Si$_2$O$_8$·2H$_2$O) can be possible minerals. Calcium carbonate strongly influences soil properties in arid and semi-arid soils. Most calcareous soils have soil a pH in the range of 7.3–8.5. When sodium is predominant in soils, soil pH is above 8.5. In most arid and semi-arid soils, calcium carbonates (calcite and dolomite) generally accumulate and are most likely to control the Ca^{2+} and Mg^{2+} solubility in these soils (Lindsay, 1979).

Trace elements can be precipitated as carbonates, sulfates, phosphates and hydroxides in arid and semi-arid environments. But most carbonates are more stable in arid and semi-arid soils than other solid phases. Cadmium hydroxide (Cd(OH)$_2$), sulfate (CdSO$_4$) and phosphates (Cd$_3$(PO$_4$)$_2$) are more soluble than carbonate (CdCO$_3$, octavite), therefore the former minerals are not stable in arid soils. In calcareous soils, CdCO$_3$ (octavite) is the main Cd mineral to control Cd^{2+} activity in soil solution. At high CO$_2$

Table 3.9. Solubility product constants of common compounds and minerals of selected major and trace elements at 25 °C[a]

Element	Compound	Formula	K_{sp}
Ca	calcium carbonate (calcite)	$CaCO_3$	3.36×10^{-9}
	calcium fluoride	CaF_2	3.45×10^{-11}
	calcium phosphate	$Ca_3(PO_4)_2$	2.07×10^{-33}
	calcium hydroxide	$Ca(OH)_2$	5.5×10^{-6}
	calcium oxalate hydrate	$CaC_2O_4 \cdot H_2O$	2.32×10^{-8}
	calcium phosphate	$Ca_3(PO_4)_2$	2.07×10^{-33}
	calcium sulfate	$CaSO_4$	4.93×10^{-5}
	calcium sulfate dihydrte	$CaSO_4 \cdot 2H_2O$	3.14×10^{-5}
	fluorapatite	$Ca_5(PO_4)_3F$	1.0×10^{-60}
	hydroxyapatite	$Ca_5(PO_4)_3OH$	1.0×10^{-36}
Mg	magnesium carbonate	$MgCO_3$	6.82×10^{-6}
	magnesium carbonate trihydrate	$MgCO_3 \cdot 3H_2O$	2.38×10^{-6}
	magnesium carbonate pentahydrate	$MgCO_3 \cdot 5H_2O$	3.79×10^{-6}
	magnesium fluoride	MgF_2	3.7×10^{-8}
	magnesium hydroxide	$Mg(OH)_2$	5.61×10^{-12}
	magnesium oxalate dihydrate	$MgC_2O_4 \cdot 2H_2O$	4.83×10^{-6}
	magnesium phosphate	$Mg_3(PO_4)_2$	1×10^{-25}
Fe	iron(II) carbonate	$FeCO_3$	3.13×10^{-11}
	iron(II) hydroxide	$Fe(OH)_2$	4.87×10^{-17}
	iron(II) sulfide	FeS	6×10^{-19}
	iron(III) arsenate	$FeAsO_4$	5.7×10^{-21}
	iron(III) hydroxide	$Fe(OH)_3$	4×10^{-38}
	iron(III) phosphate	$FePO_4$	1.3×10^{-22}
Mn	manganese(II) carbonate	$MnCO_3$	2.24×10^{-11}
	manganese(II) hydroxide	$Mn(OH)_2$	1.9×10^{-9}
	manganese(II) sulfide	MnS	2.5×10^{-13}
Zn	zinc carbonate	$ZnCO_3$	1.46×10^{-10}
	zinc carbonate monohydrate	$ZnCO_4 \cdot H_2O$	5.42×10^{-11}
	zinc hydroxide	$Zn(OH)_2$	3.0×10^{-17}
	zinc oxalate	ZnC_2O_4	2.7×10^{-8}
	zinc phosphate	$Zn_3(PO_4)_2$	9.0×10^{-33}
	zinc sulfide (sphalerite)	ZnS	2×10^{-4}
	zinc sulfide (wurtzite)	ZnS	3×10^{-2}

[a] From Lide D.R., 2001

Table 3.9. Continued

Element	Compound	Formula	K_{sp}
Cd	cadmium carbonate	$CdCO_3$	1.0×10^{-12}
	cadmium fluoride	CdF_2	6.44×10^{-3}
	cadmium hydroxide	$Cd(OH)_2$	7.5×10^{-15}
	cadmium sulfide	CdS	8.0×10^{-7}
	Cadmium phosphate	$Cd_3(PO_4)_2$	2.53×10^{-33}
Cr	chromium(II) hydroxide	$Cr(OH)_2$	2×10^{-16}
	chromium(III) hydroxide	$Cr(OH)_3$	6.3×10^{-31}
Co	cobalt(II) carbonate	$CoCO_3$	1.4×10^{-13}
	cobalt(II) sulfide	CoS	4.0×10^{-21}
	cobalt(II) hydroxide	$Co(OH)_2$	5.92×10^{-15}
	cobalt(II) phosphate	$Co_3(PO_4)_2$	2.05×10^{-35}
Cu	copper(II) carbonate	$CuCO_3$	1.4×10^{-10}
	copper(II) hydroxide	$Cu(OH)_2$	2.2×10^{-20}
	copper(II) sulfide	CuS	6×10^{-16}
	copper(II) phosphate	$Cu_3(PO_4)_2$	1.4×10^{-37}
	copper(II) arsenate	$Cu_3(AsO_4)_2$	7.6×10^{-36}
	copper(II) chromate	$CuCrO_4$	3.6×10^{-6}
	copper(I) sulfide	Cu_2S	2.5×10^{-48}
Pb	lead(II) carbonate	$PbCO_3$	7.4×10^{-14}
	lead(II) hydroxide	$Pb(OH)_2$	1.42×10^{-20}
	lead(II) sulfate	$PbSO_4$	2.53×10^{-8}
	lead(II) sulfide	PbS	3×10^{-7}
	lead(II) chloride	$PbCl_2$	1.7×10^{-5}
	lead(II) chromate	$PbCrO_4$	2.8×10^{-13}
	lead(II) fluoride	PbF_2	2.7×10^{-8}
	lead(II) arsenate	$Pb_3(AsO_4)_2$	4.0×10^{-36}
Hg	mercury(II) sulfide (red)	HgS	4×10^{-33}
	mercury(II) sulfide (black)	HgS	2×10^{-32}
	mercury(I) chloride	Hg_2Cl_2	5.0×10^{-13}
	mercury(I) sulfate	Hg_2SO_4	7.4×10^{-7}
	mercury(I) sulfide	Hg_2S	1.0×10^{-47}
Ni	nickel(II) carbonate	$NiCO_3$	1.42×10^{-7}
	nickel(II) hydroxide	$Ni(OH)_2$	5.48×10^{-16}
	nickel(II) sulfide	NiS	3×10^{-19}

[a] From Lide D.R., 2001

concentrations, octavite decreases the Cd^{2+} activity 100-fold for every pH unit increase above pH 7.5. Street et al. (1977) suggested that the solid phase $CdCO_3$ (log K_{sp}: -12.07) and $Cd_3(PO_4)_2$ (log K_{sp}: -32.61) most likely limit Cd^{2+} activities in soils, and found that $CdCO_3$ precipitated in sandy soils having low cation exchange capacity, low organic matter, and pH values > 7.0.

$$CdCO_3 \text{ (octavite)} + 2H^+ \rightleftharpoons Cd^{2+} + CO_2 + H_2O$$
$$\log K = 6.16 \quad (3.23)$$
$$CuCO_3 \text{ (c)} + 2H^+ \rightleftharpoons Cu^{2+} + CO_2 + H_2O$$
$$\log K = 8.52 \quad (3.24)$$
$$Cu_2(OH)_2CO_3 \text{ (malachite)} + 4H^+ \rightleftharpoons 2Cu^{2+} + CO_2 + 3H_2O$$
$$\log K = 12.99 \quad (3.25)$$
$$PbCO_3 \text{ (cerussite)} + 2H^+ \rightleftharpoons Pb^{2+} + CO_2(g) + H_2O$$
$$\log K = 4.65 \quad (3.26)$$
$$Pb_3(CO_3)_2(OH)_2 \text{ (c)} + 6H^+ \rightleftharpoons 3Pb^{2+} + 2CO_2 (g) + 4H_2O$$
$$\log K = 17.51 \quad (3.27)$$
$$ZnCO_3 \text{ (smithsonite)} + 2H^+ \rightleftharpoons Zn^{2+} + CO_2 (g) + H_2O$$
$$\log K = 7.91 \quad (3.28)$$

The low solubility of Cu oxide and hydroxide minerals and relatively high solubility of its carbonate cause the preferred association of Cu with the oxide phases, such as $CuFe_2O_4$, that may determine the solubility of Cu^{2+} in soil solution (Lindsay, 1979). In soils with high pH, lead carbonate ($PbCO_3$ (cerussite)) is stable, but its solubility is still higher than that of Pb phosphates.

3.5.1.2 Phosphate Dissolution/Precipitation

For the formation of metal phosphate, the conditions affecting soil solution H_2PO_4 concentration control the formation of metal phosphate. At high soil pH, H_2PO_4 concentrations are controlled by calcium phosphate. Lead phosphates have been found to be very stable in soil environments. Lead phosphates are relatively less soluble under equilibrium soil-surface conditions than oxides, hydrides, carbonates and sulfates (Ruby et al., 1994; Adriano, 2001). Since the concentration of Pb in arid soils is generally lower than that of phosphate, it is highly possible that phosphate may control Pb^{2+} solubility in soils since chloropyromorphite is always slightly more stable than cerussite (Lindsay, 1979). Other lead minerals such as the silicates, sulfates, carbonates and hydroxides are too soluble to form in soils. Various possible lead phosphates, which may occur as precipitation in arid soils, include the following reactions:

Solution Chemistry

$Pb(H_2PO_4)_2$ (c) $\rightleftharpoons Pb^{2+} + 2H_2PO_4^-$ logK = −9.85   (3.29)
$PbHPO_4$ (c) + $H^+ \rightleftharpoons Pb^{2+} + H_2PO_4^-$ logK = −4.25   (3.30)
$Pb_3(PO_4)_2$ (c) + $4H^+ \rightleftharpoons 3Pb^{2+} + 2H_2PO_4^-$ logK = −5.26   (3.31)
Soil Pb $\rightleftharpoons Pb^{2+}$ logK = −8.50   (3.32)
$Pb^{2+} + 2H_2PO_4^- \rightleftharpoons Pb(H_2PO_4)_2$ logK = 1.50   (3.33)
$Pb_4O(PO_4)_2$ (c) + $6H^+ \rightleftharpoons 4Pb^{2+} + 2H_2PO_4^- + H_2O$
   logK = 2.24   (3.34)
$Pb_5(PO_4)_3OH + 7H^+ \rightleftharpoons 5Pb^{2+} + 3H + 2PO_4^- + H_2O$
   logK = −4.14   (3.35)
$Pb_5(PO_4)_3Cl$(c) + $5CaCO_3$ (calcite) + $9H^+ \rightleftharpoons$
   $5Pb^{2+} + 5Ca_5(PO_4)_3OH$ (HA) +$5CO_2$(g) +$5H_2O$ +Cl^-
   logK = 9.19   (3.36)
$CaHPO_4 \cdot 2H_2O$ (brushite) + $H^+ \rightleftharpoons Ca^{2+} + H_2PO_4^- + 2H_2O(l)$
   logK = 0.63   (3.37)
$\beta Ca_3(PO_4)_2$ (c) + $4H^+ \rightleftharpoons 3Ca^{2+} + 2H_2PO_4^-$
   logK = 10.18   (3.38)

The Zn hydroxide, oxides and carbonates are very soluble and dissolve in soils. Brümmer et al. (1983) suggested that precipitation-dissolution reactions most likely occur in neutral and alkaline soils in controlling Zn activity in soil solution. Sadig (1991) reported that Zn^{2+} activity in arid soil solution from Saudi Arabia was under-saturated with respect to Zn_2SiO_4 and $Zn_2(PO4)_3.4H_2O$, but were within the range of $ZnFe_2O_4$. They concluded $ZnFe_2O_4$ was the most possible solid phase forming in soils and controlling Zn activity in soil solution. However, Ma and Lindsay (1990) did not find any solid-phase controlling Zn activity in soil solutions of 10 arid zone soils from Colorado.

Street et al. (1977) suggested that the solid phase $Cd_3(PO_4)_2$ (log K_{sp}: -32.61), except for $CdCO_3$ (log K_{sp}: -12.07), most likely limits Cd^{2+} activities in soils. Controlling the solubilities of trace elements in arid soils by solid phase will be discussed in detail in Chapter 5.

3.5.2 Adsorption/Desorption Process

Since most trace elements in soils are at parts per million levels, a separate compound may be not formed. Most likely, trace amounts of these trace elements and their compounds are adsorbed on the surfaces of clay minerals and various crystalline and amorphous Fe/Mn/Al oxides and hydroxides. Curtin and Smillie (1983) reported that the solubilities of Mn^{2+} and Zn^{2+} in limed soils were not consistent with the solubilities of any common minerals. The detailed mechanisms of adsorption and binding of trace elements by various solid-phase components will be discussed in later chapters. Trace elements are also desorbed from the surfaces of inorganic and organic colloids such as clay minerals and humic substances to replenish soil solution. Adsorption/desorption mechanisms may be one of the major processes controlling the solubility of most trace elements in arid and semi-arid soils. Lead and Cu are more strongly retained by many soil colloids than are Cd, Zn and Ni. On the other hand, it is difficult to distinguish adsorption from precipitation. This is especially true for arid and semi-arid soils with high soil pH. Very often two processes, adsorption and precipitation (co-precipitation), occur simultaneously. The presence of Mn oxides promotes the oxidation of Mn and Fe. This results in close association of Mn with Fe and trace elements through possible co-precipitation rather than adsorption at the oxide surface (Sposito, 1986).

Soil pH strongly affects adsorption of Cd, Zn, and Pb in soils, but less so for Cu. In addition, Ca^{2+} is the major soil solution cation of arid soils. Ca^{2+} has been shown to be important in inhibiting divalent heavy metal sorption in calcareous soils. Theoretically, the presence of Ca^{2+} in soil solution may reduce metal adsorption in arid soils. However, high pH increases overall metal adsorption in arid soils.

Dang et al. (1994) observed that the experimentally determined solubility lines for Zn^{2+} in 14 soil solutions from southern Queensland with soil pH from 7.45–8.98 and 0.08–2.07% $CaCO_3$ were not undersaturated with respect to the solubility of any known mineral form of Zn. Therefore, they suggested that Zn^{2+} activity was mainly controlled by adsorption-desorption reactions in these soils. Similar observation on solubility of Cr(VI) in arid soils was reported by Rai et al. (1989). In the absence of a solubility controlling solid phase, Cr(VI) aqueous concentrations under slightly alkaline conditions may be primarily controlled by adsorption/desorption reactions (Rai et al., 1989). Chromuim(VI) is adsorbed by iron and aluminum oxides, and kaolinite and its adsorption decreases with increasing pH.

3.5.3 Complexation of DOC and Organics

Dissolved organic carbon and plant- or microbe- produced phytosiderophores increase the solubility of most trace elements in arid and semi-arid soils. This is especially important in arid regions. High pH increases the solubility of organic matter as dissolved organic carbon in arid soils. Copper, lead and nickel have a strong tendency to form complexes, while Cd complexes are weaker. Zinc, cobalt and manganese are intermediate in this tendency. Dang et al. (1994) reported that complexation of total soluble Zn by organic and inorganic ligands constituted 40–50% of total soluble Zn in fertilized and unfertilized soil solutions from southern Queensland with soil pH from 7.45–8.98 and 0.08–2.07% $CaCO_3$. In South Australia, concentrations of Cu and Zn in water extracts at three water/soil ratios (1.5, 3, and 5) from soils with pH > 7 increase with DOC. Yin et al. (1996) reported that an increase in soil pH increased the solubility of dissolved organic matter and thus increased complexation of Hg by dissolved organic matter, resulting in decreased Hg adsorption by soils.

Dissolved organic molecules have many acidic functions (hydroxol and carbonic groups) to complex trace elements and their compounds to form soluble chelates. This is one of the reasons why solubility and bioavailability of trace elements in the rhizosphere are higher than bulk soils. At the same time, many organic acids also directly dissolve trace elements and their compounds in soils. Plant-produced phytosiderophores facilitate elements, such as Fe and Zn, uptake by plants (Zhang et al., 1991; Romheld, 1991; Hopkins et al., 1998). However, Shenker et al. (2001) did not find significant uptake of the Cd-phytosiderophores complex by plant roots.

Rhizosphere soil differs from the bulk soil in physical, chemical and biological aspects (Dinkelaker et al., 1993; Lombi et al., 1999), which affects the solubility of trace elements and overall bioavailability to plants. Fenn and Assadian (1999) reported that less $CaCO_3$ was found in plant rhizophere than in bulk samples. Solubility and bioavailability of Fe, Zn, Ni and Cd increase in the rhizosphere through the release of strong organic chelators by plant roots and microbes (Marschner and Romheld, 1996; Awad and Romheld, 2000a, b). Motaium and Babawy (1999) reported that the solubility of Fe, Mn, Cu, Zn, Cd and Co in rhizosphere was higher than bulk soils in Egyptian arid soils after 10, 40 and 80 years of irrigation with raw sewage water. The rhizosphere chemistry will be discussed further in later chapters.

3.5.4 Reduction-oxidation

Reduction-oxidation is one of the most important processes controlling solubility and speciation of trace elements in soils, especially for those elements with changeable values, such as Cr, As and Se. Within normal ranges of redox potentials and pH commonly found in soils, the two most important oxidation states for Cr are Cr(III) and Cr(VI). Cr(III) is the most stable form of chromium and less soluble and nontoxic, but Cr(VI) is mobile, soluble and toxic. The main aqueous species of Cr(III) are Cr^{3+}, $Cr(OH)^{2+}$, $Cr(OH)_3^0$ and $Cr(OH)_4^-$; and the major aqueous species of Cr(VI) are $HCrO_4^-$, $CrO_4^=$ and $Cr_2O_7^=$ (Palmer and Puls, 1994; Bartlett and Kimble, 1976). In arid and semi-arid soils, $CrO_4^=$ dominates. Cr(III) is oxidized into soluble chromate (Cr(VI)) by Mn oxides (Mn(IV)) in soils and vadose zone (Chung et al., 2001). The rate of oxidation is greater for todorokite and birnessite that contain the most quadrivalent Mn and least for lithiophorite that contains a greater proportion of trivalent Mn (Kim et al., 2002). However, James (1994) reported that alkaline soils with pH 8–10 oxidized only small amounts of added Cr(III) and did not reduce added Cr(VI), probably due to the high soil pH. In contrast, Cr(VI) in soils and vadose zones of arid and semi-arid regions is reduced both abiotically (by organic carbon) and biologically by microorganisms to insoluble Cr(III) (Oliver et al., 2003; Tokunaga et al., 2003). Added Mn^{2+} readily reduces > 50% of soluble Cr(VI) in the alkaline soils with pH 8–10 and more Cr(VI) is reduced in the soils with high concentrations of Cr(VI) at pH 10 (James, 1994). Fe(II) and reduced S can also reduce Cr(VI) to Cr(III). Ferrous iron is more effective than Mn^{2+} as a reducing agent, but acidifies the alkaline soils more than the Mn treatments do (James 1994). Addition of organic carbon and nitrate in arid soils and vadoze zones increases reduction processes (Oliver et al., 2003; Tokunaga et al., 2003; Cifuentes et al., 1996). Lactic acid is an effective reducing agent for Cr(VI) when the pH is decreased below 7.2 (James, 1994).

Arsenite can be oxidized by manganese dioxides in soils. The rate constants for the depletion of As(III) by birnessite and cryptomelane are much higher than those by pyrolusite due to the difference in the crystallinity and specific surfaces of the Mn oxides (Oscarson et al., 1983). The ability of the Mn dioxides to sorb As(III) and As(V) is related to the specific surface and the point-of-zero charge of the oxides. The one-to-one relationship between the amount of As(III) depleted and the amount of As(V) appearing in solution was reported by Oscarson and colleagues (1983).

Soil redox also strongly affects solubility of the compounds of other trace elements in arid soils. Amrhein et al. (1993) found that Fe, Mn, Ni and V in an evaporation pond soil were more soluble under reducing conditions. Han and Banin (2000) reported that after one year of saturated incubation, the solubility of Fe, Mn, Co, V, Ni, Cu and Zn in two Israeli arid soils with 0.5–23% $CaCO_3$ increased, while the solubility of Cd decreased with time. During saturated incubation, soil pH in highly calcareous arid soil containing high content of carbonates decreased. In a loessial soil from Israel, Han and Banin (1996) reported that soil pH decreased from 8.0 to 7.0–7.4 over saturated incubation. With the decrease in Eh over incubation, the parameter pe+pH also decreased from initial values of 12–13.6 to 4 after initial 7–9 days of saturation incubation.

Many biogeochemical conditions affect the solubility of trace element compounds in arid and semi-arid soils. These include soil pH, Eh and dissolved organic carbon, of which soil pH is the most important factor. Decreases in soil pH increases solubility of most of these trace elements (Cu, Cr(III), Co, Cd, Ni, Pb, Zn, Mn and Hg) and their compounds. Jeffery and Uren (1983) reported the solubility of zinc in soil decreased markedly with increasing soil pH. Increases in solubility of these compounds due to lowering of pH is attributed to both the increase in direct dissolution of these compounds and the indirect decrease in adsorption of the cations by various soil solid-phase components. Ma and Lindsay (1990) found that Zn, Pb and Cd in soil solutions of 10 arid soils from Colorado increased with decreasing soil pH. Acid rain, the release of organic acids from plant roots, decomposition of soil organic matter and the application of acid-generating fertilizers all decrease soil pH. Soil and plant respiration also affect soil pH. As the CO_2 pressure in soil is increased, the soil pH goes down. On the other hand, decreases in soil pH increase solubility of anion trace elements in soils, including As, Se, Cr(VI), and Mo.

Chapter 4

SELECTIVE SEQUENTIAL DISSOLUTION FOR TRACE ELEMENTS IN ARID ZONE SOILS

The risk from heavy metal inputs into a given soil as related to introduction into the food-chain and/or migration to ground water is determined by the partition of the added heavy metals between the solution and the solid-phase. This risk is also determined by its partition among the various components of the solid phase, including clay and organic matter surfaces (EXC), carbonates (CARB), easily reducible oxides (e.g., manganese oxides (ERO)), organic matter (OM), reducible oxides (e.g., iron oxides (RO)) and clay mineral/alumosilicate minerals (Tessier et al., 1979; Emmerich et al., 1982; Banin et al., 1990; Banin et al., 1997a).

The selective sequential dissolution method (SSD) has been developed and is widely used to study the forms, bioavailability, mobility and transformation of trace elements/heavy metals in sludge, manure, soils, sludge-amended soils and sediments (Emmerich et al., 1982; Banin et al., 1990; Shuman, 1985a, b; Sposito et al., 1982; McGrath and Cegarra, 1992; McLaren and Ritchie, 1993; Han and Banin, 1997, 1999; Chang et al., 1984). This method is based on both the solubility of individual solid-phase components as well as the selectivity and specificity of the chemical reagents. This chapter discusses the advantages and disadvantages of selective sequential dissolution protocols and their common extractants and orders. A widely used sequential dissolution procedure for arid soils is optimized by focusing on the carbonate step. Two commonly used SSD procedures for trace elements in arid zone and humid zone soils are compared.

4.1 SELECTIVE SEQUENTIAL DISSOLUTION: CHARACTERIZATION, ADVANTAGES, AND DISADVANTAGES

The basic assumption behind the sequential dissolution methods is that a particular extractant is phase or retention mode specific in its chemical attachment on a mixture of forms (D'Amore et al., 2005). The procedure should provide a gradient for the physicochemical association strength between trace elements and solid particles rather than actual speciation (Martin et al, 1987), thus giving a semi-quantitative indication for their relative availability to plants or to further migration to ground water.

There are a number of selective sequential dissolution procedures which have been developed for specific elements, matrices, regional soils, and specific purposes (Table 4.1). Additionally, various extractants are used in different sequential procedures for the similar targeted solid-phase component. The review on the extractants of individual fractions of metals in soil was made by Shuman (1991). However, in most protocols the trace elements/heavy metals in their native and waste-amended soils are divided into the following physicochemical forms:

1) Simple or complexed ions in solution in equilibrium with exchangeable and other phases.

2) Exchangeable ions (EXC), sometimes including ions nonspecifically adsorbed and specifically absorbed on the surface of various soil components, such as carbonate, organic matter, Fe, Mn, Si, and Al oxides, and clay minerals. This part is controlled by adsorption-desorption processes.

3) Organically bound matter (OM). Heavy metals/trace elements may be bound in living organisms, detritus, and organic matter of the soil. The organically bound trace elements or heavy metals are affected by the production and decomposition of organic matter.

4) Occluded or co-precipitated by carbonate (CARB). This fraction would be susceptible to changes of soil pH.

5) Metals-bound to iron and manganese oxides, including amorphous and crystalline oxides, which appear as nodules, concretions and coatings on particles. This fraction is reduced or oxidized by changes of Eh, including three divisions: metals bound to Mn oxides or easily reducible oxides (ERO), metals bound to amorphous Fe oxides (AmoFe), and metals bound to crystalline Fe oxides (CryFe) (the last two divisions are sometimes collectively called reducible oxides fraction (RO)).

6) Residual. This fraction mainly contains primary and secondary minerals, which hold elements within their crystal structure. This fraction also contains trace elements remained from the extraction of all previous fractions (e.g., humin bound). These metals/trace elements are not expected to be released into soil solutions over a reasonable time span under conditions normally encountered in nature.

The terms of the fractions are more likely to be operationally, rather than chemically, defined. However, each extractant in the sequentially selective procedure mostly targets one major solid-phase component. Therefore, trace elements bound to the targeted fraction can be either

Selective Sequential Dissolution

Table 4.1. Some selective sequential dissolution procedures employed to fractionate trace elements in soils[a,b]

No.	References	Fractions					
		Soluble	Exchangeable	Carbonate bound	Fe/Mn oxides (Amorphous/crystalline Fe oxide)	Organically bound	Residual
1	Tessier et al., 1979		1M $MgCl_2$ (1)	1M NaOAc/HOAc (2)	0.04M $NH_2OH.HCl$/ 25% HOAc (3)	30% H_2O_2/ 0.02M HNO_3 (4)	HF/$HClO_4$ (5)
2	Shuman, 1985a, b		1M $MgCl_2$ (1)		0.1M $NH_2OH.HCl$ (3), 0.2M $(NH_4)_2C_2O_4/H_2C_2O_4$ (4) 0.2M $(NH_4)_2C_2O_4/H_2C_2O_4$/ ascorbic acid (5)	0.7M NaOCl (2)	HF/HNO_3/HCl (6-8) (Sand, silt and clay)
3	Sposito et al., 1982		0.5M KNO_3/H_2O (1)	0.05M EDTA (3)		0.5M NaOH (2)	4M HNO_3 (4)
4	Emmerich et al., 1982 McLaren and Crawford, 1973	0.05M $CaCl_2$ (1)	2.5% HOAc (2)			1M $K_4P_2O_7$ (3)	HF
5	McGrath and Cegarra, 1992		0.1M $CaCl_2$ (1)	0.05M EDTA (3)		0.5M NaOH (2)	5% HCl (4)
6	Ma and Rao, 1997b	H_2O (1)	1M $MgCl_2$ (2)	1M NaOAc-HOAc (3)	0.04M $NH_2OH.HCl$ + 25%HOAc (4)	H_2O_2 (5)	HF/HCl/HNO_3 (6)
7	Han and Banin, 1996		1M NH_4OAc (1)	1M NaOAc-HOAc (2)	0.04 M $NH_2OH.HCl$ + 25% HOAc (3, 6 with heating)	H_2O_2 (4)	4M HNO_3 (7)
	Han et al., 2001b				0.2 M $(NH_4)_2C_2O_4/H_2C_2O_4$ (5) 0.1M $NH_2OH.HCl$ + 25% HOAc (3)		
8	Han et al., 1995		1M $MgCl_2$ (1)	1M NaOAc-HOAc (3)	0.1M $NH_2OH.HCl$ (4) 0.2M $(NH_4)_2C_2O_4/0.2M\ H_2C_2O_4$ (6) 0.04M $NH_2OH.HCl$ + 25%HOAc (7)	0.1M $Na_4P_2O_4$ (2) H_2O_2 (5)	HF/$HClO_4$ (8)

[a] Number in parenthesis is the order in the sequential dissolution procedure.
[b] From Han and Banin, 1995, Copyright (1995), with permission from Taylor & Francis US.

referred to as the targeted solid-phase component or the extractant itself; both can be found in the literature. In this book, trace element fractions among various solid-phase components of soils are named in terms of the targeted solid-phase components. The SSD techniques are, in general, easy to apply, inexpensive, and require minimal data analysis (D'Amore et al., 2005).

Functionally, one cannot remove all of a targeted solid-phase component with only one extractant and without experiencing some degree of interaction on or with other components. No selective dissolution scheme can be considered completely accurate in distinguishing between different forms of an element (i.e., various organic-inorganic solid-phase components). The common criticism of sequential dissolution extractions is the lack of selectivity and sufficiency of extractions of some solid-phase components (Biester and Scholz, 1997; Nirel and Morel, 1990). Biester and Scholz (1997) reported that sequential dissolution extraction did not completely extract organically bound Hg with H_2O_2 and observed that mercury found in the residual step consisted of Hg^0, humic-bound Hg or mercury sulfide. This may be true of those trace elements with a strong interaction with organic matter, especially humin in soils. Additionally, there may be redistribution and re-adsorption during sequential dissolution extraction (Rendell et al., 1980; Kheboian and Bauer, 1987). However, re-adsorption observed during sequential extraction may be due to the use of either large spikes or simple model materials (Belzile et al., 1989). Re-adsorption of metals (Cd, Cu, Pb and Zn) during sequential extractions has been reported to be minimal (Belzile et al., 1989; Kim and Fergusson, 1991). Also, as discussed below, various extractants with different selectivities and specificities are used in different sequential procedures. These make comparisons of distribution of trace elements/heavy metals in soils or waste-amended soils obtained by different sequential procedures more difficult.

Despite these shortcomings, common to any chemical extraction procedure, sequential dissolution techniques still furnish more useful information on metal binding, mobility and availability than can be obtained with only a single extractant.

4.2 OPTIMIZATION OF THE SELECTIVE SEQUENTIAL DISSOLUTION PROCEDURES FOR TRACE ELEMENTS IN ARID SOILS – FOCUSING ON THE CARBONATE STEP

In many SSD procedures, one of the important steps, which is still problematic, is "carbonate"-extraction. This chemical component is an abundant constituent of soils, particularly in the arid regions. It contains a number of trace elements in a relatively mobile state. Thus, a correct estimation of its elemental content is of great importance.

The carbonate-bound trace elements are usually extracted by NaOAc-HOAc (Tessier et al., 1979; Hickey and Kittrick, 1984; Banin et. al., 1990; and Han et al., 1992) or by using chelating agents, such as EDTA (Sposito et al., 1982; Emmerich et al., 1982; Miller and Mcfee, 1983; Chang et al., 1984; Knudtsen and O'Connor, 1987; McGrath and Cegarra, 1992). Generally, EDTA lacks the required specificity and selectivity for carbonate phase extractions and tends to extract metals from a number of soil components.

The protocol involving NaOAc-HOAc at pH 5 was first proposed and used by Jackson (1958) to remove carbonates from calcareous soils to analyze soil cation exchange characteristics (Grossman and Millet, 1961). Other researchers used HOAc for the extraction of metals from sediments and soils (Nissenbaum, 1972; Mclaren and Crawford, 1973). Tessier et al. (1979) first used the NaOAc-HOAc solution at pH 5 to dissolve the carbonate fraction from sediments. Since then, the NaOAc-HOAc buffer has been widely used as a specific extractant for the carbonate phase in various media (Tessier et al., 1979; Hickey and Kittrick, 1984; Rapin et al., 1986; Mahan et al., 1987; Han et al., 1992; Clevenger, 1990; Banin et al., 1990). Despite its widespread use, this step is not free from difficulties, and further optimization is required in its application. Questions arise with regard to this step in the elemental extraction from noncalcareous soils, the dissolution capacity and dissolution rates imposed by the buffer at various pHs, and the possibility that different carbonate minerals may require different extraction protocols (Grossman and Millet, 1961; Tessier et al., 1979).

The following sections summarize the studies on the dissolution technique for the carbonate fraction from arid and semi-arid soils with different amounts and types of carbonate minerals. In addition, the selectivity and effectivity of the NaOAc-HOAc extraction technique at varying pHs to extract the carbonate phase, and only the carbonate phase, from soils is examined (Han and Banin, 1995).

4.2.1 Dissolution Capacity of NaOAc-HOAc Solutions at Various pHs

Dissolution of carbonate by NaOAc-HOAc solutions at varying pHs from arid-zone soils, as indicated by X-ray diffraction, are presented in Fig. 4.1. Arid soils from Israel contained varying contents of $CaCO_3$ (from 13.7–68.1%). X-ray diffraction showed that calcareous soils used in this study mainly contained calcite (very strong peak at d = 3.03 Å, and weak peaks at 2.29 Å and 2.10 Å), and only soils K2 and J3 had a small amount of dolomite (very weak peak at d = 2.89 Å) (Fig. 4.1).

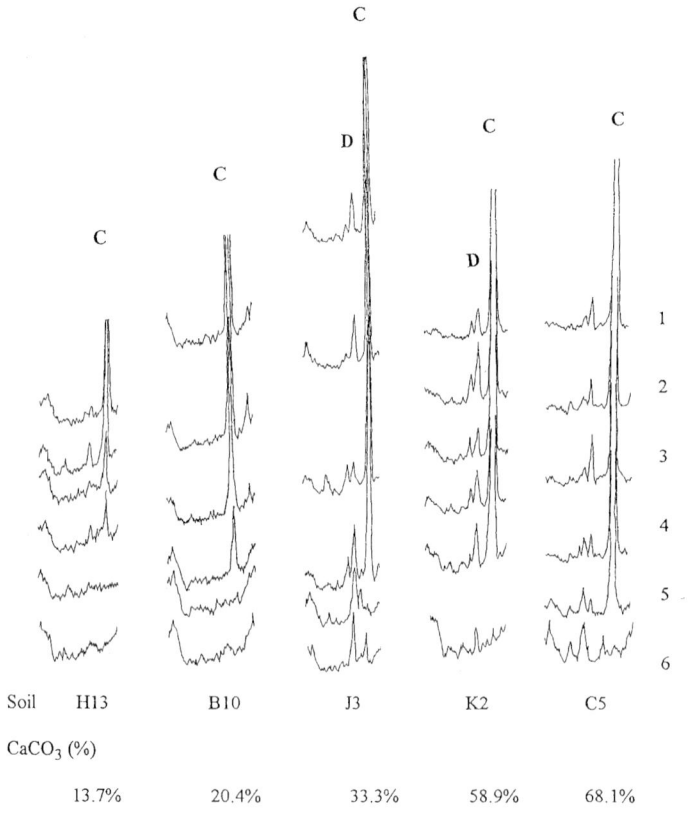

Figure 4.1. Removal of carbonate from Israeli arid soils as indicated by the X-ray diffractograms after extraction of the carbonate fraction by NaOAc-HOAc solutions at various pHs for 16 hours. C: calcite: d = 3.04 Å, and D: dolomite, d = 2.89 Å. Number 1, 2, 3, 4, 5, and 6 indicate non-treated soil (No. 1), treatments (No. 2-6) with NaOAc-HOAc solutions at pH 7.0, 6.0, 5.5, 5.0 and 4.0, respectively (after Han and Banin, 1995. Reprinted from Commun Soil Sci Plant Anal, 26, Han and Banin A., Selective sequential dissolution techniques for trace metals in arid-zone soils: The carbonate dissolution step, p 563, Copyright (1995), with permission from Taylor & Francis US)

The X-ray diffraction results showed that peaks of calcite disappeared after extraction of the carbonate (CARB) fraction by the NaOAc-HOAc solution at pH 5.0 from the soils with 10–30% of $CaCO_3$, such as soils H13, B10 and J3. This indicates that the buffer solution at pH 5.0 is able to completely dissolve calcite from these arid-zone soils (Fig. 4.1). The buffer solution at pH 5.5, however, could dissolve almost all of the calcite only from the soils with 10–20% of carbonate (soils H13 and B10). Even this buffer solution at pH 6.0 dissolved most of calcite from the soil with 10% of carbonate (soil H13). The buffer solution at pH 4.0 thoroughly dissolved all the calcite from the soils with 60% and 70% of $CaCO_3$. The buffer solutions at pH 5.0, 5.5, 6.0 and 7.0, however, did not dissolve all of the calcite from these two soils. Soil extractions (as Ca) also showed the same trend (Fig. 4.2).

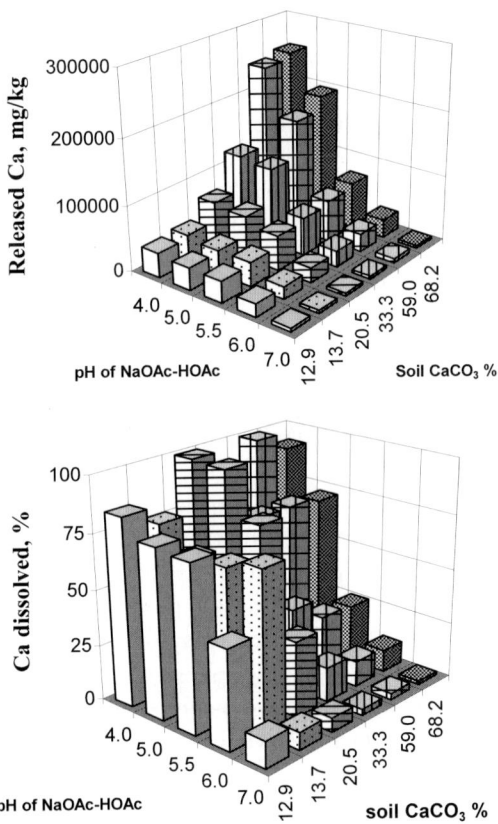

Figure 4.2. Dissolution of Ca from Israeli arid soils by NaOAc-HOAc solutions at various pHs after the extraction of the exchangeable fraction (after Han and Banin, 1995. Reprinted from Commun Soil Sci Plant Anal, 26, Han and Banin A., Selective sequential dissolution techniques for trace metals in arid-zone soils: The carbonate dissolution step, p 568, Copyright (1995), with permission from Taylor & Francis US)

The graph obtained by X-ray diffraction (Fig. 4.1) showed that the NaOAc-HOAc solution at pH 5.0 could not completely dissolve the trace quantities of dolomite from the soils during the first 16 hours of extraction. Tessier et al. (1979), however, reported that the peak generated by dolomite after extraction by the buffer at pH 5.0 disappeared from sediment.

4.2.2 Kinetics of Dissolution of $CaCO_3$ with NaOAc-HOAc Solutions

The kinetics of dissolution of pure $CaCO_3$ and soil $CaCO_3$, as indicated by the volume of CO_2 released and Ca dissolved during extraction, are presented in Fig. 4.3. It shows that dissolution of both pure and soil $CaCO_3$ by the NaOAc-HOAc solutions at various pHs reached a plateau after two hours. This indicates that a certain acid dose reacts completely with the proper content of soil carbonate within two hours. Tessier et al. (1979) reported that after five hours of leaching sediments, there was no increase in the calcium concentration, thus indicating that it is unnecessary to allow 16 hours for extraction of the CARB fraction, as was originally done in this sequentially selective dissolution procedure.

4.2.3 Dissolution of Major and Trace Elements by NaOAc-HOAc at Various pHs

Calcium can be used as an indicator for selective carbonate dissolution. Its concentration in the extracts is limited either by its content in the carbonate phase (at low pHs and/or low carbonate content of the soil) or by the acid capacity of the extracting solution (at higher pHs and/or higher carbonate content in the soil). The interplay of these factors is clearly shown in Fig. 4.2.

The effective dissolution capacity of the NaOAc-HOAc buffer solutions at various pHs in the CARB step of the current SSD protocol is estimated from the maximum quantity of Ca released (Fig. 4.2). When the buffer solution is at pH 5.0, almost all $CaCO_3$ can be dissolved from calcareous soils with about 45–50% $CaCO_3$ at this step. For soils with higher carbonate contents, a second dose of the buffer solution must be added to complete the dissolution step.

The specificity of the NaOAc-HOAc solutions for the carbonate phase may be measured by other major elements released during extraction. For example, only limited attack of the alumosilicate components in the soil by this buffer solution was observed (Table 4.2). When the buffer solution is at pH 5.0 and above, the dissolution of these major elements is very limited.

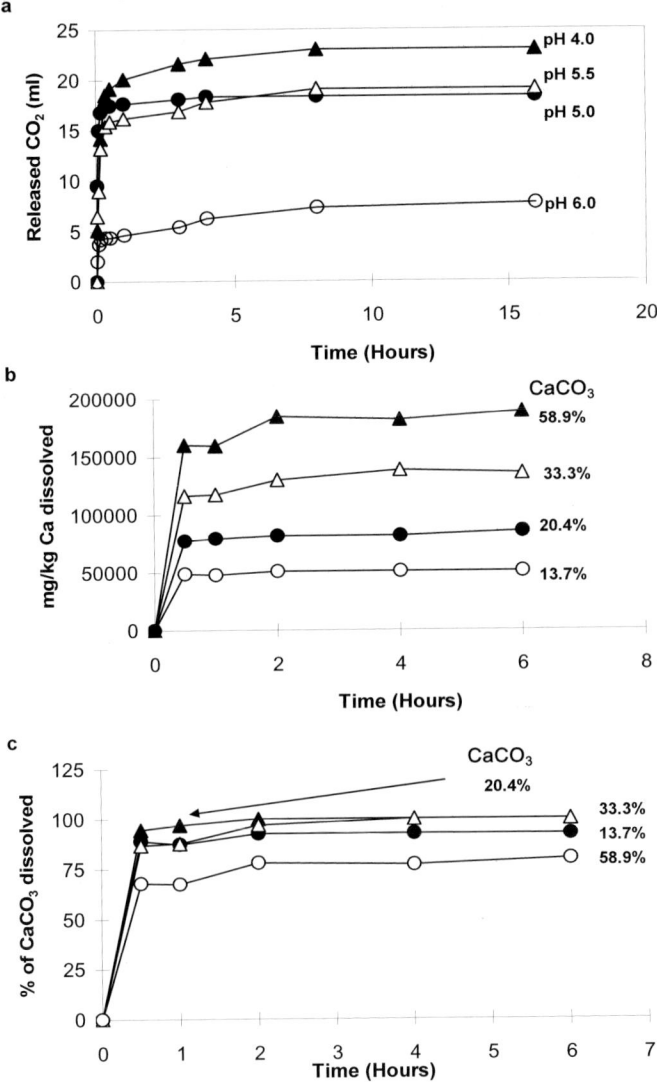

Figure 4.3. Dissolution kinetics of $CaCO_3$ by NaOAc-HOAc solutions at varying pHs. a: Dissolution of pure $CaCO_3$ by NaOAc-HOAc (0.1 g $CaCO_3$, 25 mL buffer solution), b and c: Dissolution of soil $CaCO_3$ by NaOAc-HOAc at pH 5.0 (1.0 g soil, 25 ml buffer solution) (after Han and Banin, 1995. Reprinted from Commun Soil Sci Plant Anal, 26, Han and Banin A., Selective sequential dissolution techniques for trace metals in arid-zone soils: The carbonate dissolution step, p 565, Copyright (1995), with permission from Taylor & Francis US)

Table 4.2. Dissolution of some major elements extracted by NaOAc-HOAc solutions at various pHs after removing the exchangeable fraction[a]

Soil (CaCO$_3$, %)	pH of Solution	Ca mg/kg	Ca % of total	Mg mg/kg	Mg % of total	Fe mg/kg	Fe % of total	Mn mg/kg	Mn % of total	Al mg/kg	Al % of total	K mg/kg	K % of total	Si mg/kg
S1 12.9	4.0	37391	84.5	810	12.9	30.6	0.20	58.4	24.4	133	1.30	1298	47.2	218
	5.0	33811	76.4	795	12.7	0.00	0.00	43.6	18.3	23.4	0.20	0	0.00	117
	5.5	33189	74.9	564	9.00	0.00	0.00	37.0	15.5	0.00	0.00	0	0.00	83
	6.0	19727	44.6	359	5.70	0.00	0.00	14.3	5.95	0.00	0.00	0	0.00	53
	7.0	5319	12.0	62	1.00	0.00	0.00	1.8	0.70	0.00	0.00	0	0.00	18
H13 13.7	4.0	41979	76.6	635	6.00	60.7	0.20	147	18.2	247	0.80	1365	22.3	445
	5.0	36168	66.0	513	4.90	0.00	0.00	63.1	7.80	11.3	0.00	272	4.40	232
	5.5	36803	67.2	217	2.10	0.00	0.00	23.0	3.00	0.00	0.00	0	0.00	83
	6.0	19695	71.7	262	2.50	0.00	0.00	20.2	2.50	0.00	0.00	0	0.00	95
	7.0	4344	7.9	112	1.10	0.00	0.00	2.9	0.40	0.00	0.00	0	0.00	37
B10 20.5	4.0	70453	106.0	664	7.40	95.0	0.30	73.3	11.8	521	1.70	85	1.60	662
	5.0	65692	99.3	561	6.30	0.00	0.00	25.3	4.10	79.6	0.26	0	0.00	329
	5.5	52704	79.7	448	5.10	0.00	0.00	24.0	3.90	0.00	0.00	0	0.00	169
	6.0	22582	34.1	236	2.60	0.00	0.00	4.3	0.70	0.00	0.00	0	0.00	83
	7.0	4223	6.4	90	1.00	0.00	0.00	0.0	0.00	0.00	0.00	0	0.00	29
J3 33.3	4.0	123495	77.6	1494	13.8	65.2	0.40	136	42.8	123	0.90	1082	25.3	308
	5.0	117220	73.7	1166	10.8	0.00	0.00	90.6	28.5	5.22	0.04	191	4.50	150
	5.5	57572	36.2	552	5.10	0.00	0.00	52.0	16.60	0.00	0.00	0	0.00	92
	6.0	24480	15.4	272	2.50	0.00	0.00	16.3	5.20	0.00	0.00	0	0.00	57
	7.0	4542	2.9	74	0.70	0.00	0.00	1.8	0.60	0.00	0.00	0	0.00	43
K2 59.0	4.0	243612	103.4	1837	10.9	181.0	1.30	181	60.5	86.0	0.55	624	15.0	535
	4.5	237408	100.8	1554	9.20	143.0	1.03	143	47.8	19.00	0.12	11	0.26	339
	5.0	173561	73.7	709	4.20	88.0	0.63	88.0	29.4	0.00	0.00	0	0.00	142
	5.5	62706	26.6	416	2.50	31.0	0.22	31.0	10.4	0.00	0.00	0	0.00	127
	6.0	27375	11.6	281	1.70	11.0	0.08	11.0	3.68	0.00	0.00	0	0.00	148
	7.0	6910	2.9	144	0.90	1.50	0.01	1.5	0.50	0.00	0.00	0	0.00	95
C5 68.2	4.0	252072	92.4	1026	20.9	86.0	0.62	86.0	49.7	103	0.59	639	42.2	347
	4.5	293303	107.5	923	18.8	74.0	0.54	74.0	42.8	28.0	0.16	101	6.68	230
	5.0	195206	71.6	576	11.8	38.0	0.28	38.0	22.0	0.00	0.00	0	0.00	115
	5.5	69620	25.5	201	4.10	25.0	0.18	25.0	14.5	0.00	0.00	0	0.00	63
	6.0	27591	10.1	93	1.90	21.0	0.15	21.0	12.1	0.00	0.00	0	0.00	69
	7.0	4100	1.5	35	0.70	6.00	0.04	6.0	3.47	0.00	0.00	0	0.00	48

[a] From Han and Banin, 1995, Copyright (1995), with permission from Taylor & Francis US.

Table 4.3. Dissolution of some trace elements extracted by NaOAc-HOAc at various pHs after removing the exchangeable fraction (mg/kg)[a]

Soil	pH of NaOAc-HOAc	Trace elemetns								
		Ti	Cd	Cu	Pb	Zn	Ni	Cr	V	Co
S1	4.0	0.02	0.85	0.73	0.00	0.00	0.22	0.44	1.68	0.43
	5.0	0.03	0.68	0.00	0.00	0.00	0.32	0.22	0.51	0.66
	5.5	0.00	0.64	0.15	0.00	0.00	0.34	0.09	0.69	0.40
	6.0	0.00	0.50	0.00	0.00	0.00	0.00	0.00	0.54	0.30
	7.0	0.00	0.22	0.00	0.00	0.00	0.00	0.06	0.48	0.36
H13	4.0	0.01	0.92	21.94	0.00	3.55	1.95	0.85	2.81	0.82
	5.0	0.00	0.63	1.50	0.00	0.00	0.78	0.20	0.63	0.18
	5.5	0.00	0.47	0.00	0.00	0.00	0.00	0.00	0.31	0.10
	6.0	0.00	0.46	0.25	0.00	0.00	0.31	0.00	0.60	0.00
	7.0	0.00	0.17	0.14	0.00	0.00	0.00	0.00	0.26	0.07
B10	4.0	0.01	1.02	1.30	0.00	0.19	1.42	0.89	2.04	0.21
	5.0	0.00	0.72	0.05	0.00	0.00	0.58	0.30	0.43	0.00
	5.5	0.00	0.78	0.00	0.00	0.00	0.34	0.00	0.49	0.15
	6.0	0.00	0.43	0.00	0.00	0.00	0.00	0.00	0.27	0.04
	7.0	0.04	0.14	0.00	0.00	0.00	0.00	0.00	0.00	0.13
J3	4.0	0.57	1.28	3.21	10.70	6.87	6.79	1.00	2.13	0.85
	5.0	0.36	0.99	1.79	4.90	7.08	6.21	0.47	1.04	0.37
	5.5	0.00	0.79	0.00	0.00	0.00	0.36	0.03	0.87	0.15
	6.0	0.00	0.44	0.00	0.00	0.00	0.00	0.00	0.51	0.11
	7.0	0.00	0.15	0.00	0.00	0.00	0.00	0.00	0.19	0.27
K2	4.0	0.55	1.93	7.54	13.58	16.77	13.88	6.43	3.07	1.41
	4.5	0.61	1.47	4.56	8.38	11.28	10.75	4.91	0.67	0.76
	5.0	0.03	1.24	1.13	0.00	0.00	1.90	2.59	1.13	0.44
	5.5	0.00	0.80	0.74	0.00	0.00	0.76	0.69	0.99	0.33
	6.0	0.00	0.45	0.20	0.00	0.00	0.53	0.16	0.63	0.21
	7.0	0.00	0.19	0.12	0.00	0.00	0.00	0.00	0.70	0.19
C5	4.0	0.68	1.75	8.07	12.06	13.65	8.36	11.23	2.42	1.07
	4.5	0.59	1.46	4.49	8.98	12.94	8.21	8.73	1.10	0.78
	5.0	0.57	1.19	2.56	8.00	8.78	6.90	4.87	1.05	0.27
	5.5	0.00	0.76	0.03	0.00	0.00	0.19	0.95	0.81	0.04
	6.0	0.00	0.48	0.00	0.00	0.00	0.00	0.16	0.55	0.13
	7.0	0.00	0.17	0.00	0.00	0.00	0.00	0.00	0.27	-

[a]From Han and Banin, 1995. Copyright (1995), with permission from Taylor & Francis US.

Grossman and Millet (1961) found that the free Fe-oxide concentration in noncalcareous soils was unchanged after contact with this buffer for nine weeks. Other researchers have shown that acetic acid at a concentration of 2.5% and pH 2.5 led to a partial attack of Fe and Mn oxides (Nissenbaum, 1972; Mclaren and Crawford, 1973; Tessier et al., 1979). Tessier et al. (1979) also indicated that this buffer solution at pH 5.0 was minimal in the attack of silicate minerals and sulfide.

The contents of trace elements extracted by the buffer solutions depend upon the solution's acid capacity in dissolving carbonate from soils. Trace elements dissolved by the buffer solution increased with decreasing pH of the buffer solution (Table 4.3). Release of trace elements by the buffer solutions at pH 6.0 was much smaller from calcareous soils with more than 30% of $CaCO_3$. The dissolution of trace elements by the buffers paralleled with the dissolution of Ca and Mg. The correlation coefficients between Ca and trace elements were as follows: Cd (0.92), Pb (0.87), Zn (0.90), Ni (0.90), Cr (0.91), V (0.54) and Co (0.70); and between Mg and trace elements were Cd (0.88), Pb (0.80), Zn (0.79), Ni (0.87), Cr (0.58), V (0.69) and Co (0.80), (all with n = 32).

4.2.4 Effects of pH of NaOAc-HOAc Solutions on the Subsequent Fractions

Extraction of the CARB fraction by the NaOAc-HOAc solution at various pHs resulted in the larger differences obtained in the subsequent fractions, i.e., ERO fraction, OM fraction, and to some extent, the RO fraction (Figs. 4.4 and 4.5). Extraction of the CARB fraction by NaOAc-HOAc solutions at pH 6.0 and 7.0 usually led to higher contents of major and trace elements in the ERO fraction than that by the buffer solutions at pH 4.0 and 5.0. This was especially obvious for Ca and Mg in calcareous soils. Metal extraction from the OM and RO fractions after extraction of the CARB fraction by the NaOAc-HOAc solutions at pH 7.0 and 6.0 were also higher than those extracted by the buffer solutions at pH 4.0 and 5.0, as shown in Figs. 4.4 and 4.5.

The effects of the pH of the NaOAc-HOAc solutions on the subsequent fractions are related to the partitioning patterns of elements in soils. Calcium and Cd in the calcareous soils are predominately present in the CARB fraction (Banin et al., 1990). Cadmium and Ca in the CARB fraction of the soils studied accounted for 40–50% and 75–99%, respectively. Even NaOAc-HOAc solutions at pH 7.0 extracted 3–6% and

7–10% of Ca and Cd from soils, respectively. The pH of the buffer solution for the CARB fraction strongly affected the subsequent fractions of the elements with high to intermediate preference for the CARB fraction, such as Ca, Mg, Cd, Pb, Zn, and Cu (Figs. 4.4 and 4.5, Cd and Pb data not shown). However, the pH of the buffer solution for the CARB fraction did not significantly affect the subsequent fractions of the elements with a majority in the RO and RES fractions, such as Fe (Fig. 4.5).

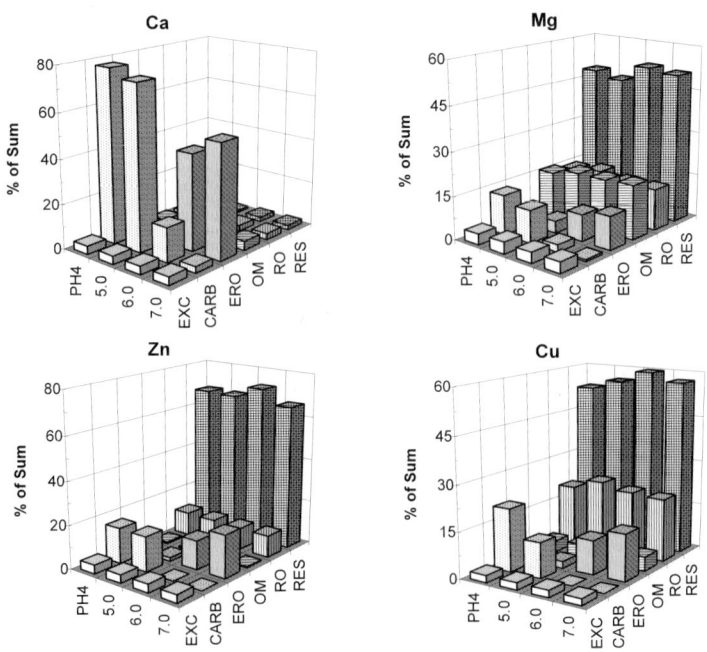

Figure 4.4. Effects of pH of NaOAc-HOAc solutions for the CARB fraction on the subsequent fractions (Soil $CaCO_3$: 33.3%) (after Han and Banin, 1995. Reprinted from Commun Soil Sci Plant Anal, 26, Han and Banin A., Selective sequential dissolution techniques for trace metals in arid-zone soils: The carbonate dissolution step, p 572, Copyright (1995), with permission from Taylor & Francis US)

In summary, sodium acetate buffer solutions at pH 5.5 and at the soil to solution ratio of 1:25 only extract the carbonate from calcareous soils with 10–20% of carbonate, and, at pH 5.0, all of the carbonate from soils with 30–50% of carbonate is dissolved. A second extraction with a fresh buffer solution is required for soils with more than 50% carbonate. The dissolution kinetics of carbonate showed that six hours of extraction are generally sufficient for complete carbonate dissolution. The part of the carbonate fraction not dissolved at the carbonate fraction step is mainly released at the easily reducible oxides fraction step, and to some extent, at the organic matter fraction and reducible oxides fraction steps, leading to gross misinterpretation of the elemental partitioning in arid zone soils.

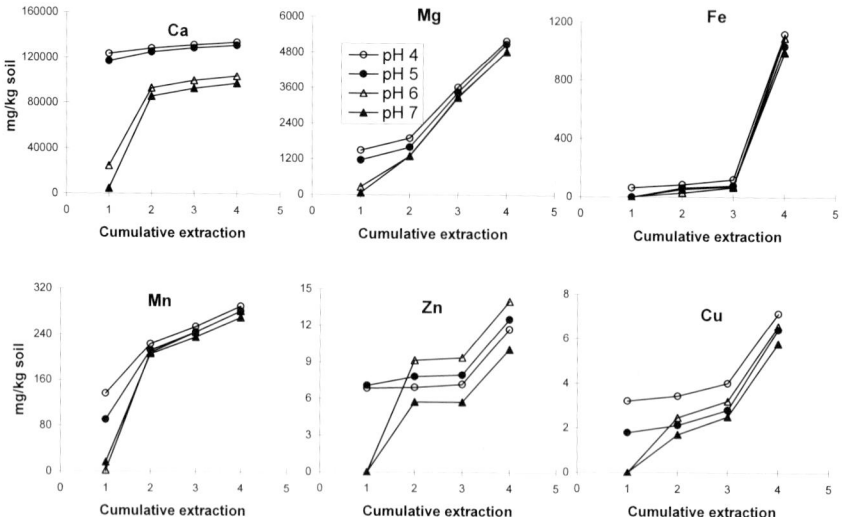

Figure 4.5. Cumulative sums of extractable elements in soil J3 ($CaCO_3$: 33.3%) in consecutive steps of the selective sequential dissolution procedure as a function of the cumulative steps. 1: CARB (carbonate); 2: CARB (carbonate) + ERO (easily reducible oxide); 3: CARB + ERO + OM (organic matter); 4: CARB + ERO + OM + RO (reducible oxide) (after Han and Banin, 1995. Reprinted from Commun Soil Sci Plant Anal, 26, Han and Banin A., Selective sequential dissolution techniques for trace metals in arid-zone soils: The carbonate dissolution step, p 573, Copyright (1995), with permission from Taylor & Francis US)

4.3 COMPARISONS OF THE TWO SELECTIVE SEQUENTIAL DISSOLUTIONS PROTOCOLS

The two representative selective sequential dissolution procedures (Bonn and Rehovot procedures) were employed to comparatively study the distribution of trace elements in Israeli arid soils (Table 4.4). The results of both experiments have been fully reported by Banin et al. (1995) and Han and Banin (1995).

An optimal selective sequential dissolution procedure should satisfy the following requirements:

1. Selectivity/specificity. Each step should extract the desired phase to the maximum extent and attack the other phases to the minimum extent.

Table 4.4. Comparisons of two selective sequential dissolution procedures[a]

Step No.	SSD	Targeted Solid-phases	Reagents	Conditions
1	Rehovot	Soluble and Exchangeable	1M NH_4NO_3	pH 7, solution/soil ratio: 25, shake for 30 min
	Bonn	Mobile	1M NH_4NO_3	pH, not buffered (for soils with pH 4.5, buffered to pH 4.5, solution/soil ratio: 25, shake for 24 h)
2	Rehovot	Carbonate	1M NaOAc-HOAc	pH 5, shake for 6 hours
	Bonn	Easily mobilizable	1M NH_4OAc	pH 6, shake for 24 h
3	Rehovot	Mn oxides	0.04M $NH_2OH.HCl$ + 25%HOAc	pH2, shake for 30 min
	Bonn	Mn oxides	0.1M $NH_2OH.HCl$ + 1M NH_4OAc	pH6, shake for 30 min
4	Rehovot	Organic matter	H_2O_2 + 0.01M HNO_3 + 1 M NH_4NO_3	pH 2, at 80° C for 3 hours shake for 10 min
	Bonn	Organic matter	0.025M NH_4-EDTA	pH 4.6, shake for 90 min
5/ 5+6	Rehovot	Fe oxides	0.04M $NH_2OH.HCl$ + 25%HOAc	pH 2, at 90° C for 3 h
	Bonn	Poorly crystalline	0.2M Tamm's solution	pH 3.25 shake in darkness for 4 h
		Crystalline Fe oxide	0.2M Tamm's solution + 0.1M Vtc.	pH 3.25 boiling for 30 min
7	Rehovot	Clay minerals/ residual	4M HNO_3	Microwave at 90% power, 100 psi for 30 min
	Bonn	Clay minerals/ residual	$HClO_4$ + HNO_3	
8	Rehovot	Total	4M HNO_3	Microwave at 90% power, 100 psi for 30 min
	Bonn	Total	$HClO_4$ + HNO_3	

[a] Rehovot and Bonn procedures are from Han and Banin, 1996, 1997; and Zeien and Brummer, 1991.

2. Relative simplicity and rapidity. A procedure has to be easy to practice, not time or labor intensive, and enable simple and accurate analysis of all elements.

Various sequential dissolution protocols have been developed by different research groups in order to accommodate their types of soils, experimental conditions, and objectives. This makes it difficult to compare the results with different procedures. We compared two SSD procedures for humid zone and arid zone soils, developed by German and Israeli soil scientists, respectively, based on aggressiveness of extractants, their specificity and selectivity, completeness of phase-extraction by each extractant from defined phases and their effects on subsequent fractions. We also appraised the applicability as well as the limitations of each procedure under different conditions.

The cumulative sums of selected major and trace metals extracted by the two SSD procedures from representative arid-zone soils are shown in Fig. 4.6. As can be seen from the figure, the Rehovot procedure is stronger in attacking desired fractions, such as the carbonate bound, Mn oxide bound and organically bound fractions. Extraction of certain major elements, indicating selectivity, specificity and completeness of extraction of given soil components, was found to differ between the two procedures. Calcium and Mg were more completely extracted from the carbonate fraction in arid zone soils by the Rehovot procedure. Calcium and relevant trace elements bound in the carbonate fraction, which were not completely dissolved by the Bonn procedure at this step, were released at the following steps, such as the ERO, OM or RO fractions.

On the basis of this comparison study, at present, it is still difficult to adopt a "universal" selective sequential dissolution procedure, which may be used everywhere and be suitable for all soils with diversified physical, chemical and mineralogical properties. The application of the SSD procedure must consider individual soil characteristics, such as soil type and properties. The two typical SSD procedures were developed to address soils formed in two climates. The Rehovot procedure was developed to be suitable for the calcareous soils in arid and semi-arid zone soils, whereas the Bonn procedure was created to primarily handle the acid and neutral soils in humid zones. In general, the Bonn procedure appears to be unsuited for calcareous soils in arid and semi-arid zones. The Rehovot procedure has limitations in handling acid and neutral soils, especially forest soils with higher content of organic matter.

There are very limited comparative studies on fractionations and distribution of trace elements in soils extracted by various selective sequential dissolution protocols. Sutherland and Tack (2003) compared fractionation of Cu, Pb and Zn in reference soils using three selective

Selective Sequential Dissolution

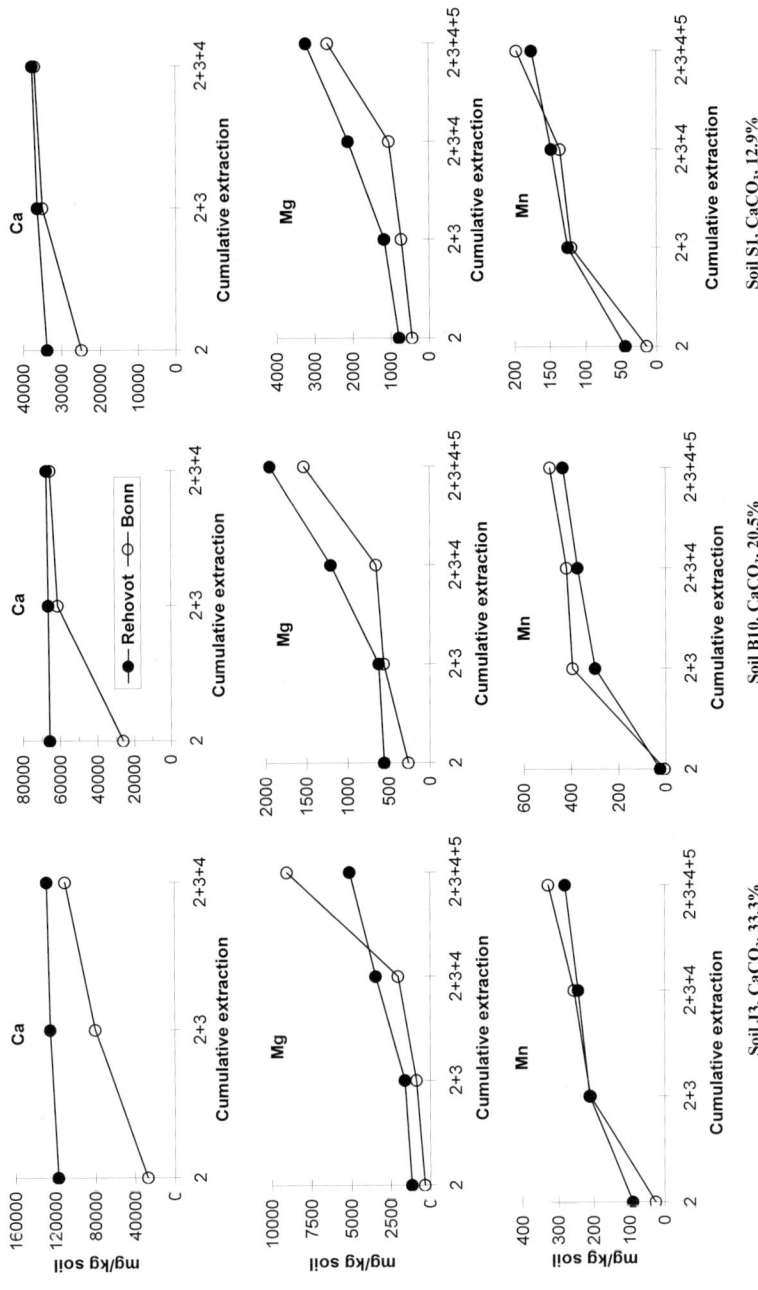

2: Carbonate/Mobilizable, 3: Easily reducible oxide/Mn oxide, 4: Organic matter, 5: Reducible oxide/Fe oxide

Figure 4.6. Cumulative sums of extractable Ca, Mg, and Mn in arid soil J3, B10, and S1 in consecutive steps of two sequential dissolution procedures.

sequential protocols (Ho and Evans, 1997; Tessier et al., 1979, and Hall et al., 1996). They found that difference between schemes were greatest for Pb, moderate for Cu and least for Zn.

4.4 CURRENT AVAILABLE SSD PROTOCOLS FOR FRACTIONATION OF TRACE ELEMENTS IN ARID AND SEMI-ARID SOILS

Below is a review of the chemical extractants and their extraction order, in sequence, in the selective sequential dissolution procedures (Table 4.5).

4.4.1 The Chemical Extractants for Different Fractions in Solid-phase Components

4.4.1.1 Exchangeable Fraction (EXC)

The exchangeable trace elements/heavy metals (EXC) are usually extracted with a neutral salt solution, such as 0.5 M KNO_3 (Emmerich et al., 1982; Sposito et al., 1982; Miller and Mcfee, 1983; Chang et al., 1984), 0.01M $Ca(NO_3)_2$ (Miller and Mcfee, 1983; Levy et al., 1992, Bell et al., 1991), 0.1 M $CaCl_2$, (Mclaren and Crawford, 1973; McGrath and Cegarra, 1992), 1M NH_4NO_3 (Banin et al, 1990; Han et al., 2004b), 1M NH_4OAc (Singh and Abrol, 1986; Levy et al., 1992), 1M $MgCl_2$ (Tessier et al., 1979; Lindau and Hossner, 1982; Han et al., 1990a, 1993, Hickey and Kittrick, 1984), $Mg(NO_3)_2$ (Mandal et al., 1992) and NaOAc (Feijtel et al., 1988). However, NH_4OAc (pH 7.0) and NaOA could attack carbonate (Tessier et al., 1979; Han and Banin, 1995). Additionally, the exchangeable fraction can be further divided into the nonspecifically (with $CaCl_2$ or NH_4OAc) and specifically exchangeable forms (with $Pb(NO_3)_2$, $MgCl_2$, and 2.5% HOAC) (Iyengar et al., 1981; Miller et al., 1986; Mclaren and Crawford, 1973). This fraction can extract water-soluble trace elements and heavy metals from soils. Sometimes, H_2O is directly used to extract the soluble fraction of trace elements from soils (Sposito et al., 1982; Emmerich et al., 1982; Chang et al., 1984; Miller et al., 1986; Feijtel et al., 1988).

Table 4.5. Extractants for trace elements bound to various solid-phase components in soils

No.	Fraction		Extractants
1	Soluble		H_2O
2	Exchangeable		0.5M KNO_3, 0.01M $Ca(NO_3)_2$, 0.1M $CaCl_2$, 1M NaOAc, 1M NH_4OAc, 0.5M KNO_3, 1M NH_4NO_3, 1M $MgCl_2$
3		Nonspecifically	0.1 M $CaCl_2$, 1M NH_4OAc
4		Specifically	1M $MgCl_2$, $Pb(NO_3)_2$, 2.5%HOAc
5	Carbonate		1M NaOAc-HOAc, 0.05 M EDTA
6	Organic		30% H_2O_2, 0.7M NaOCl, 0.11M $Na_4P_2O_7$, 0.1M $K_4P_2O_7$, 0.5M NaOH, 0.1M $Na_4P_2O_7$ + 0.1M NaOH, 0.025M $Cu(OAc)_2$
7	Fe, Mn Oxides		0.04/0.25M $NH_2OH.HCl$ + 25%HOAc, 0.1M $NH_2OH.HCl$, 0.2M $(NH_4)_2C_2O_4$ - 0.2M $H_2C_2O_4$, $Na_2S_2O_4$
8		Mn Oxide	0.1M $NH_2OH.HCl$
9		Amorphous Fe Oxides	0.2M $(NH_4)_2C_2O_4$/ 0.2M $H_2C_2O_4$ (dark), 0.25M $NH_2OH.HCl$ + 0.25M HCl
10		Crystalline Fe Oxides	0.2M $(NH_4)_2C_2O_4$/0.2M $H_2C_2O_4$ + 0.1M ascorbic acid, Hot 0.04M $NH_2OH.HCl$ + 25% HOAc
11	Residual		HF-$HClO_4$, 4M HNO_3, HCl-HNO_3

4.4.1.2 Carbonate Bound Fraction (CARB)

The carbonate bound fraction is, in many cases, extracted with 1M NaOAc-HOAc (Tessier et al., 1979; Han et al., 1990a, 1992; Hickey and Kittrick, 1984) and 0.05M EDTA (Sposito et al., 1982; Emmerich et al., 1982; Miller and Mcfee, 1983; Chang et al., 1984; McGrath and Cegarra, 1992). Grossman and Millet (1961) reported that organic carbon and free iron concentrations in noncalcareous soil samples were unchanged after contact with NaOAc-HOAc buffer for nine weeks. However, other studies have demonstrated that lower pH solutions lead to a partial attack on Fe and Mn oxides (Mclaren and Crawford, 1973). Tessier et al. (1979) found that dolomite was fully dissolved (disappearance of X-ray peak of dolomite) from sediments following treatment with this solution while attack of silicate minerals, sulfide, and organic matter were minimal. In contrast, NaOAc-HOAc can extract some elements from noncalcareous soils and may extract elements easily bound to organic matter (Jiang, 1983; Han, 1998).

4.4.1.3 Readily Reducible Oxide Bound Fraction (ERO)

The readily reducible oxide fraction is customarily extracted with $NH_2OH \cdot HCl$ (Chao, 1972; Miller and Mcfee, 1983; Lindau and Hossner, 1982; Han et al., 1990a, 1992; Feijtel et al., 1988; Banin et al., 1990). Trace elements/heavy metals extracted in this fraction are mostly from Mn oxides (Shuman, 1982). This acid might attack some organic matter, resulting in underestimating the organically bound fraction. However, after extraction of the exchangeable fraction, this attack is less serious.

4.4.1.4 Organically Bound Fraction (OM)

The organically bound trace elements are customarily extracted by treatment with 30% H_2O_2 (Tessier et al., 1979; Lindau and Hossner, 1982; Han et al., 1990a, 1992; Hickey and Kittrick, 1984), $K_4P_2O_7$, or $Na_4P_2O_7$ (Mclaren and Crawford, 1973; Miller and Mcfee, 1983; Miller et al., 1986; Levy et al., 1992), 0.1M $Na_4P_2O_7$ + 0.1M NaOH (Levy et al., 1992), 0.7 M NaOCl (Shuman, 1985a, b; Mandal et al., 1992), 0.5 M NaOH (Sposito et al., 1982; Emmerich et al., 1982; Chang et al., 1984; McGrath and Cegarra, 1992) and $Cu(OAc)_2$ (Sedbery and Reddy, 1976; Singh and Abrol, 1986; Mandal and Mandal, 1986; Han et al., 1992). However, H_2O_2 does not completely oxidize all of the organic matter. It could also oxidize sulfide minerals to a large extent, but the attack on the major silicate phases is minimal. This is noteworthy because the more efficient methods for destroying organic matter lack specificity – in the sense that these reagents partially attack the silicate minerals (Tessier et al. 1979). Thus, H_2O_2, as an oxidant, is selected to obtain a compromise between the complete oxidation of organic matter and the alteration of silicate material (Tessier et al., 1979). Shannon and White's (1991) recent research showed that during oxidation of organic materials in sediments, H_2O_2 extracted 69% and 104% of the Fe and S, respectively, added as FeS; and 92% and 80% of the Fe and S, respectively, added as pyrite (FeS_2). About 77% of the total carbonate of the sediments was selectively extracted in the oxidation step of the organic material (Shannon and White, 1991). Bermond and Benzineb (1991) found that the limitation for hydrogen peroxide was due to cation transfer. Shuman (1982, 1985a) pointed out that H_2O_2 may attack Mn oxide while $Na_4P_2O_7$ dissolves iron oxide. He suggested that the organically bound fraction for trace elements was more labile and should, therefore, be taken early in the sequence. He found that NaOCl was more selective and specific for organic matter than H_2O_2 and completely destroyed organic matter (Shuman, 1985a). NaOCl did not attack Mn and Fe oxides (Shuman, 1985a). Moreover, some researchers further divided the organically-bound form into the chelated (weakly bound to organic matter) and insoluble (strongly bound to organic matter) organic bound fractions on the basis of their availability to plants (Han et al., 1990a, 1992; Khalid et al., 1981). Han et al. (1990a, 1992) reported that 0.1 M $Na_4P_2O_7$ and 30% H_2O_2 were used to sequentially extract weakly organic bound trace elements and insoluble organically bound fractions from arid and semi-arid soils in China, respectively.

4.4.1.5 Reducible Oxide Bound Fraction (RO)

The reducible oxide fraction is extracted by more aggressive extraction with 0.04M $NH_2OH \cdot HCl$+25%HOAc (Feijtel et al., 1988), 0.2M $(NH_4)_2C_2O_4$–0.2 M $H_2C_2O_4$ (Mclaren and Crawford, 1973) and citrate-bicarbonate-dithionite (CBD) (Miller and Mcfee, 1983). Trace element concentrations are consistently lower in the extractant of dithionite-citrate (CBD) than with hydroxylamine hydrochloride-acetic acid ($NH_2OH \cdot HCl$ + 25%HOAc); the differences being greatest for those trace metals which form the most insoluble sulfide salts (Gupta and Chen, 1975; Tessier et al., 1979). Badri and Aston (1981) pointed out that the extraction of sediments with $NH_2OH \cdot HCl$ was more efficient than extraction with $Na_2S_2O_4$ for determining the reducible oxide fraction, which should follow the removal of the organically bound fraction by 30% H_2O_2. The iron oxide could be further divided into an amorphous oxide fraction (extracted with 0.2 M

$(NH_4)_2C_2O_4$ + 0.2 M $H_2C_2O_4$) and a crystalline iron oxide fraction (dissolved with ammonia oxalate buffer + 0.1 M ascorbic acid, 0.04 M $NH_2OH \cdot HCl$+25% HOAc or oxalate reagent at $85^\circ C$ under ultraviolet irradiation) (Shuman, 1985a; Miller et al., 1986; Mandal et al., 1992; Han et al., 1990a). Shuman (1982) found that $NH_2OH \cdot HCl$ was specific for Mn oxide, oxalate buffer in the dark specific for amorphous Fe oxide and $NH_2OH \cdot HCl$+ascorbic acid specific for crystalline Fe oxide. Chao and Zhou (1983) recommended 0.25 M $NH_2OH \cdot HCl$+0.25 M HCl (at $50^\circ C$, for 30 min) as the most desirable extractant for amorphous iron oxides based on the minor dissolution of crystalline iron oxides, 30 min extraction time, low slope of the time series dissolution curve and the close agreement with Tamm's extraction. However, Shuman (1982) found that $NH_2OH \cdot HCl$ and $NH_2OH \cdot HCl$-acetic acid extracted very low amounts of Fe, and oxalate extracted minor Al and high amounts of Zn and Cu from soils.

4.4.1.6 Residual Fraction (RES) and Total (TOT)

The residual fraction and the total of trace elements are usually digested with HF or $HF-HClO_4$ (Mclaren and Crawford, 1973; Sedbery and Reddy, 1976; Tessier et al., 1979; Sposito, et al., 1982; Han et al., 1990a, 1992, 1993; Han and Zhu, 1992). Also, there are many researchers who used 4M HNO_3 or $HCl-HNO_3$ (McIntosh, 1978; Miller and Mcfee, 1983; Banin et al., 1990; and McGrath and Cegarra, 1992; Han et al., 2001b, 2004b). However, even though these reagents could extract most of the mobile trace elements, they could not dissolve all of the total elements, especially Cr and Pb, and some of the major elements.

4.4.2 The Order of Extraction Steps

The order of extraction in selective sequential dissolution protocols is important. If the carbonate and easily reducible oxide fractions are extracted before the organically-bound fraction, then the trace elements found in organically-bound form will probably be decreased (Hirsch and Banin, unpublished data). In a preliminary study, Han (1998) observed that the extracts of NaOAc solution at pH 5 (for the CARB fraction) from soils with higher organic matter became brown colored. This implies some dissolution of organic matter. Hence, it would appear that the fraction of weakly bound to organic matter should be extracted before the extraction of the CARB fraction. Han et al. (1990a, b, 1991, 1992, 1993, 1995) first extracted the soluble organically-bound fraction with $Na_4P_2O_7$ before the

CARB fraction, followed by the insoluble organically-bound fraction with H_2O_2. Shuman (1982) also pointed out that the extractant that reduces oxides will probably extract a considerable amount of micronutrients from the organic matter; therefore, he recommended that the organically-bound fraction should be extracted with NaOCl first. However, in highly calcareous soils, this is not always feasible. Miller et al. (1986) showed that if the organically-bound fraction was extracted with $K_4P_2O_7$ first, before the readily reducible oxide fraction, it extracted more trace elements, and he suggested that $NH_2OH \cdot HCl$ for readily reducible oxides should be used before the $K_4P_2O_7$ step.

The sulfides of trace elements in soils and sediments are also of importance in controlling the availability and mobility of trace elements, especially for land disposal of sulfide-rich sediments or anaerobic digested sludge. Due to the oxic nature in arid soils, most of the sulfur is present as sulfate; thus, this problem may not be pressing. In most current SSD schedules, the majority of the sulfide forms are included in the organic bound or residual fractions.

4.4.3 Other Specific Sequential Fractionation

Many selective sequential fractionation protocols have been developed for special elements, such as arsenic, selenium, and manganese. Wenzel et al. (2001) developed arsenic fractionation procedures in soils, including non-specifically sorbed (0.05M $(NH_4)_2SO_4$)), specifically-sorbed (0.05M $NH_4H_2PO_4$), amorphous and poorly-crystalline hydrous oxides of Fe and Al (ammonium oxalate buffer in the dark at pH 3.25) and well-crystallized hydrous oxides of Fe and Al and residual phases (HNO_3 + H_2O_2). They found that carbonate bound As in calcareous soils as extracted by 1M NaOAc-HOAc is very negligible; therefore, they did not include the carbonate bound As fraction in the sequential procedures. This procedure did not include organically-bound As since this fraction is not as important as Fe and Al oxide-bound As. But in many solid wastes or waste-amended soils, organically-bound As may occur. Chao and Sanzolone (1989) proposed a Se fractionation protocol consisting of soluble (0.25M KCl), ligand exchangeable (0.1M KH_2PO_4), acid extractable (4M HCl), oxidative acid decomposable ($KClO_3$ + concentrated HCl) and strong mixed–acid digestable (HF + HNO_3 + $HClO_4$). Warden and Reisenauer (1991) employed a sequential extraction procedure for Mn from the plant-availability viewpoint, which consisted of readily soluble Mn, weakly absorbed Mn, carbonate bound Mn, specifically absorbed Mn and Mn oxide-bound fractions. It should be recognized, however, that additional Mn-pools may be present in the organic matter, iron oxides, and secondary and primary minerals in soils, and that these may also be partly available to plants under a specific condition.

Besides chemical extraction procedures, the physical separation of trace element species also has been proposed, which will be discussed in the next chapter. The advantages of utilizing a physical separation approach are the limited attacks on the inherent speciation of an element, such as by use of ion exchange resin or by gel chromatography. The residual form may be further divided into sand, silt and clay sizes (Singh et al, 1986; Shuman, 1985a). Additionally, the combination of chemical fractionation with physical separation, such as infrared spectrometry, scanning electron microscopy, X-ray diffraction, energy dispersive X-ray analysis and electron spectroscopy for chemical analysis, may also be helpful in determining the forms of the trace elements in the waste-amended soils, including the minerals and organic functional groups (Hirsh and Banin, unpublished data; Essington and Mattigod, 1991). However, application of these methods is usually dependent on the initial characterization of the fractions by chemical protocols. More complete review on selective sequential dissolution for trace elements in soils was made by Shuman (1991).

Chapter 5

BINDING AND DISTRIBUTION OF TRACE ELEMENTS AMONG SOLID-PHASE COMPONENTS IN ARID ZONE SOILS

Carbonates, organic matter, Fe and Mn oxides, and clay minerals play important roles in controlling overall reactivity of trace elements in soils and sediments. This chapter addresses the interaction of trace elements with carbonates, organic matter, Fe and Mn oxides and clay minerals. Analytical techniques for trace element speciation in solid-phase and their distribution among various solid-phase components in arid and semi-arid soils are reviewed. Solubilities of trace elements in solid phases and their mineralogical characteristics in arid and semi-arid soils also are discussed.

In addition, the removal of organic matter and Fe oxides from soils and sediments is common practice as a pretreatment for soils prior to physical, chemical and biological analyses. The effects of the removal of these components on physicochemical and surface chemical properties of soils will be discussed as well.

5.1 REACTIONS OF TRACE ELEMENTS WITH ORGANIC MATTER IN SOILS

Soil organic matter is one of the most important components governing solubility and bioavailability of trace elements in soil-plant systems (Ross, 1994). The organic matter bound fraction is the most abundant solid-phase component for Cu and Cr (III) in newly contaminated arid soils (Han and Banin, 1997, 1999) (Table 5.1). The organic matter bound fraction is also the most abundant fraction for Cu and Cr in non-arid soils receiving Cu(II), Cr(III) and Cr(VI) salts (Han et al., 2001b, 2004b). The major functional sites for metal bonding are oxygen-containing ligands, including carboxyl, phenol, alcohol and carbonyl groups (Ross, 1994), but acidic humic acid functional groups are principal binding sites for trace elements (Davies et al., 2001). Trace element bindings in humic substances are either inner sphere complexes (for Cu and Pb) or outer sphere complexes (for most other divalent heavy metals such as Zn). Electron paramagnetic resonance studies show that Cu(II) is bound in inner sphere complexes mostly to sites containing O and, to some extent, N sites as well. X-ray spectroscopy studies found binding of Hg^{2+} to O sites of humic substances at high metal/C ratios and to reduced S sites with high specificity at low Hg/C ratios, controlling Hg chemistry in soils and natural waters (Bloom et al., 2001). Ionic strength and pH strongly affect metal-humic reactions. In general, humic metal complexes become more dissociated as the ionic strength of the solution increases, but increases in pH decrease protonation of humic materials, thus increasing the stability of the metal humic complexes. However, the net stability of metal humic complexes also depends upon hydrolysis and complexation reactions with other inorganic ligands. The stability of Cu-humic complexes increases with increasing pH in the lower pH region, and may decrease with increasing pH in the higher pH regions. Low rates of metal release by humic acid solids under basic solution conditions are due to tight binding and slow diffusion of metal cations through the electrostricted water of aqueous humic acid gels (Davies et al., 2001). Metal affinities for humic acid (pH 4–6) are in the order: Cu > Pb >> Cd > Zn (Stevenson, 1977) and for fulvic acid (pH 5): Cu > Pb > Zn (Schnitzer and Skinner, 1967).

Table 5.1. Percentages of trace metals bound in soil organic matter in two Israeli arid soils treated with metal nitrates[a]

Soil		Cd	Cu	Pb	Zn	Ni	Cr
		% of the Total					
Sandy	Average	10.5	28.1	14.7	12.1	22.6	64.6
pH: 7.15	Standard deviation	13.4	3.7	4.9	2.9	3.7	3.9
$CaCO_3$: 0.5%	Maximum	38.9	37.4	27.2	17.7	30.5	71.6
	Minimum	0.0	22.1	9.7	8.5	17.5	59.5
Loessial	Average	1.8	3.7	0.3	2.7	16.8	48.2
pH: 8.03	Standard deviation	2.8	1.3	0.6	0.9	4.5	5.5
$CaCO_3$: 23.2%	Maximum	9.2	6.7	2.0	4.5	25.4	60.0
	Minimum	0.0	0.6	0.0	1.2	9.3	40.6

[a]The values were the results of measurements during one year of incubation at wetting-drying cycle regime. Cd, Cu, Pb, Zn, Ni, and Cr(III) were added as nitrates at the levels of 0.31–0.52, 30–51, 37.7–62, 93–155, 21–34, and 44–73 mg/kg for sandy soil, and 0.54–0.90, 45–75, 50–85, 125–210, 65–110, and 90–150 mg/kg for loessial soil, respectively.

Ahnstrom and Parker (2001) studied Cd reactivity in metal-contaminated soils using a coupled stable isotope dilution-sequential extraction procedure. They found that in uncontaminated arid soil and in soils contaminated with sewage sludge and smelter emission, considerable tracer (stable isotope Cd) had accrued in the NaOCl-oxidizable fraction (mostly organically bound fraction in these soils) by 24 hours. This indicates that the labile portion of the organically bound fraction has a $t_{1/2}$ on the order of hours. But in a Pb-Zn mined soil, where refractory ZnS sphalerite is present, $t_{1/2}$ may be a few days.

Addition of organic matter such as compost, poultry litter, and sewage sludge increases trace elements adsorption, especially in the light-textured soils (Shuman, 1999). Most organic materials increase Zn adsorption, but those with high soluble C decrease Zn adsorption (Shuman, 1999). Addition of biosolids increases Cd adsorption in a range of soils from California to Washington State (Li et al., 2001). The inorganic fraction of biosolid-amended soils contributes to the increased Cd adsorption either due to added inorganic adsorptive phases (e.g., Fe, Mn and Al oxides, silicates, carbonates or phosphates) or due to the increased adsorptive characteristics of the inorganic phases present in the soil (Li et al., 2001). Almas et al. (2000a) reported that the addition of organic matter containing dissolved organic acids (pig manure) increased the solubility of Cd and Zn in soils by the formation of organo-metallic complexes. Almas et al. (2000b) further found that the addition of organic matter (pig manure) reduced the rate of Cd and Zn transfer from the reversibly sorbed fraction into the irreversibly sorbed fraction.

Losi et al. (1994) found that organic matter-amended soils removed higher amounts of Cr(VI) from irrigated water. Chromium concentrations and plant uptake were less in organic matter-amended soils than in soils without organic matter amendment. Bolan et al. (2003) reported that amendment of biosolid composts increased Cr in the organic-bound fraction and effectively reduced the phytotoxicity of Cr(VI). Addition of organic matter accelerated in situ Cr(VI) reduction (Tokunaga et al., 2003).

Soil organic matter exists as a coating or a thin layer on the mineral surface (Bailey et al., 2001). Removing these organic matter coatings results in a change of adsorption of trace elements in soils. Removal of organic matter with hydrogen peroxide has been reported to decrease or increase metal sorption capacity or cation exchangeable capacity (CEC) (Shuman, 1988; Petruzelli et al., 1981; Sequi and Aringhieri, 1977; Mangaroo et al., 1965; and Hinz and Selim, 1999). A sorption capacity decrease is observed more often than an increase. Li et al. (2001) reported that the removal of organic carbon from a range of biosolid-amended soils (from California to Washington State) caused a reduction of Cd adsorption. Sequi and Aringhieri (1977) reported that the electronegative surface area decreased and the electropositive and external surface area increased after H_2O_2 treatment. In addition, the removal of organic matter with hypochlorite was reported to decrease the clay dispersivity (Goldberg et al., 1990).

Removal of organic matter has been reported to both increase and decrease heavy metal adsorption. Extraction of soil organic matter with sodium hydroxide increases Zn adsorption (Mangaroo et al, 1965). Cavallaro and McBride (1984a) reported that the removal of organic matter in soil clays with sodium hypochlorite either increased or had little effect on Zn and Cu adsorption. However, Petruzelli et al. (1981) reported that Pb adsorption capacity, as indicated by the Langmuir isotherm, was reduced after the removal of soil organic matter with H_2O_2. Shuman (1988) found that adsorption of Zn by soils with organic matter removed by sodium hypochlorite conformed to the Langmuir isotherm. He concluded that organic matter removal lowered Zn adsorption capacity and decreased Zn-bonding energy. The removal of organic matter with hydrogen peroxide reduced Cu adsorption on all sizes of clay fractions (Wu et al., 1999). The fine clay (< 0.02 µm) exhibited higher Cu retention than did the coarse (0.2–2 µm) and medium clay (0.02–0.2 µm). Copper is preferentially sorbed on organic matter associated with the coarse clay fractions of the soil.

More recently, Hinz and Selim (1999) used the thin disk flow method and found that the removal of organic matter resulted in the doubling and quadrupling of Zn retention. They suggested that these treatments increased the number of specific sites for Zn adsorption. These observations were further confirmed by Kingery et al. (1999) using traditional batch experiments. The maximum adsorption of Zn on the soil with organic matter removed was about 2 times that of the nontreated soils. Maximum adsorption capacity of Zn on the soil after removing organic matter (1162 mg kg^{-1}, 17.8 µmol kg^{-1}) was higher than the untreated soil (278 mg kg^{-1}, 4.3 µmol kg^{-1}). The observation with the traditional batch technique is consistent with the findings with a thin disk flow method by Hinz and Selim (1999) as stated above.

Zinc sorbed in the untreated soil after desorption was less than that in the soil with organic matter removed (Kingery et al., 1999). This shows that the Zn maintained in the soil solid phase after desorption was proportional to Zn adsorption of the soil solid-phase. A strong hysteresis between Zn desorption and Zn adsorption curves was found in the untreated and treated soils. This indicates that some sorbed Zn is in the non-exchangeable form and on the specific adsorption sites on soil solid-phase components.

5.2 REACTIONS OF TRACE ELEMENTS WITH IRON, MANGANESE, AND ALUMINUM OXIDES IN SOILS

Iron and Mn oxides and hydroxides play an important role in trace element binding in soils. Iron oxides or hydroxides accumulate trace elements through the specific adsorption mechanism. Singh and Gilkes (1992) reported that major proportions of the Co, Cr, Cu, Mn, Ni and Zn in 39 soils from South-Western Australia were concentrated with the iron oxides. In Israeli arid soils, total Co and Cu concentrations strongly correlate with total Fe and Mn in soils, respectively (Figs. 5.1 and 5.2). Oxides and hydroxides bind transition and heavy metal ions by direct coordination to surface oxygen anions, i.e., chemisorption. The chemisorption process is not wholly reversible and is less pH-reversible than the sorption reaction. A higher pH promotes hydrolysis and precipitation of Cu^{2+} at sites of chemisorbed Cu^{2+} (McBride, 1991). The clustering of metal hydroxides on the surface of oxides may be the first step of heterogeneous nucleation of metal precipitation on the surface.

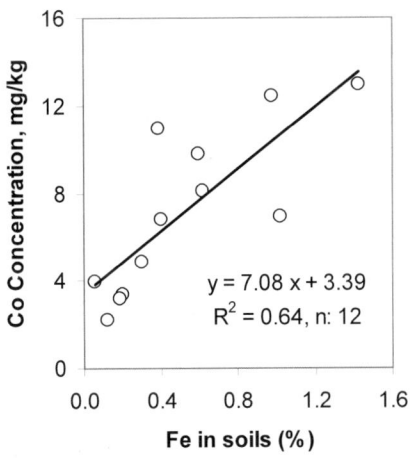

Figure 5.1. Correlation between Co and Fe contents in Israeli arid soils

Metal sorption on Fe/Al oxides is an inner sphere complexion. The formation of a surface-metal bond releases protons for every metal ion adsorbed. Heavy metal sorbed on Fe oxides can be exchanged only by other metal cations having a similar affinity or by H^+ (McBride, 1989). Metal adsorption on Fe oxides is an initial rapid adsorption reaction, followed by slow diffusion (Barrow et al., 1989). Metal ions (Ni^{2+}, Zn^{2+} and Cd^{2+}) slowly diffuse into the solid (oxides, such as goethite), reducing the reversibility of adsorption reactions with time (Gerth and Brummer, 1981). The meso- (2–30 nm width) and micropores (< 2 nm width) in goethite crystals are observed by atomic force microscopy. Therefore heavy metals, phosphate and other ions could slowly diffuse into goethite particles (Fischer et al., 1996). Metal ion adsorption into goethite involves three steps: surface adsorption, diffusion into the mineral and fixation at positions within the mineral. Ahnstrom and Parker (2001) observed the slow migration of ^{111}Cd into the reducible oxide bound fraction in uncontaminated arid soil and soils contaminated with sewage sludge, smelter emission and Pb-Zn mine soil. This indicates that reaction half-time is in the range from weeks to years. Compared to reducible oxides, the Cd isotopic exchange reaction is much faster in the organically bound and carbonate (sorbed) fractions.

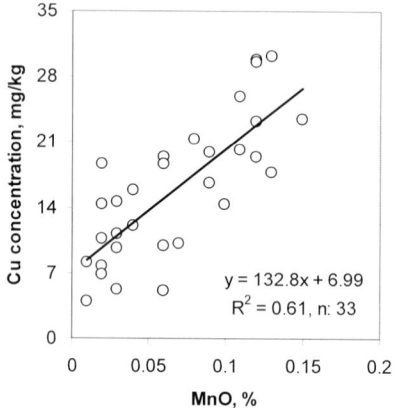

Figure 5.2. Correlation between Cu and MnO contents in Israeli arid soils

The high energy binding sites on hydrous manganese oxide are an exchange sorption on the negatively charged oxide sites by displacement of surface and diffuse layer cations and protons, which bind Cd more strongly than Zn (Zasoski and Burau, 1988). However, at pH 6 and 8, Zn is more strongly bound by Mn oxides than Cd. The high surface charge density and great proportion of Zn hydroxy solution phase species at high pH may favor Zn sorption (Zasoski and Burau, 1988). Competitive adsorption of Cd and Zn was found on Mn oxides (Zasoski and Burau, 1988). Birnessite-hausmannite and pyrolusite coatings also increase Zn sorption on quartz sand, principally in nonexchangeble forms (Stahl and James, 1991).

Iron and Mn oxides are two of the most important solid-phase components controlling distribution and availability of trace elements in arid soils. Han and Banin (1997, 1999) reported that the Fe oxide bound fraction (both amorphous Fe oxides and crystalline Fe oxides) was one of the most abundant fractions for Zn, Cr, Pb and Ni in arid soils receiving soluble salts (Table 5.2). Iron and Mn oxides bound fractions accounted for 30–40%, 20–40%, and 20–30% of added Cu, Zn and Ni in contaminated arid soils, respectively (Table 5.2). In newly polluted arid soils, heavy metals in the amorphous Fe oxide bound fraction are higher than those in the crystalline Fe oxide bound fraction, while with time the latter increases (Han and Banin, 1995, 1997, 1999). Han et al. (1995) reported that in eight representative calcareous soils from arid and semi-arid North and Northwestern China containing 0.1–14.7% of $CaCO_3$ with pH from 7.2–8.1, Zn bound to the Fe oxides accounted for 20–30% of total Zn, which was higher than Zn bound to organic matter (< 5%) and carbonate (< 5%) fractions.

The adsorption on Fe and Mn oxides is higher than that on carbonate and clay minerals. Brummer et al. (1983) studied Zn adsorption capacity of several soil materials in $CaCO_3$ buffered systems. They found that amorphous Fe/Al oxides and Mn oxides had much higher adsorption capacity (1190–1540 μ mol g^{-1}) than bentonite (44 μ mol g^{-1}) and $CaCO_3$ (0.44 μ mol g^{-1}). Han et al. (1995) reported that the binding capacity of Zn in Fe oxides (139.7 ± 62 mg Zn/mol Fe) was much higher than $CaCO_3$ (2.7 ± 4.5 mg Zn/mol $CaCO_3$) in Chinese arid soils. Similarly, the binding capacity of Zn by Mn oxides was also higher than the carbonate. In general, the Mn oxides (hausmannite and cryptomelane) adsorb more heavy metals such as Co and Cd than the Fe oxides (goethite and ferrihydrite) (Backes et al., 1995). The metals are also much less readily desorbed from the Mn than the Fe oxides, and Cd is more readily desorbed than Co from both the Mn and Fe oxides (Backes et al., 1995). The orders of metal affinities for amorphous Fe oxides are Pb > Cu > Zn > Ni > Cd > Co (Kinniburgh et al., 1976). The metal affinities for geothite (FeOOH) and haematite are Cu > Pb > Zn > Co > Ni > Mn, and Pb > Cu > Zn > Co > Ni > Mn, respectively (McKenzie, 1980). Soil Fe oxides preferentially adsorb Pb, Cu and Zn compared to Cd, Ni and Co.

Heavy metal adsorption by Fe oxide and hydroxides in arid soils containing $CaCO_3$ can occur at lower pH and increases with higher pH (Ross, 1994). Han et al. (1995) reported that in calcareous soils from arid regions of China, the binding capacity of Zn by Fe oxides in calcareous soils (139.7 ± 62 mg Zn/mol Fe) was significantly higher than that in acid soils (9.3 ± 5.9 mg Zn/mol Fe) and neutral soils (57.5 ± 16.4 mg Zn/mol Fe). Zinc binding by iron oxides (including amorphous and crystalline Fe oxides) is exponentially related to soil pH (Y, mg Zn/mol Fe, = 0.674 x $10^{0.30 pH}$). (Figure 5.3) (Han et al., 1995).

Table 5.2. Percentages of trace metals bound in the Mn and Fe oxides in two Israeli arid soils treated with metal nitrates[a]

Components	Soil		Trace Elements (% of the Total)					
			Cd	Cu	Pb	Zn	Ni	Cr
Fe and Mn oxides	Sandy	Average	3.7	37.2	27.9	20.1	18.8	11.7
	pH: 7.15	Maximum	27.1	45.8	37.4	25.6	24.1	15.5
	CaCO$_3$: 0.5%	Minimum	0.0	27.8	17.5	16.4	11.9	7.9
	Loessial	Average	6.0	34.2	13.0	36.7	29.3	18.1
	pH: 8.03	Maximum	33.0	48.0	18.6	49.9	43.9	29.2
	CaCO$_3$: 23.2%	Minimum	0.0	26.1	9.2	26.8	19.9	12.9
Mn oxides	Sandy	Average	2.0	10.7	14.3	13.0	15.2	4.0
		Standard deviation	3.1	1.6	2.2	1.2	1.8	1.3
		Maximum	9.9	12.8	17.7	15.3	18.6	6.6
		Minimum	0.0	7.1	10.4	11.1	10.1	2.1
	Loessial	Average	2.9	16.1	12.8	23.7	18.3	7.6
		Standard deviation	3.8	2.0	2.1	3.4	3.7	1.9
		Maximum	11.4	21.5	16.6	32.3	27.5	11.3
		Minimum	0.0	13.0	9.2	17.6	12.4	4.7
Fe oxides	Sandy	Average	1.7	26.5	13.6	7.1	3.6	7.7
		Standard deviation	3.7	3.8	2.7	1.4	0.9	0.8
		Maximum	17.2	33.0	19.7	10.3	5.4	8.9
		Minimum	0.0	20.6	7.1	5.3	1.8	5.8
	Loessial	Average	3.1	18.1	0.3	13.0	11.1	10.5
		Standard deviation	7.1	3.3	0.6	2.4	2.3	2.1
		Maximum	21.6	26.5	2.0	17.6	16.4	17.9
		Minimum	0.0	13.0	0.0	9.3	7.5	8.2

[a]The values were the results of measurements during one year of incubation at wetting-drying cycle regime. Cd, Cu, Pb, Zn, Ni, and Cr(III) were added as nitrates at the levels of 0.31–0.52, 30–51, 37.7–62, 93–155, 21–34, and 44–73 mg/kg for sandy soil, and 0.54–0.90, 45–75, 50–85, 125–210, 65–110, and 90–150 mg/kg for loessial soil, respectively.

The aging of oxides decreases solubility and extractability of trace elements adsorbed. Aging decreases desorption of Cd and Co from Fe oxides but has less effect on desorption from Mn oxides (Backes et al., 1995). Long-term aging (up to two years) induces the transformation of an initially noncrystalline alumina to more ordered products including gibbsite (Martinez and McBride, 2000). Copper movement is towards the surface of the coprecipitate with increased aging time. Copper is initially evenly distributed within the alumina, but segregates at or near the alumina surface forming CuO and/or clusters after long-term reaction with alumina (Martinez and McBride, 2000).

Boron is adsorbed on Fe/Al oxides (goethite and gibbsite) via an inner-sphere mechanism with shifts in zero point of charge (Goldberg et al., 1993). Boron adsorption on Fe/Al oxides increases from pH 3 to 6, exhibiting a peak at pH 6 to 8.5, and then decreases from pH 8.5 to 11. The constant capacitance model successfully describes B adsorption on goethite, gibbsite, and kaolinitic soils.

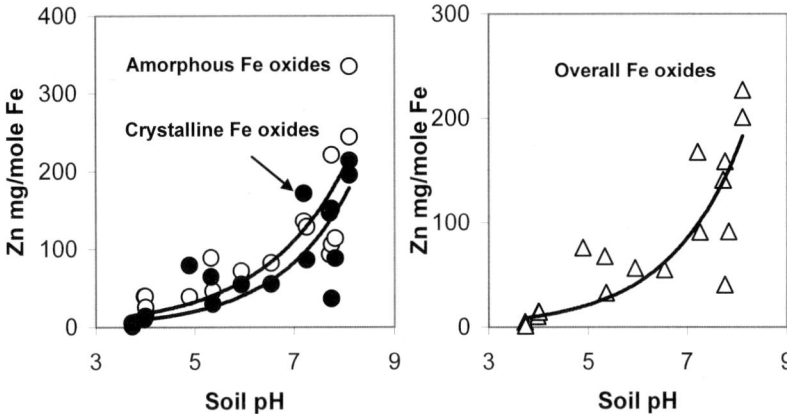

Figure 5.3. Effects of soil pH on the Zn amount bound to the Fe oxide fraction (amorphous/crystalline Fe oxide and overall Fe oxide bound fractions) in soils from China with pH 3.73-8.1 and 0–14.7% $CaCO_3$ (after Han et al., 1995. Reprinted from Geoderma, 66, Han F.X., Hu A.T., Qi H.Y., Transformation and distribution of forms of zinc in acid, neutral and calcareous soils of China, p 128, Copyright (1995), with permission from Elsevier)

Arsenate is readily adsorbed to Fe, Mn and Al hydrous oxides similarly to phosphorus. Arsenate adsorption is primarily chemisorption onto positively charged oxides. Sorption decreases with increasing pH. Phosphate competes with arsenate sorption, while Cl, NO_3 and SO_4 do not significantly suppress arsenate sorption. Hydroxide is the most effective extractant for desorption of As species (arsenate) from oxide (goethite and amorphous Fe oxide) surfaces, while 0.5 M PO_4 is an extractant for arsenite desorption at low pH (Jackson and Miller, 2000).

Cobalt is strongly adsorbed by Mn oxides. There are close relationships between Co and the easily reducible fraction of Mn (Mn oxides) in soils (Jarvis, 1984) and will be in detail discussed in next chapter. Cobalt is frequently accumulated in Mn nodules in soils (Mckenzie, 1975). It was suggested that the Co^{2+} ion was first sorbed, then slowly oxidized to Co^{3+} and became incorporated into the surface layers of the crystal lattice, releasing the Mn^{2+} ion into the solution (Burns, 1976; Mckenzie, 1975). X-ray photoelectron spectroscopy showed that Co^{3+} was present on the surface of birnessite after the sorption of the Co^{2+} ions took place (Murray and Dillard, 1979). Traina and Donor (1985) suggested that the Mn release during Co^{2+} sorption resulted not only from the oxidation of Co^{2+} to Co^{3+}, but also from a direct exchange of Co^{2+} for Mn^{2+} produced during the redox reaction. Cobalt adsorption by Mn oxides at pH 6.0 is larger than that by the Fe oxides. Increasing the initial sorption period substantially decreases the proportion of desorbed Co from goethite but has much smaller effects with the Mn oxides (Backes et al., 1995). Ainsworth et al. (1994) found that Co was incorporated into the hydrous iron oxide structure via recrystallization after adsorption. The major part of Co adsorption in soil is due to the clay fraction and iron oxides (Borggaard, 1987). McLaren et al. (1986b) pointed out that soil-derived oxides sorbed by far the greatest amounts of Co, while clay minerals and non-pedogenic iron and manganese oxides sorbed relatively little Co.

Iron and Mn oxides occur in soils as coating or precipitation on the surfaces of layer silicate minerals. Robert and Terce (1989) pointed out that gels or coatings of Fe compounds on clay surfaces have important metal cation adsorption properties. Removing the coatings of Fe oxides decreases adsorption of heavy metals. Iron oxides/hydroxides are customarily removed by dithionite-citrate-bicarbonate (DCB), oxalate buffer and hydroxylamine hydrochloride as discussed in the previous chapter. Oxalate buffer solution extracts mostly amorphous Fe oxides. Trace element concentrations are consistently lower in the extracts of DCB than of hydroxylamine hydrochloride-acetic acid (25%) ($NH_2OH \cdot HCl$) (Gupta and Chen, 1975; Tessier et al., 1979). The differences are greatest for those trace metals which form the most insoluble sulfide salts. Badri and Aston (1981) pointed out that $NH_2OH \cdot HCl$ more efficiently extracted Fe oxides from sediments than dithionite ($Na_2S_2O_4$). Chao and Zhou (1983) recommended 0.25 M $NH_2OH \cdot HCl$ + 0.25 M HCl (at $50°C$, for 30 min) as the most desirable extractant for amorphous iron oxides due to its minor dissolution of crystalline iron oxides and close agreement with oxalate buffer extraction. However, Shuman (1982) found that $NH_2OH \cdot HCl$ and $NH_2OH \cdot HCl$-acetic acid extracted very low amounts of Fe, and oxalate extracted minor Al and high amounts of Zn and Cu from soils. He concluded that $NH_2OH \cdot HCl$ is specific for extraction of Mn oxides, while oxalate buffer (in the dark) is specific for amorphous Fe oxide, and $NH_2OH \cdot HCl$ is specific for crystalline Fe oxide. In addition, crystalline Fe oxides are also extracted with ammonia oxalate buffer + ascorbic acid, $NH_2OH \cdot HCl$ + 25% HOAc or oxalate reagent (at $85°C$, under ultraviolet irradiation) (Shuman, 1985a; Miller et al., 1986; Mandal et al., 1992; Han et al., 1990a).

The effects of the removal of Fe oxides by DCB on soils have been studied. In contrast to organic matter destruction, the removal of Fe oxide by dithionite commonly increases CEC more often than it decreases CEC (Shuman, 1976; Cavallaro and McBride, 1984a; Shuman, 1988; Ghabru et al., 1990; and Hinz and Selim, 1999). The zero point charge (ZPC) value of the soil is related to the surface coatings of Fe oxide, and the removal of Fe oxides by DCB lowers the ZPC of the soil (Elliott and Sparks, 1981). The removal of oxides shifts positive zeta potentials toward more negative values (Cavallaro and McBride, 1984b). Iron removal results in conversion of the surface from repulsive to attractive toward metal cations and increases electrostatic attraction for metals (Elliott et al., 1986). Another possible mechanism is by exposure of additional exchange sites through the removal of bound cationic iron species and elimination of the hydroxy-polymeric Fe and Al compounds physically obstructing such sites. Treatment of dithionite increases reactive surface area (Ghabru, et al., 1990) and reduces the structural Fe of clay minerals (Ericsson, et al., 1984; Ghabru, et al., 1990). In addition, the removal of amorphous Fe oxides with oxalate solution and/or crystalline oxides with dithionite increases the clay dispersivity (Goldberg et al., 1990).

Cavallaro and McBride (1984a) observed that the removal of Fe oxides from two clay soils reduced Zn adsorption. Shuman (1976) reported that the removal of Fe oxides resulted in an increase or decrease in Zn adsorption, but later in another similar study (1988) he found that the removal of either amorphous or crystalline Fe oxides increased Zn adsorption capacity and decreased Zn-bonding energy. The author explained that adsorption sites on the Fe oxide coatings were not as numerous as those released when the coatings were removed. Elliott et al. (1986) observed that DCB extraction of Fe oxides from two subsoils of the Atlantic Coastal Plain increased heavy metal adsorption. Wu et al. (1999) found that Cu adsorption on the fine clay fraction increased after dithionite treatment with possible exposure of much more high-affinity sites for Cu on the fine clay.

Hinz and Selim (1999) used a thin disk flow method to study the effects of the removal of both organic matter and iron oxides on Zn adsorption. They found that such a treatment caused doubling and quadrupling of Zn retention. Significant increases in Zn adsorption occurred at low input Zn concentrations. This suggests that more specific sites become available as a result of the treatments for the removal of both organic matter and Fe oxides. Kingery et al. (1999), in subsequent batch experiments, verified that Zn adsorption significantly increased after the removal of Fe oxides by DCB. In the batch experiments, the iron oxide/hydroxide fraction was removed with the dithionite-citrate method (Jackson, 1958) after the removal of organic matter with hydrogen peroxide. The maximum adsorption of Zn on the soil with removing organic matter and iron oxides was about three times that on the soil with only removing organic matter and about five times that on the untreated soil. The soil, after removing both organic matter and iron oxides, possessed the highest maximum adsorption capacity (2173 mg kg^{-1}, 33.2 µmol kg^{-1}), while the untreated soil had the lowest maximum adsorption capacity (278 mg kg^{-1}, 4.3 µmol kg^{-1}). The Zn maximum adsorption capacity on the untreated and treated soils measured by the traditional batch technique were much in agreement with that measured with the thin disk flow method by Hinz and Selim (1999). Zinc sorbed in the whole soil after desorption was less than that in the soil with the removal of organic matter, which was even less than that in the soil without both organic matter and Fe oxides.

The effects of the removal of organic matter and iron oxides on Zn adsorption on soils are also influenced by Zn concentration. At low concentrations (5–10 mg L^{-1}, initial concentration), both treated soils (removed organic matter and iron oxides) behaved similarly. At high Zn concentration, however, treated soils behaved differently. When the initial Zn concentration was between 5 and 10 mg kg^{-1}, adsorption of Zn by soils without organic matter and without both organic matter and iron oxides were 2–2.5 times that of the untreated soil. With an increase in initial Zn concentration, the soil without both iron oxides and organic matter adsorbed more Zn than the soil without organic matter. This indicates that the available sites for Zn decrease with increases in the initial Zn concentration.

The effects of pH on Zn adsorption are clearly shown in the treated soils with removing organic matter and iron oxides (Figure 5.4). Removing organic matter from the soil resulted in an increase in pH of about 1 unit (Kingery et al., 1999). This slight increase in pH did not increase the CEC, but instead slightly decreased the CEC. The increase in Zn adsorption capacity of soils after removing the organic matter as discussed above may be due to the increase in the proportion of hydrated Zn ions, resulting in increasing adsorption (Chairidchai and Ritchie, 1990). A large increase in soil pH of about 2 units, after removing both the organic matter and iron oxides, may result in an increase in both the proportion of hydrated Zn ions and CEC. The CEC in the whole soil (Kingery et al., 1999) was 1.52 ± 0.13 cmol$_c$ kg^{-1}, while the CEC increased up to 3.79 ± 0.05 cmol$_c$ kg^{-1} in the soil with removing organic matter and iron oxides. The CEC in the soil without organic matter was similar to the untreated soil, 1.40 ± 0.04 cmol$_c$ kg^{-1}. The zinc adsorption capacity of the whole soil and the soil with organic matter and Fe oxides removed increased with the pH of the solution (Fig. 5.4). In effect, at pH > 6–6.5, Zn adsorption on whole soils was higher than soils with the removal of organic matter and Fe oxides, and Zn adsorption on the soils without organic matter and without organic matter and Fe oxides were in a similar range.

Adsorption kinetics of Zn on the untreated soil and the soils with the removal of organic matter and Fe oxides has been studied in detail by Hinz and Selim (1999) using a thin disk flow method. They reported that Zn sorption was highly concentration-dependent. When Zn concentration was low (2.5×10^{-7} M), the sorption/desorption was more time-dependent, while at a higher concentration (2.6×10^{-5} M), it was instantaneous and mostly equilibrated. For the soil with the removal of organic matter and Fe oxides, stronger sorption and desorption kinetics behavior was exhibited compared to the untreated soil. They suggested that diffusion into soil minerals or surface-controlled reactions may cause such behavior. Han and Banin (1997, 1999) found that heavy metal retention in arid soils was characterized by an initial fast fixation process, followed by a slow long-term process. Redistribution of heavy metals among varying solid-phase components is mainly responsible for the slow long-term process in soils, which may involve the slow diffusion of metals from the surface into the interior minerals or oxides (Brummer et al., 1988).

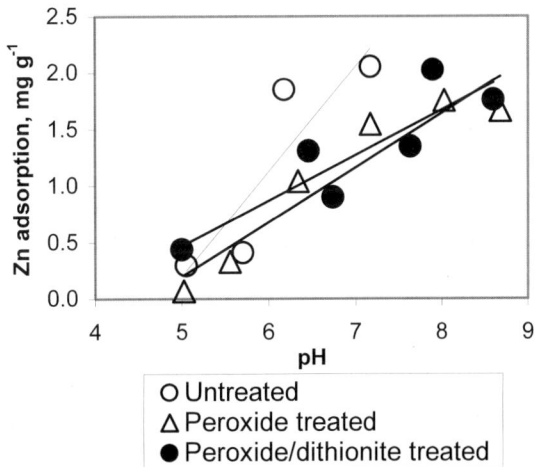

Figure 5.4. Effects of pH on Zn adsorption on an untreated and treated soils with peroxide and dithionite

The adsorption/desorption kinetics of Zn on soils with the removal of organic matter and Fe oxides were studied by Kingery et al. (1999) using traditional batch experiments on the same soil used by Hinz and Selim (1999). Adsorption of Zn on the three soils was kinetically characterized by three kinetic rates: a very fast initial retention process within 30 min, an intermediate retention process from 30 min to 8 hours, followed by the very slow process (Fig. 5.5). About 50% of Zn was adsorbed on the three soils within the initial first 30 min. Initially, Zn desorption increased with time during the first 20 minutes from both the untreated soils and the soils without organic matter and Fe oxides. However, after 20 minutes of adsorption, net Zn adsorption from the whole soil and the soil with the removal of organic matter and Fe oxides decreased with time. This indicates that some re-adsorption takes place after Zn is desorbed from soils. Re-adsorption is much more obvious in the soil without organic matter and Fe oxides than in whole soil. This study implies that a desorption experiment including any solution extraction may be highly characterized by desorption and extraction kinetics. Overall equilibrium of the extraction and desorption on soils and sediments would lead to some underestimation of targeted solid-phase components. This is especially important for sequential selective dissolution procedures. The time for each step may be the maximum of dissolution of that targeted solid-phase, but the minimum extent of re-adsorption.

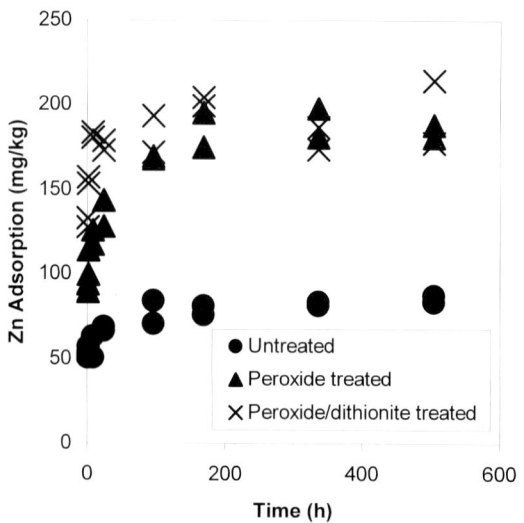

Figure 5.5. Kinetics of Zn adsorption on a soil and soil with removals of organic matter and Fe oxides (1 g soil reacted with 20 ml of 10 mg/L Zn solution concentration)

In summary, the removal of organic matter and Fe oxides significantly changes the physicochemical and surface chemical properties of soils. Thus, this pretreatment affects the overall reactivity of heavy metals in soils. The removal of organic matter and Fe oxides may either increase or decrease heavy metal adsorption. The mechanisms responsible for the changes in metal adsorption in soils with the removal of organic matter and Fe oxides include increases in pH, surface area, CEC and electrostatic attraction, decreases in the ZPC, shifts of positive zeta potentials toward more negative values and exposure of additional exchange sites. A strong hysteresis in the Zn desorption and Zn adsorption curves is found in the untreated soils. Zinc adsorption/desorption kinetics in the soils with the removal of organic matter and Fe oxides is similar to that in the untreated soils. Zinc maximum adsorption capacity on the untreated soil and the soils with the removal of organic matter and Fe oxides obtained with a traditional batch technique is in agreement with that obtained with a thin disk flow method.

5.3 REACTIONS OF TRACE ELEMENTS WITH CLAY MINERALS

Layer silicate minerals have a high selectivity of trace transition and heavy metals and greater irreversibility of their adsorption. Some chemisorbing sites such as –SiOH or AlOH groups may be at clay edges and form hydroxyl polymers at the mineral surface. Another possible reason for the high selectivity may be hydrolysis of the metal and strong adsorption of the hydrolysis ion species.

Clay may promote hydrolysis of the metal at low pH, but also inhibit hydrolysis at high pH (McBride, 1991). At a higher pH, clay prevents complete hydrolysis of the metal due to the affinity of the charged polymeric metal ions for the silicate surface. This keeps the metal from becoming a separate hydroxide phase.

Both organic and inorganic ligands such as Cl and dissolved organic carbon (fulvic acid and carboxylic acids) decrease metal adsorption. In the arid soils with higher pH, folic acids increase the solubility of metals such as Cu and Zn. The interaction between the transition of heavy metals and silicate surfaces was reviewed by McBride (1991).

The presence of hydroxyaluminum- and hydroxyaluminosilicate polymer in interlayered montmorillonite greatly promotes the adsorption of Cd, Zn, and Pb (Saha et al., 2001). The adsorption selectivity sequences of montmorillonite (Pb > Zn > Cd) and interlayered montmorillonite (Pb >> Zn >> Cd) resemble the metal selectivity on amorphous Fe and Al hydroxides (Saha et al., 2001). On montmorillonite, the metals are predominantly adsorbed on the permanent charge sites in an easily replaceable state. However, a substantial involvement of the edge OH^- groups of montmorillonite in specific adsorption of the metals is also observed, especially at higher pH (Saha et al., 2001).

Compared to the importance of Fe and Mn oxides as metal adsorbing surfaces, the primary role of layer silicates is as a substrate on which Fe and Mn oxides precipitate and coat. This is especially true in arid and semi-arid soils. Iron and Mn oxides bind much more Zn in calcareous soils than in carbonate soils and the binding capacity increases with soil pH. As discussed earlier, coatings of Fe oxide/hydroxides on clay surfaces have important effects on metal cation adsorption properties.

The migration of heavy metals into mineral lattices is very slow. Ahnstrom and Parker (2001) observed the slow migration of ^{111}Cd into the residual fraction in arid soils with a weeks-to-years reaction half-time. Theoretically, the residual fraction is comprised of very refractory Cd bound in the lattices of aluminosilicate minerals. Lattice diffusion, a process necessary for isotopic exchange, can require years.

Boron is adsorbed on montmorillonite via an outer-sphere adsorption mechanism (Goldberg et al., 1993). This mechanism is also observed for B adsorption on arid-zone soils. The maximum boron adsorption on montmorillonite and two arid-zone soils occurs near pH 9, while B adsorption on kaolinite increases from pH 3–6, peaks at pH 6–8.5 and decreases from pH 8.5–11. The constant capacitance model is able to describe B adsorption on kaolinite and kaolinitic soils and to predict B adsorption behavior on arid soils using the average set of B surface complexation constants, but fails to describe B adsorption on montmorillonite and montmorillonitic soil (Goldberg et al., 1993; Goldberg, 1999). In addition, As(V), As(III) and Se(IV) anions have been reported to be adsorbed by kaolinite and montmorillonite, strongly depending upon the solution pH (Frost and Griffin, 1977). Montmorillonite adsorbs more As(V), As(III) and Se(IV) anions than kaolinite due to a higher edge surface area.

5.4 REACTIONS OF TRACE ELEMENTS WITH CARBONATE

Trace elements can be adsorbed on the surface of calcite, influencing their solubility in calcareous soils of arid and semi-arid zones. The carbonate bound fraction is the major solid-phase component for many trace elements (Cd, Pb, Zn, Ni and Cu) in arid and semi-arid soils, especially in newly contaminated soils (Table 5.3). In Israeli arid soils treated with metal nitrates, the carbonate bound fraction is the largest solid-phase component (60–80%, 50–60%, 40–60%, 30–40%, and 25–36% for Cd, Pb, Zn, Ni, and Cu respectively). Divalent metallic cations at low aqueous concentrations first associate with calcite via adsorption reactions. Then they may be incorporated into the calcite lattice as a co-precipitate by recrystallization (Franklin and Morse, 1983; Kornicker et al., 1985; Davis et al., 1987; Zachara et al., 1988; Reeder and Prosky, 1986; Pingitore and

Eastman, 1986; Reeder and Grans, 1987). Carbonates in soils have direct (through their surface interactions) or indirect (through their pH effect on other soil constituents) effects on metal solubility (Papadopoulos and Roweli, 1988). The $CaCO_3$ surface has a high affinity for Cd at low Cd concentrations, and a solid solution between $CdCO_3$ and $CaCO_3$ is formed on the surface (Papadopoulos and Roweli, 1988). Equilibrium is attained in several minutes for a low initial Cd concentration. However, at a high aqueous concentration, discrete metal carbonate or basic–metal carbonate precipitates may form on the calcite surface (McBride, 1979, 1980; Glasner and Weiss, 1980; Kaushansky and Yariv, 1986; Papadopoulos and Roweli, 1988, Zachara et al., 1989). Papadopoulos and Roweli (1988) found that at a high Cd concentration, precipitation of $CdCO_3$ on calcite predominated, and the precipitation was a slow process. Zachara et al. (1991) reported that at higher initial metal concentrations (10^{-5} to 10^{-4} mol/L), equilibrium solutions were saturated with a discrete solid phase of the sorbates: $CdCO_3(s)$, $MnCO_3(s)$, $Zn_5(OH)_6(CO_3)_2(s)$, $Co(OH)_s(s)$ and $Ni(OH)_2(s)$. This indicates that aqueous concentrations of trace elements may be governed by solubilities of discrete carbonates. It was suggested that Cd was adsorbed by $CaCO_3$ on the surface and then slowly diffused into the inside of the particle. Finally, $CdCO_3$ and $CaCO_3$ formed a solid-solution (Stipp et al., 1992). Divalent metal ion sorption is dependent upon aqueous Ca concentration (Zachara et al., 1991). Zachara et al. (1988) reported that Zn adsorption occurred via exchange with Ca^{2+} in a surface-adsorbed layer on calcite. Their exchange constants vary with surface coverage (Zachara et al., 1991).

Table 5.3. Percentages of trace metals bound in carbonates in two Israeli arid soils treated with metal nitrates

Soil		Trace Elements (% of the Total)					
		Cd	Cu	Pb	Zn	Ni	Cr
Sandy	Average	60.7	24.8	48.1	56.2	41.0	10.8
pH: 7.15	Standard deviation	14.9	4.4	5.4	3.9	4.4	2.6
$CaCO_3$: 0.5%	Maximum	93.4	32.7	57.1	62.4	48.4	15.3
	Minimum	26.6	18.1	37.5	47.1	32.0	5.9
Loessial	Average	79.6	36.2	57.7	38.6	33.0	8.8
pH: 8.03	Standard deviation	14.1	5.6	6.0	5.7	9.4	3.4
$CaCO_3$: 23.2%	Maximum	100.0	46.0	69.5	50.2	50.3	15.2
	Minimum	51.2	25.4	42.1	27.5	17.6	3.8

Greater adsorption of trace metals is found at higher pH and $CO_2(g)$ concentrations. Sites available for Zn^{2+} sorption are less than 10% of the Ca^{2+} sites on the calcite surface, and Zn adsorption is independent of surface charge. This indicates a surface complex with a covalent character (Zachara et al., 1991). Furthermore, the surface complex remains hydrated and labile because Zn^{2+} is rapidly exchangeable with Ca^{2+}, Zn^{2+} and $ZnOH^-$. At the dolomite-solution interface, the carbonate(CO_3)-metal (Ca/Mg) complex dominates surface speciation at pH > 8, but at pH 4–8, hydroxide (OH) – metal (Ca/Mg) dominates surface speciation (Pokrovsky et al., 1999). Calcite has an observed selectivity sequence: $Cd > Zn \geq Mn > Co > Ni \geq Ba = Sr$, but their sorption reversibility is correlated with the hydration energies of the metal sorbates. Cadmium and Mn dehydrate soon after adsorption to calcite and form a precipitate, while Zn, Co and Ni form surface complexes, remaining hydrated until the ions are incorporated into the structure by recystallization (Zachara et al., 1991).

Carbonate removal decreases the adsorption capacity of calcareous soils for trace elements such as Zn (Martinez and Motto, 2000). Solubility curves of the calcareous soil after removal of carbonates resemble those of noncalcareous soils. Solubility values in calcareous soils are within those found for noncalcareous soils. Zinc is completely desorbed after adsorption on pure $CaCO_3$ surfaces by lowering the pH of the solution (Zachara et al., 1989).

5.5 ANALYTICAL TECHNIQUES FOR TRACE ELEMENT SPECIATION IN SOLID-PHASE

In general, there are two categories of techniques used to study solid phase speciation of trace elements in soils or solid mixtures. These include chemical extraction methods and instrumental techniques. Among chemical extraction methods, sequential selective dissolution techniques have been commonly used for chemical partitioning of environmental solid samples. The characterization, advantages, disadvantages and current sequential selective dissolution procedures have been discussed and reviewed in detail in Chapter 4. In contrast, instrumental techniques have been used to study *in situ* solid-phase speciation of trace elements and heavy metals in soils and other solid mixtures. These techniques generally exploit the interaction of energy probes (electromagnetic fields, i.e., photons, charged particles, atoms and neutrons) with the sample to be analyzed (D'Amore et al., 2005). When energy probes interact with samples, they produce signals reflecting their interaction. A histogram of the intensities of the energy produced within a sequence of narrow energy bands by a sample probe is called a spectrum or spectral lines when discrete amounts of energy are exchanged between the probe and the sample (D'Amore et al., 2005).

Mass spectrometric techniques, electron techniques (imaging, spectroscopy and diffraction), ion probes (proton induced X-ray emission), X-ray methods, vibrational spectroscopy, UV-visible and luminescence spectroscopy and magnetic spectroscopy have been used to study the surface chemistry of trace elements by the removal or sputtering of molecular ions and ion fragments from those surfaces to identify them. The mass spectrometric techniques include secondary ion mass spectroscopy, fast atom bombardment, ion microprobe mass analyzer and laser microprobe mass analyzer. The electron techniques are represented by scanning electron microscopy (SEM), transmission electron microscopy (TEM), electron energy loss spectroscopy (EELS), scanning electron microscopy-energy dispersive X-ray analysis (SEM-EDX) and auger electron spectroscopy. Proton induced X-ray emission and Rutherford backscattering spectrometry are examples of ion probes. X-ray methods consist of X-ray photoelectron spectroscopy, X-ray diffraction, X-ray fluorescence microprobe and auger electron spectroscopy. Vibrational spectroscopy includes surface enhanced Raman spectroscopy and laser Raman microanalysis. The most often used types of magnetic spectroscopy are nuclear magnetic resonance, electron paramagnetic resonance and Mossbauer spectroscopy (D'Amore et al., 2005).

However, there are several factors limiting the widespread application of these instrumental techniques for determining solid-phase speciation of trace elements in soils. These are: 1) the presence of trace amounts of most trace elements and heavy metals in soils; 2) complicated matrix composition with possible interference; 3) the presence of both organic and inorganic forms of most trace elements; 4) the number of possible compounds which are greater than the number of elements; 5) the mixtures of both known and unknown forms (D'Amore et al., 2005); and 6) the expense of these instruments. All of these factors frequently contribute to less accessibility to and less sensitivity of the instrumental techniques and matrix interference. Therefore, the focus will be made below on the distribution of trace elements in solid phase components by sequential selective dissolution techniques. Throughout the rest of this chapter, the discussion will be organized by the targeted solid-phase component instead of by the trace element extractant.

5.6 DISTRIBUTION OF TRACE ELEMENTS AMONG SOLID-PHASE COMPONENTS IN ARID AND SEMI-ARID SOILS

The addition of trace elements to arid soils, in either form of addition, is known to change the original distribution patterns of trace elements in soils. Examples include land-applied sewage sludge, reclaimed sewage for irrigation, fertilizer application, municipal and industrial refuse and atmospheric fallout. As will be reviewed, the distribution of zinc and copper in native soils and soils amended with sewage sludge has been well documented in the literature. Much less information has been found for Cd, Cr, Ni and Pb. In general, most trace elements in native uncontaminated arid and semi-arid soils are mainly present in the iron oxide bound and clay mineral bound fractions, while in freshly contaminated arid and semi-arid soils, they are predominately in the carbonate and organic matter bound fractions, which increase after contamination. The distribution of heavy metals among solid phase fractions of soils is affected by climate, soil formation processes, elemental make-up of the parent material, soil properties (e.g., pH and Eh, texture, mineralogical composition, organic matter and carbonate contents) as well as the chemical properties of the elements. The pathways of transformation of trace elements and their transfer fluxes among various solid-phase components in Israeli arid soils contaminated with metal salts will be discussed as a case study in detail in the next chapter.

In Californian soils, the application of sludge (over four years) tends to increase the organically bound Cu, the carbonate bound Cd, Pb, and to some extent, Zn and Ni (Sposito et al., 1982). After being treated with sludge for seven years, the carbonate and organically bound fractions become the most prevalent solid phase for Cu, Ni and Zn, but distribution patterns of Cd, Cr and Pb are not significantly affected by the amounts of sludge added (Chang et al., 1984). In calcareous soils of Israel after prolonged irrigation with treated sewage effluents for up to 28 years, the distribution of some heavy metals is characterized by less than 0.2% in the water soluble fraction (Banin et al., 1981). Hydroxylamine hydrochloride in 25% acetic acid extractable Cd, Cr, Cu, Ni and Pb are 9–20%, 5–11%, 16–23%, 10–14%, and 1–3.9%, respectively. The initial distribution, transformation and redistribution of Cd, Pb, Ni, Cu, Cr(III) and Zn in metal salts contaminated Israeli soils are discussed in detail in Chapter 7.

5.6.1 Cadmium

Cadmium in native arid soils predominates in the carbonate fraction. In 45 uncontaminated Israeli soils, Cd is mainly present in the carbonate bound fraction (43.5% ± 22.3%), followed by the easily reducible oxide bound fraction (22% ± 19.5%) (Table 5.4) (Banin et al., 1997a). The

Table 5.4. Distribution of trace elements among solid-phase components[a] in 45 Israeli arid soils

Element		Solid-phase components (mg/kg)						
		Exchangeable	Carbonate	Easily Reducible Oxide	Organic Matter	Reducible Oxides	Residual	Sum
Cd	Average	0.01	0.16	0.09	0.04	0.04	0.04	0.36
	Standard deviation	0.02	0.24	0.12	0.07	0.05	0.04	0.33
	Maximum	0.13	1.56	0.66	0.39	0.25	0.17	2.13
	Minimum	0.00	0.01	0.00	0.00	0.00	0.00	0.07
Zn	Average	0.2	4.0	4.1	0.6	11.7	35.4	56.1
	Standard deviation	0.9	4.6	4.3	1.2	7.2	22.3	28.6
	Maximum	5.3	17.6	17.5	6.0	31.5	114.4	144.2
	Minimum	0.0	0.0	0.0	0.0	2.6	3.5	7.9
Ni	Average	0.1	2.9	2.5	2.7	7.2	15.6	31.0
	Standard deviation	0.1	3.5	2.2	2.6	4.5	11.4	18.5
	Maximum	0.5	18.4	9.4	11.1	21.7	63.4	98.7
	Minimum	0.0	0.0	0.0	0.0	0.6	1.3	3.9
Cr	Average	0.0	0.9	0.8	4.4	3.6	31.4	41.0
	Standard deviation	0.0	1.3	1.2	3.2	2.2	26.9	32.2
	Maximum	0.1	5.3	5.4	12.2	12.5	156.2	181.8
	Minimum	0.0	0.0	0.0	1.1	0.9	4.0	7.5
Cu	Average	0.1	1.2	0.9	0.3	4.0	11.6	18.2
	Standard deviation	0.2	1.2	1.0	0.3	2.1	9.7	11.7
	Maximum	0.6	6.6	3.3	1.3	8.2	60.5	73.9
	Minimum	0.0	0.0	0.0	0.0	0.6	1.5	4.4
Co	Average	0.1	0.5	1.2	0.3	2.6	4.5	9.2
	Standard deviation	0.1	0.3	1.4	0.2	1.5	2.8	4.9
	Maximum	0.2	0.9	4.0	0.8	5.0	9.7	16.8
	Minimum	0.0	0.0	0.0	0.0	0.0	0.0	3.0
Mn	Average	1.6	86.3	167	39.2	72.3	83.9	450
	Standard deviation	2.0	65.5	172	38.1	81.4	50.3	299
	Maximum	9.6	337	708	182	391	200	1271
	Minimum	0.0	3.8	3.4	0.0	3.6	11.7	64

[a]Exchangeable, carbonate, easily reducible oxide, organic matter, reducible oxide, and residual fractions were extracted by 1 M NH_4NO_3, 1 M NaOAc-HOAc, 0.04 M $NH_2OH \cdot HCl$+25%HOAc, 30% H_2O_2, hot 0.04 M $NH_2OH \cdot HCl$ + 25%HOAc and 4 M HNO_3, respectively.

organically bound and reducible oxide bound fractions account for 7.4% ± 11.3% and 11.4 ± 13.3%, respectively, while exchangeable Cd is 1.8 ± 4.3%.

In arid soils from California, U.S., 73% of the Cd exists in the carbonate fraction and only 13% in the residual fraction (Emmerich et al., 1982). In the delta of the Guadalquiver River of southwestern Spain, more than 50% of Cd is present in the carbonate fraction, and the Cd in the exchangeable fraction increases in polluted soils (Ramos et al., 1994).

In waste amended soils, Cd is primarily present in the exchangeable and carbonate fractions. In Californian soils, most of the Cd is in the carbonate bound (46%) and residual fractions (42%) (Emmerich et al., 1982). In the calcareous soils amended with sewage sludge of Southeast Spain, a high percentage of Cd is soluble and exchangeable (Moral et al., 2005). Cadmium in the carbonate and exchangeable fractions increases, and the percentage of the residual fraction decreases with time during annual applications of sludge. Sposito et al. (1983) and Chang et al. (1984) found that the carbonate bound Cd fraction increased with time in arid soils amended with sludge. In western Australian soils incubated at the field capacity regime, added Cd is redistributed from soluble to less soluble fractions with time, depending upon the type of soil, soil pH and rate of application. In the sandy soil, Cd is redistributed from the water soluble to the exchangeable fraction; in the lateritic podzolic soils from the exchangeable to the organic matter fraction, especially at higher pH; and in soils dominated by goethite from the exchangeable to the oxide bound and the residual fractions with time (Mann and Ritchie, 1994). However, Cd in soils contaminated with $PbHAsO_4$ and in battery breaking sites is predominately in the residual fraction (> 90%), while in smelter sites, Cd is mostly in the exchangeable, carbonate and residual fractions.

5.6.2 Chromium

In uncontaminated and sludge-amended arid soils, Cr is predominately bound to the residual phase (Chang et al., 1984). In 45 Israeli arid and semi-arid soils, Cr is primarily bound to the residual fraction (74.4% ± 7%), followed by the reducible oxide bound and organically bound fractions (each accounting for 10–11.6%) (Banin et al., 1997a) (Table 5.4). Chromium in the carbonate and easily reducible oxide bound fractions accounts for small percentages (each around 2%). Exchangeable Cr is only 0.03%. In the sewage sludge-amended calcareous soils of Southeast Spain, the residual fraction is the main Cr fraction (Moral et al., 2005).

5.6.3 Copper

Copper is mostly present in the residual fraction and organically bound fraction in arid soils. Sixty five to eighty-two percent of Cu in Californian soils is in the residual fraction and 12–32% is in the organically bound fraction (Emmerich et al., 1982). In 45 Israeli arid and semi-arid soils, Cu is predominately present in the residual fraction (60.4% ± 11.2%), followed by the reducible oxide bound fraction (24.5% ± 10.2%) (Table 5.4). The carbonate and easily reducible oxide bound fractions account for 7.4 ± 6.2% and 5.6 ± 5.9%, respectively. However, the organically bound fraction and exchangeable Cu in these native soils are 1.6% ± 1.9% and 0.67 ± 1.2%, respectively. Banin et al. (1990, 1997a) found that the percentages of organically bound and the carbonate bound Cu in uncontaminated Israeli soils were very small. Most of Cu in Indian calcareous soils with 14–40% $CaCO_3$ occurs in the residual (61–76%), the organically bound (7–12%), and the oxides bound fractions (12–22%) (Singh et al., 1989); whereas, in orchard soils, there are higher levels of Cu in the organically bound fraction (40%) and some exchangeable Cu (1.3%), and less in the residual fraction (42.7%) (Prasad et al., 1984). In uncontaminated arid soils of China, Cu is primarily in the residual fraction, followed by the carbonate and the oxide bound fractions (Jin et al., 1996). In the delta of the Guadalquiver River of Southwestern Spain, Cu is primarily present in the residual fraction and is associated with the organic matter fraction with small amounts in the crystalline Fe-oxide fraction (Ramos et al., 1994).

In sludge-amended Californian soils, Cu is predominately present in both the organically bound (35–54%) and carbonate (25–36%) fractions, followed by the residual fraction (22–30%) (Emmerich et al., 1982; Chang et al., 1984). Addition of sludge increases the organically bound, and to some extent, the carbonate bound fractions. In the sludge-amended calcareous soils of Southeast Spain, the predominant phase is the residual Cu (Moral et al., 2005). Copper in Colorado soils contaminated with mine tailings is mainly associated with the organic fraction (Levy et al., 1992).

5.6.4 Nickel

Nickel is found mainly in the residual and the reducible oxide bound fractions in arid soils. More than 80% of Ni in Californian soils is present in the residual fraction (Emmerich et al., 1982). In 45 Israeli arid soils, Ni is primarily in the residual fraction (48 ± 10.2%), followed by the reducible oxide bound fraction (24.2 ± 8.9%) (Table 5.4) (Banin et al., 1997a). A small portion of Ni is evenly distributed among the organically bound, the easily reducible oxide bound, and the carbonate bound fractions (each accounting for 8–11.6%). Exchangeable Ni is only 0.47 ± 0.95%. In uncontaminated arid soils of China, Ni is predominately in the residual fraction, followed by the carbonate, the organic matter and the oxide bound fractions (Jin et al., 1996). In Chinese loess soils or soils from the arid North region, Ni mainly exists in the residual (62–76%) and the Fe-Mn oxides bound fractions (13.8–26%), compared to only 4.1–6.4% in the organically bound fraction and 2.6–6.2% in the carbonate fraction (Wang and Qu, 1992).

In Californian soils amended with sewage sludge for seven years, Ni is mostly present in the residual (64%), the organically bound (12%) and the carbonate fractions (18%) (Chang et al., 1984). Nickel in the carbonate fraction is found to increase with time in arid zone soils amended with sludge (Knudtsen and O'Connor, 1987). In the sludge-amended calcareous soils of Southeast Spain, the residual and the carbonate bound Ni fractions are the major solid-phase (Moral et al., 2005).

5.6.5 Lead

Lead has been found to reside mainly in the carbonate and the residual fractions in arid soils. In Californian soils, most of the Pb is in the carbonate fraction (55%), compared to 20% in the residual fraction (Sposito et al., 1982). In Israeli arid soils, Pb is predominately in the residual and the crystalline Fe oxide bound fractions (Han and Banin, 1999). In the delta of the Guadalquiver River of Southwestern Spain, Pb is primarily present in the Fe-Mn oxide bound fraction, followed by the carbonate fraction, while the organic and the exchangeable fractions are small (Ramos et al., 1994). In uncontaminated arid soils of China, Pb is mainly in the residual fraction, followed by the carbonate and the oxide bound fractions (Jin et al., 1996).

In the sludge-amended arid soils, Pb is also primarily present in the carbonate fraction. Lead in sludge-amended Californian soils is mostly present in the carbonate fraction (80–90%) and less in the residual fraction (18%) (Sposito et al., 1982). Lead in the carbonate fraction in sludge-amended soils increases. Similarly, Misra et al. (1990) reported that the transformation of Pb into DTPA-nonextractable forms was relatively slow during 30–60 days of flooding-incubation. Lead in mine-contaminated Colorado soils is predominately present in the Fe and Mn oxide fractions (Levy et al., 1992). In the sludge-amended calcareous soils of Southeast Spain, the residual and the carbonate bound Pb are the major solid-phase (Moral et al., 2005). In a long-term $PbHAsO_4$ pesticide contaminated orchard soil from Washington State, Pb is mainly present in the carbonate and the Fe-Mn oxide bound fractions (Ma et al., 1997b).

5.6.6 Zinc

Most of the zinc in native arid soils occurs in the residual and the reducible oxides (iron oxides) fractions. In 45 Israeli arid soils, Zn is mainly bound in the residual fraction (61.3 ± 16.6%), followed by the reducible oxide bound fraction (23 ± 11.4%) (Table 5.4) (Banin et al., 1997a). About 12–14% of Zn is present in the carbonate bound and the easily reducible oxide bound fractions. The organically bound fraction and the exchangeable Zn fraction are very low, accounting for 1.1 ± 2.0% and 0.48 ± 2.2%, respectively. In Californian soils, Zn is predominately present in the residual fraction (92–96%) (Emmerich et al., 1982; Chang et al., 1984). In calcareous soils of China, 60–68% of the Zn is in the residual fraction and 15–30% is bound to crystalline iron oxides (Table 5.5), while in acid soils 80–93% of the Zn resides in the residual fraction (compared with 65–84% in neutral soils) and 7–12% in the crystalline iron oxide fraction (compared with 10–24% in neutral soils) (Han et al., 1992). Similarly, in calcareous paddy soils of China, Zn is predominately in the residual fraction (97%–99%) (Han and Zhu, 1992). More than 47% of land in India is deficient in Zn because 80–96% of their Zn is in the residual fraction (Singh et al., 1987). In the delta of the Guadalquiver River of Southwestern Spain, Zn is primarily present in the Fe-Mn oxide bound fraction, followed by the carbonate fraction, while amounts in the organic and the exchangeable fractions are small (Ramos et al., 1994).

The application of sludge increases Zn in the carbonate, the organically bound, and the exchangeable fractions in arid soils. In sludge-amended Californian soils Zn is predominately in the carbonate bound fraction (46%) and the residual fraction (36%), and less in the organically bound fraction (19%) (Emmerich et al., 1982). In Israeli soils, Zn is mostly in the organic bound, the carbonate and the easily reducible oxide bound fractions. With time, more Zn is transferred into the carbonate and easily reducible oxide fractions (Banin et al., 1990). In sludge-amended calcareous soils of Southeast Spain, the residual and the reducible oxides bound Zn are the major solid-phase (Moral et al., 2005). In calcareous soils of China after seven months of application of Zn as sulfate, Zn is redistributed in the residual (30–60% of the added soluble Zn), the soluble organically bound (10–25%), the iron oxide bound (20–30%) and the carbonate bound (only 2–13%) fractions (Han et al., 1995, 1992). Zinc in the carbonate fraction increases in the amended soils. In Indian calcareous paddy soils, Zn is transported into the amorphous-iron oxide > crystalline iron oxide > complexed Zn > residual > exchangeable fractions after three rice crops (Singh and Abrol, 1986). In Chinese upland soils, especially in calcareous soils, the amorphous iron oxide Zn decreases, the crystalline iron oxide fraction increases, and the soluble organically bound Zn fraction is transferred into the insoluble organic matter fraction with time (Han et al., 1995). In mine tailings-contaminated soils, Zn has been shown to occur mainly in the Fe and Mn oxide fraction (Levy et al., 1992). In smelter sites and battery breaking sites, Zn is mainly in the residual fraction, followed by the Fe-Mn oxide and the carbonate fractions (Ma and Rao, 1997a).

Table 5.5. Distribution of Zn among solid-phase components in some Chinese arid soils[a]

Soil	pH	Solid-phase components (mg/kg)						
		Soluble organically complexed	Carbonate	Mn oxide	Insoluble Organic	Amorphous Fe oxide	Crystalline Fe oxide	Residual
Eluvial soil	7.82	0.07	0.1	0.14	1.46	3.35	18.1	88.17
Eluvial soil	7.75	0.83	0.45	0.65	0.66	3.39	19.6	NA[b]
Black soil on old deposits	7.20	0.37	0.04	0.43	2.07	1.19	12.9	29.42
Rendizina	7.75	2.22	1.89	1.22	2.07	1.71	10.6	26.58
Meadow cinnamon soil	7.71	1.09	0.62	1.34	2.05	3.44	14.7	74.55
Cinnamon soil	7.21	1.06	0	0.5	2.16	2.31	133	67.7
Inland salt soil	8.20	1.33	0.55	0.93	0.08	5.03	16.1	NA
Grey desert soil	8.10	0.63	0	0.86	0.74	1.57	18.2	62.8
Grey cinnamon soil	6.71	5.4	1.07	1.08	4.47	4.24	6.9	90.2

[a]Data extracted from Han et al., 1995. Exchangeable (below detection limit of Atomic Absorption Spectroscopy), soluble organically complexed, carbonate, Mn oxide, insoluble organic, amorphous Fe oxide, crystalline Fe oxide, and residual fractions were extracted by 1N $MgCl_2$, 0.1M $Na_4P_2O_7$ + 1N Na_2SO_4, 1M NaOAc-HOAc, 0.1M $NH_2OH·HCl$, 30% H_2O_2, $H_2C_2O_4$-$(NH_4)_2C_2O_4$, 0.04M $NH_2OH·HCl$+25%HOAc, and $HClO_4$-HF, respectively. [b]NA: Not available.

5.6.7 Manganese

Manganese in uncontaminated Israeli arid soils is predominantly in the easily reducible oxide fraction (35–40% of the total-HNO_3 Mn), followed by the carbonate fraction (18–25%), and the residual fraction (14–25%) (Han and Banin, 1996) (Table 5.4). Manganese in the organic fraction amounts to 9–12% and in the reducible oxide fraction to 5–11%. The exchangeable fraction of Mn in the soils is very low. In the sludge-amended calcareous soils of Southeast Spain, the residual and the carbonate bound Mn fractions are the major solid-phase (Moral et al., 2005). In comparison, the Mn in fine-textured soils from the southeastern United States is primarily in the organic and the easily reducible oxide fractions (Shuman, 1985b). In acidic soils from England, the Mn mostly resides in the easily reducible oxide fraction (Jarvis, 1984), with only a very small percentage found in the exchangeable fraction. The percentages of the exchangeable, the carbonate, the organic matter, and the easily reducible oxide bound fractions in the Israeli sandy soil are higher than those in the loessial soil. This is partly related to differences in the abundance of various soil components, which compete with each other for the element. Paradoxically, carbonate and organic matter in the Israeli sandy soil contain relatively more Mn than in the loessial soil, even though the carbonate and organic matter are lower in the sandy soil. This may be because of relatively low amounts of alumosilicate minerals in the sandy soil. In general, it is noted that Mn does not tend to reside in the residual fraction of soils which are mainly constituted of alumosilicate minerals (Banin et al., 1990). This is attributed to the following facts: the solubility of Mn silicates is high; Mn inclusion as an isomorphous substitution in layer silicates is not favored because Mn^{4+} is too highly charged to be included in tetrahedral and octahedral oxygen packing, and Mn^{2+} is not frequently present in oxidized environments.

5.6.8 Cobalt

Cobalt in the native Israeli arid-zone soils is predominantly in the residual fraction (40–55% of the total-HNO_3 Co), followed by the easily reducible oxide fraction (20–35%) and the reducible oxide fraction (10–25%) (Table 5.4). Much smaller content is found in the other fractions (the organic bound fraction, 3–10%; the carbonate fraction, < 8% and the exchangeable fraction, < 1%). This shows that a large fraction of the soil Co in the native Israeli arid soils is incorporated in primary and secondary alumosilicate minerals and in oxides. In uncontaminated arid soils of China, Co is predominately in the residual fraction, followed by the oxide and the carbonate bound fractions (Jin et al., 1996). In the sludge-amended calcareous soils of Southeast Spain, the residual, the carbonate and the reducible oxides bound Co are the major solid-phase (Moral et al., 2005). Cobalt is known to replace divalent metals in the octahedral layers of clay minerals as an isomorphous substitution, and it may also be incorporated in Mn oxides and Fe oxides (Burns, 1976; Mckenzie, 1975; Ainsworth, et al., 1994). In comparison, Co in the soils of southern Scotland is mostly in the residual fraction (44–77%), with small amounts in the organic and the easily reducible oxide (Mn oxide) fractions (3–26%) and in the Fe oxide fraction (14–42%) (McLaren et al., 1986a). It appears that in arid zone soils a higher proportion of Co is present in the easily reducible oxide (Mn oxide) bound fraction and a lower proportion is in the organic matter fraction compared to the humid-zone soils.

The data for the six Israeli arid soils show that Co contents in the easily reducible oxide, the organic matter, the reducible oxide and the residual fractions and the total Co are linearly related to the Mn contents in each respective fraction. Correlation coefficients between the respective Co and Mn fractions are significant at the 5% probability level (Fig. 5.6). The Co/Mn molar ratios are 0.0186, 0.0177, 0.0075, 0.0205, 0.0364 in the total, the easily reducible oxide, the organic matter, the reducible oxide and the residual fractions, respectively. Therefore, Co is more preferentially incorporated in the reducible oxide (Fe oxide) fraction than Mn. Much of the Co in arid soils is included in the easily reducible oxide (Mn oxide) fraction,

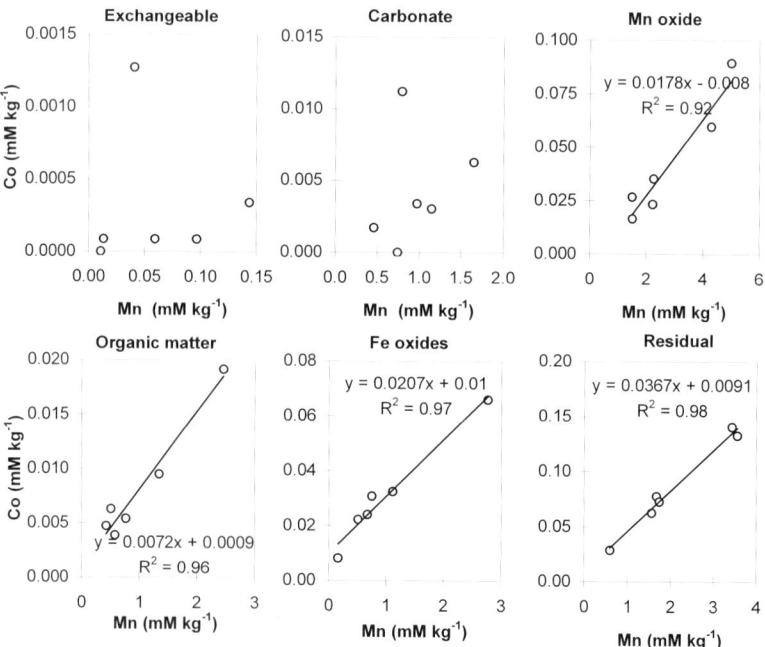

Figure 5.6. Relationship between Co and Mn contents extracted from solid-phase of six Israeli arid-zone soils with sequential dissolution procedures (after Han et al., 2002b. Reprinted from J Environ Sci Health, Part A, 137, Han F.X., Banin A., Kingery W.L., Li Z.P., Pathways and kinetics of transformation of cobalt among solid-phase components in arid-zone soils, p 184, Copyright (2002), with permission from Taylor & Francis)

which is relatively labile (under reducing conditions) in soils. In addition, Co^{3+} (the product of Co reacted with Mn oxides as well as a result of weathering) has been proven to occur in arid and semi-arid soils (Cai and Liu, 1991). In the extracts of 0.02 M EDTA + 0.5 M HOAc + 0.5M NH_4OAc from 17 arid and semi-arid soils of China, Co^{2+} and Co^{3+} are in the range of 0.04–0.80 and 0.72–6.42 mg/kg, respectively. The amount of Co^{3+} increases with increasing amorphous Mn oxides in soils (r = 0.81**, n = 14). However, in general, Co^{2+} is the majority of soil Co (Cai and Liu, 1991).

5.6.9 Other Trace Elements

Woolson et al. (1971) used a modified soil P procedure to study distribution of As in U.S. soils. The mass of the residual As in both uncontaminated and soils contaminated with arsenic pesticides from Washington and Oregon is found as the Fe oxide bound As (0.1 N NaOH extractable), followed by the Al oxide and the Ca bound As (extracted by NH_4F and H_2SO_4, respectively).

The distribution of trace elements and heavy metals in waste amended-arid soils is influenced by soil properties (organic matter, oxide, and carbonate contents, soil texture etc.), soil condition (soil pH and Eh) and the loading levels of waste (i.e., sewage sludge) (Banin et al., 1990; Sposito et al. 1983). Banin et al. (1990) reported that an Israeli sandy soil amended with sludge contained more Cu in the carbonate and the easily reducible oxide fractions than a clay soil, but the clay soil had higher Zn in the reducible oxide fraction and less in the carbonate fraction. They also found that Cd in soils at a low sludge rate (1%) was mainly in the reducible oxide (35–45%) and the easily reducible oxide (25–30%) fractions, whereas at a higher sludge rate (10%), most of the Cd was in the carbonate (20–30%) and the easily reducible oxide (20–30%) fractions. The effects of levels of sewage sludge on metal redistribution in soil may be, in part, related to dilution. Sposito et al. (1982) found that in sludge-treated Californian soils at a high rate of application (90 tons/ha/yr), the predominant forms of the metals were the organic fraction for Cu, the residual fraction for Ni and the carbonate fraction for Zn, Cd and Pb.

5.7 SOLUBILITY OF SOLID PHASES AND THEIR MINERALOGICAL CHARACTERISTICS CONTROLLING DISTRIBUTION OF TRACE ELEMENTS AMONG SOLID PHASE COMPONENTS IN ARID SOILS

General physical and chemical properties and mineralogical characteristics of selected trace elements (Cd, Cu, Cr, Ni, Pb, Zn, Co and Mn) controlling their distribution and solubility in arid and semi-arid soils are briefly discussed below. The properties of relevant major and trace metals are presented in Table 5.6 and Table 5.7. The eight trace elements are divided, in general, into four groups according to these general properties. This division is compatible with their atomic, physical and chemical properties and mineralogical characteristics: Group I contains Cd and Pb; Group II contains Ni, Zn and Cu; Group III contains Cr; and Group IV contains Co and Mn.

Table 5.6. Some basic atomic properties of studied and related elements[a]

Element	Atomic Number	Electronic Configuration	Electronegativity	Crystal Ionic Radius A°
Al	13	$[Ne]3s^23p$	1.5	0.5
Ca	20	$[Ar]4s^2$	1	0.99
Cd	48	$[Kr]4d^{10}5s^2$	1.7	0.97
Co	27	$[Ar]3d^74s^2$	1.8	0.74(+2[b]), 0.63(+3)
Cr	24	$[Ar]3d^54s$	1.6	0.64(+3), 0.52(+6)
Cu	29	$[Ar]3d^{10}4s$	2.0 (+2)	0.72 (+2)
Fe	26	$[Ar]3d^64s^2$	1.8 (+2), 1.9 (+3)	0.76(+2), 0.64(+3)
Mg	12	$[Ne]3s^2$	1.2	0.65
Mn	25	$[Ar]3d^54s^2$	1.5	0.80(+2), 0.62(+3), 0.54(+4)
Ni	28	$[Ar]3d^84s^2$	1.8	0.72
Pb	82	$[Xe]4f^{14}5d^{10}6s^26p^2$	1.8	1.2
Zn	30	$[Ar]3d^{10}4s^2$	1.6	0.74

[a] Data from Lange's Handbook of Chemistry, edited by Dean, 1973.
[b] Number in parenthesis is the valence.

Table 5.7. Solubility equilibria of selected minerals of related elements[a]

Element	Reaction (Oxides)	logK°	Reaction (Hydroxides)	logK°
Al	$0.5\gamma\text{-}Al_2O_3$ (c) $+ 3H^+ \rightleftharpoons Al^{3+} + 1.5H_2O$	11.49	$\gamma\text{-}Al(OH)_3$ (gibbsite) $+ 3H^+ \rightleftharpoons Al^{3+} + 3H_2O$	8.04
	$0.5\alpha\text{-}Al_2O_3$ (corundum) $+ 3H^+ \rightleftharpoons Al^{3+} + 1.5H_2O$	9.73	$Al(OH)_3$ (amorp) $+ 3H^+ \rightleftharpoons Al^{3+} + 3H_2O$	9.66
Ca	CaO (lime) $+ 2H^+ \rightleftharpoons Ca^{2+} + H_2O$	32.95	$Ca(OH)_2$ (portlandite) $+ 2H^+ \rightleftharpoons Ca^{2+} + 2H_2O$	22.80
Fe	$1/2\alpha\text{-}Fe_2O_3$ (hematite) $+ 3H^+ \rightleftharpoons Fe^{3+} + 3/2H_2O$	0.09	$Fe(OH)_3$ (soil) $+ 3H^+ \rightleftharpoons Fe^{3+} + 3H_2O$	2.70
			$\alpha\text{-}FeOOH$ (goethite) $+ 3H^+ \rightleftharpoons Fe^{3+} + 2H_2O$	-0.02
			$\gamma\text{-}FeOOH$ (lepidocrocite) $+ 3H^+ \rightleftharpoons Fe^{3+} + 2H_2O$	1.39
Mg	MgO (periclase) $+ 2H^+ \rightleftharpoons Mg^{2+} + H_2O$	21.74	$Mg(OH)_2$ (brucite) $+ 2H^+ \rightleftharpoons Mg^{2+} + 2H_2O$	16.84
Cd	CdO (monteponite) $+ 2H^+ \rightleftharpoons Cd^{2+} + H_2O$	15.14	$\beta\text{-}Cd(OH)_2$ (c) $+ 2H^+ \rightleftharpoons Cd^{2+} + 2H_2O$	13.65
Cu	CuO (tenorite) $+ 2H^+ \rightleftharpoons Cu^{2+} + H_2O$	7.66	$Cu(OH)_2$ (c) $+ 2H^+ \rightleftharpoons Cu^{2+} + 2H_2O$	8.68
Mn	$\beta\text{-}MnO_2$ (pyrolusite) $+ 4H^+ + 2e^- \rightleftharpoons Mn^{2+} + 2H_2O$	41.89	$\gamma\text{-}MnOOH$ (manganite) $+ 3H^+ + e^- \rightleftharpoons Mn^{2+} + 2H_2O$	25.27
	$\delta\text{-}MnO_{1.8}$ (birnessite) $+ 3.6H^+ + 1.6e^- \rightleftharpoons Mn^{2+} + 1.8H_2O$	35.38	$Mn(OH)_2$ (pyrochroite) $+ 2H^+ \rightleftharpoons Mn^{2+} + 2H_2O$	15.19
	Mn_3O_4 (hausmannite) $+ 8H^+ + 2e^- \rightleftharpoons 3Mn^{2+} + 4H_2O$	63.03		
Pb	PbO (yellow) $+ 2H^+ \rightleftharpoons Pb^{2+} + H_2O$	12.89	$Pb(OH)_2$ (c) $+ 2H^+ \rightleftharpoons Pb^{2+} + H_2O$	8.16
Zn	ZnO (zincite) $+ 2H^+ \rightleftharpoons Zn^{2+} + H_2O$	11.16	$Zn(OH)_2$ (amorp) $+ 2H^+ \rightleftharpoons Zn^{2+} + H_2O$	12.48

[a] Data from Lindsay, 1979. (Log K° at 25°C)

Table 5.7. Continued[a]

Element	Reaction	logK°	Reaction	logK°
	Carbonate		Others	
Ca	$CaCO_3$ (calcite) $+2H^+ \rightleftharpoons Ca^{2+} + CO_2(g) + H_2O$	9.74	$CaSO_4 \cdot 2H_2O$ (gypsum) $\rightleftharpoons Ca^{2+} + SO_4^{2-} + 2H_2O$	-4.64
Fe	$FeCO_3$ (siderite) $+2H+ \rightleftharpoons Fe^{2+} + CO_2(g) + H_2O$	7.92	Fe_2SiO_4 (fayalite) $+4H^+ \rightleftharpoons 2 Fe^{2+} + H_4SiO_4$	19.76
Mg	$MgCO_3$ (magnesite) $+ 2H^+ \rightleftharpoons Mg^{2+} + CO_2(g) + H_2O$	10.69	$MgSiO_3$ (clinoenstatite) $+2H^+ +2H_2O \rightleftharpoons Mg^{2+} + H_4SiO_4$	11.42
	$CaMg(CO_3)_2$ (dolomite) $+ 4H^+ \rightleftharpoons Ca^{2+} + Mg^{2+} + 2CO_2(g) + H_2O$	18.46		
Cd	$CdCO_3$ (octavite) $+ 2H^+ \rightleftharpoons Cd^{2+} + CO_2(g) + H_2O$	6.16	$Cd_3(PO_4)_2$ (c) $+ 4H^+ \rightleftharpoons 3 Cd^{2+} + 2H_2PO_4^-$	1.00
			$CdSiO_3$ (c) $+ 2H^+ + H_2O \rightleftharpoons Cd^{2+} + H_4SiO_4$	7.63
Cu	$CuCO_3$ (c) $+ 2H^+ \rightleftharpoons Cu^{2+} + CO_2(g) + H_2O$	8.52	α-$CuFe_2O_4$ (cupric ferrite) $+ 8H^+ \rightleftharpoons Cu^{2+} + 2Fe^{3+} +4 H_2O$	10.13
	$Cu_2(OH)_2CO_3$ (malachite) $+ 4H^+ \rightleftharpoons 2Cu^{2+} + CO_2(g) + 3H_2O$	12.99		
Mn	$MnCO_3$ (rhodochrosite) $+ 2H^+ \rightleftharpoons Mn^{2+} + CO_2(g) + 2H_2O$	8.08	$MnSiO_3$ (rhodonite) $+ 2H^+ +H_2O \rightleftharpoons Mn^{2+} + H_4SiO_4$	10.25
Pb	$PbCO_3$ (cerussite) $+ 2H^+ \rightleftharpoons Pb^{2+} + CO_2(g) + H_2O$	4.65	$Pb_3(PO_4)_2$ (c) $+ 4H^+ \rightleftharpoons 3 Pb^{2+} + 2H_2PO_4^-$	-5.26
			$PbSiO_3$ (c) $+ 2H^+ + H_2O \rightleftharpoons Pb^{2+} + H_4SiO_4$	5.94
Zn	$ZnCO_3$ (smithsonite) $+ 2H^+ \rightleftharpoons Zn^{2+} + CO_2(g) + H_2O$	7.91	$ZnFe_2O_4$ (franklinite) $+ 8H^+ \rightleftharpoons Zn^{2+} + 2Fe^{3+} +4 H_2O$	9.85
			Zn_2SiO_4 (willemite) $+ 4H^+ \rightleftharpoons 2Zn^{2+} + H_4SiO_4$	13.15

[a] Data from Lindsay, 1979. (Log K° at 25°C)

5.7.1 Group I: Cd and Pb

Cadmium and lead in the native and amended calcareous soils are predominately present in the carbonate fraction. These two elements in arid soils tend to bind to calcium carbonate surfaces and form surface precipitates of metal-carbonate. The low solubility of $CdCO_3$, the similarity of the ionic radii of Cd to Ca, and similar electron configuration and unit crystal structure as carbonate cause preferential sorption of Cd on the calcite surfaces in the arid soil and smooth transition to its surface precipitation in the carbonate phase of the soils. Cadmium is adsorbed by $CaCO_3$ on the surface and then slowly diffuses into the inside particles, and finally $CdCO_3$ and $CaCO_3$ form a solid-solution (Stipp et al., 1992). The relatively high solubility of Cd oxide and hydroxide (Table 5.7) decreases the affinity of Cd to the oxide phase. The large ionic size of Cd and Pb prevents the incorporation of both of them as an isomorphous substitutent in layer silicates. Since lead has a large ionic radius (120 pm), it tends to associate predominately with calcium carbonate grains in the arid soils. The reaction forms a solid solution of $PbCO_3$ in $CaCO_3$, either on the surface or inside the grains. Thus Cd and Pb in arid soil tend to reside mainly in the carbonate phase, and due to low affinity of Cd to all of the other more stable phases, Cd appears in the exchangeable phase, especially when the carbonate content of the soil is low. This distribution pattern is responsible for the high mobility of Cd in soils and sediments and its high risk of introduction into the food-chain and contamination of underground water. Results by Han and Banin (1997, 1999) and Han et al. (2001a, c) suggest that the moisture regime of soils does not change the status of Cd in the arid soils. Due to their high affinity with carbonate, Cd and Pb in the metal salted amended soils quickly approached the quasi-equilibrium characteristic in the native soils. This will be discussed in detail in the next chapter.

5.7.2 Group II: Zn, Cu, and Ni

Zinc, Cu and Ni have similar ionic radii and electron configurations (Table 5.6). Due to the similarity of the ionic radii of these three metals with Fe and Mg, Zn, Cu and Ni are capable of isomorphous substitution of Fe^{2+} and Mg^{2+} in the layer silicates. Due to differences in the electronegativity, however, isomorphous substitution of Cu^{2+} in silicates may be limited by the greater Pauling electronegativity of Cu^{2+} (2.0), whereas Zn^{2+} (1.6) and Ni^{2+} (1.8) are relatively more readily substituted for Fe^{2+} (1.8) or Mg^{2+} (1.3) (McBride, 1981). The three metals also readily coprecipitate with and form solid solutions in iron oxides (Lindsay, 1979; Table 5.7).

The solubility ordering of the carbonates is Ni (NiCO$_3$, Ksp: 6.6 x 10^{-9}) > Cu (CuCO$_3$, 1.4 x 10^{-10}) > Zn (ZnCO$_3$, 1.4 x 10^{-11}) (Dean, 1973; Table 5.7). This makes the presence of some Zn in the carbonate phase more likely than for Cu and Ni. Addition of excess free zinc initially leads to increased association to the carbonate phase and only a limited binding to oxides (Fe/Mn oxide) and silicate (residual) phases. Eventually, however, incorporation of Zn in the silicate and oxide phases, due to formation of Zn$_2$SiO$_4$ (willemite) and ZnFe$_2$O$_4$ (franklinite), will be responsible for the redistribution and transformations of added Zn in the amended soils (Lindsay, 1979).

On the other hand, the low solubility of Cu oxide and hydroxide minerals and relatively high solubility of its carbonate cause the preferred association of Cu with the oxide phases, building increasingly thicker surface layers and smoothly transposing from sorption to surface precipitation on oxides. It was suggested that CuFe$_2$O$_4$ may determine the solubility of Cu^{2+} in soil solutions (Lindsay, 1979). However, the tetrahedral coordination of Fe^{3+} required for this structure may inhibit its formation under soil conditions. McBride (1981) suggested that Cu^{2+} was in chemisorbed or occluded forms on iron oxides rather than as a separate phase, or possibly a formation of Cu^{2+}-O-Fe^{3+} or Cu-O-Al bonds. However, chemisorption is likely to occur only at the edges on crystalline mineral grains. Co-precipitation of Cu^{2+} in aluminum and iron hydroxides occurs readily. Furthermore, since Cu^{2+} substitution inhibits crystallization and produces high surface area oxides, the deficit created by substitution of Al^{3+} or Fe^{3+} by Cu^{2+} must be balanced by the adsorbed cations. Cu^{2+} substitution is usually located at or near surfaces (McBride, 1981). In fact, Cu^{2+} is the most strongly adsorbed of all the divalent transition and heavy metals on Fe and Al oxides and oxyhydroxides. In contrast, the affinity of synthetic Mn oxides to specifically adsorb Cu^{2+} is even stronger than that of Fe or Al oxides (McKenzie, 1980). Nevertheless, the higher abundance of iron relative to manganese in soils may give Cu a better opportunity to associate with iron-rich secondary mineral phases in soils despite a very strong affinity of manganese oxides for Cu. It is much less likely that Cu will be retained by soil carbonates, or form solid solution in soil calcium carbonate due both to the large ionic size difference between Cu and Ca and the relatively high solubility of Cu carbonate. However, Cu^{2+} may substitute more freely for Mg^{2+} (0.72 °A) in magnesium carbonate or Fe^{2+} (0.78 °A) in iron carbonate (McBride, 1981). Excess soluble Cu added to soils will tend to temporarily associate with the oxide phases rather than the carbonate phases.

Copper also binds very strongly as an inner-sphere complex with organic matter while other divalent transition metals such as Ni^{2+} and Co^{2+}

are probably bound by outer-sphere complexation with the metal retaining its inner hydrated sphere (McBride, 1981). The contribution of organic matter to the binding of Cu will be higher in soils poor in the clay minerals or other active components, and will be highest when the predominant clay mineral is kaolinite and lowest when montmorillonite is the main clay mineral (Stevenson and Fitch, 1981). Copper incorporation in layer silicates is due to its coordination number of six, leading to isomorphous substitution in the octahedral sites of phyllosilicates, but the process is very slow.

Results by Han and Banin (1997, 1999) and Han et al. (2001a, c) reported that in the native arid soils, Zn, Ni and to some extent Cu, were predominately present in the residual, the reducible oxide, and to some extent, the carbonate fractions. After addition of metal salts, the metals are rapidly bound to the carbonate fraction. Later, metals are further transferred from the carbonate fraction into the more stable fractions, such as the easily reducible oxide, the organic matter, the reducible oxide bound and the residual fractions. Because cupric ferrite ($CuFe_2O_4$) has lower solubility than frankinite ($ZnFe_2O_4$) at the same conditions (i.e., pH), earlier incorporation of added Cu into Fe oxides than of added Zn might be preferred. The affinity of Cu for various Fe, Al, and Mn oxides is greater than Zn, which is, in turn, greater than Ni (Ross, 1994). All of these facts explain the observation that the Cu in the amended arid soils approaches the quasi-equilibrium faster than Ni and Zn, and these three metals are slower than Cd and Pb, as observed by Han and Banin (1997, 1999) and Han et al. (2001a, c).

5.7.3 Group III: Cr

Chromium has a similar electron configuration to Cu, because both have an outer electronic orbit of 4s. Since Cr^{3+}, the most stable form, has a similar ionic radius (0.64 A°) to Mg (0.65 A°), it is possible that Cr^{3+} could readily substitute for Mg in silicates. Chromium has a lower electronegativity (1.6) than Cu^{2+} (2.0) and Ni (1.8). It is assumed that when substitution in an ionic crystal is possible, the element having a lower electronegativity will be preferred because of its ability to form a more ionic bond (McBride, 1981). Since chromium has an ionic radius similar to trivalent Fe (0.65°A), it can also substitute for Fe^{3+} in iron oxides. This may explain the observations (Han and Banin, 1997, 1999; Han et al., 2001a, c) that the native Cr in arid soils is mostly and strongly bound in the clay mineral structure and iron oxides compared to other heavy metals studied. On the other hand, humic acids have a high affinity with Cr (III) similar to Cu (Adriano, 1986). The chromium in most soils probably occurs as Cr (III) (Adriano, 1986). The chromium (III) in soils, especially when bound to organic matter, remains reduced and resistant to oxidation for considerable lengths of time. Thus, as Han and Banin's (1997, 1999) results show, when soluble Cr is added to arid soils, added Cr^{3+} is initially and immediately bound to the organic matter fraction. With time, however, it is transferred into the more stable fractions. Due to its slow conversion into the reducible oxide and the residual fractions, Cr in the amended arid soils mostly departs from and remains removed from the quasi-equilibrium; however, with time it still very slowly approaches the quasi-equilibrium.

5.7.4 Group IV: Mn and Co

Since the solubility of Mn silicates is high, and further, its inclusion as an isomorphous substitution in layer silicates is not favored (Mn^{4+} is too highly charged to be included in tetrahedral and octahedral oxygen packing, and Mn^{2+} is not often present in oxidized environments), Mn does not appear in the residual phase of arid soils (Banin et al., 1990). On the other hand, Mn has much higher concentrations in soils than other heavy metals leading to the formation of separate minerals. The complex solubility interplay between the oxides and oxyhydroxides of Mn under varying redox conditions affects frequent transformations among the easily reducible oxide and the reducible oxide phases. Since $MnCO_3$ has a solubility similar to MnOOH at the redox and CO_2 concentration prevailing in soils, Mn is frequently transformed between the oxide phase and the carbonate phase. In soils with a high total Mn, low carbonates and low biological activity, Mn tends to be present in various Mn-oxide minerals such as pyrolusite (β-MnO_2; Mn^{4+}) and, possibly, manganite (γ-MnOOH; Mn^{3+}) in several forms including concretions, pans, coatings and mottles. Nevertheless, in arid soils high in carbonates or low in oxide surfaces, Mn associates with the carbonate phase forming a solid solution with calcite or pure rhodochrosite ($MnCO_3$, Mn^{2+}) (Banin et al., 1990; Lindsay, 1979). Results by Han and Banin (1996) show that Mn in native arid soils is mainly present in the easily reducible oxide (Mn oxide) and the carbonate fractions. However after saturation, Mn is mainly transferred from the Mn oxides into the Mn carbonate and the exchangeable fractions (Han and Banin, 1996).

Cobalt is capable of substitution for Fe^{2+} and other transition metals in the phyllosilicates due to the similarity of ionic radii. On the other hand, cobalt (Co^{2+}) is specifically adsorbed by Mn and Fe oxides, and concentrations of Co sorbed by Mn oxides are much greater than those by Fe oxides (Backes, et al., 1995). Traina and Donor (1985) suggested that the Mn release during Co^{2+} ion sorption resulted not only from the oxidation of

Co^{2+} to Co^{3+}, but also from a direct exchange of Co^{2+} for Mn^{2+} produced during the redox reaction. The cobalt in arid soils, as indicated by Han et al. (2002b), mainly occurs in the residual and the Mn oxide (easily reducible oxide) fractions. Furthermore, after water saturation, the Co is transferred mainly from the Mn oxide fraction into the carbonate and exchangeable fraction. This will be discussed in detail in the next chapter.

Chapter 6

TRANSFER FLUXES OF TRACE ELEMENTS IN ARID ZONE SOILS – A CASE STUDY: ISRAELI ARID SOILS

Kinetic analyses of trace element sorption processes indicate that metal binding reactions take place almost instantaneously (Brümmer et al., 1988). In pure mineral systems, a slow process has been identified which associates with solid-state diffusion from external to internal binding sites. For example, it is thought to take place in meso- and micro-pores in geothite (Fischer et al., 1996). In arid soils, however, the slow process observed is largely controlled by both the redistribution of trace metals among various solid-phase components and the slow diffusion of surface sorbed or precipitated metals into particles (Han, 1998). Solid-phase components have differing metal binding intensities, and hence, strongly affect the overall kinetics of trace elements in polluted soils (Han et al., 2003a). The better understanding of the effects of metal species, loading levels, time of reaction and soil management on transfer fluxes of trace elements or heavy metals in soils is essential to remediate metal polluted soils. Transfer fluxes of seven trace elements (Cd, Cu, Cr, Ni, Zn, Mn and Co) between the solution and solid phase and among the major solid-phase components in Israeli arid soils are discussed in this chapter.

The present case study included both laboratory incubation experiments and a field study. Soluble metal salts and three soil moisture regimes were employed in the laboratory incubation experiments, while the field study used sewage sludge as a source for trace elements. The soil moisture regimes employed in the incubation experiments included the saturation paste, field capacity and wetting-drying cycling moisture regimes. The wetting-drying cycle consisted of the first two days at the field capacity regime, followed by subsequent 12 days of air-drying at $25°C$ in each cycle. The relevant soil properties are presented in Table 6.1. Metal nitrates (Cd, Cu, Cr, Ni, Pb and Zn) were added as salt powder to two representative arid soils (sandy and loessial soils) at various loading levels from 0.5T to 5T. 'T' is the total concentration of trace elements in native soils (Tables 6.2–6.3). The wet soils were placed in a constant-temperature room and incubated at

$25 \pm 1\,°C$. On the basis of determined moisture content changes, deionized water was supplied to the remaining soil to keep it in the respective water regime.

Table 6.1. The relevant soil properties

Soil	Experiment	Metal treatment	pH	CaCO$_3$ %	OM (%)	CEC (meq/100g)	SSA (m^2/g)	clay %	silt %	sand %
Sandy E20[a]	Incubation	Metal salts	7.15	0.5	1.0	4.5	29.0	6.0	2.0	92.0
Loessial S20[a]	Incubation	Metal salts	8.03	23.2	1.4	7.6	75.0	15.0	23.0	62.0
Sandy	Field study	Sewage sludge	7.40	1.1	1.6	NA[b]	NA	16.0	7.0	77.0

[a] E20, a sandy soil and S20, a loessial soil were used for all incubation studies at three moisture regimes
[b] NA: not available

Table 6.2. The original metal concentration (HNO$_3$-extractable) in native soils and levels of addition to the two Israeli arid soils incubated under the saturated paste regime condition

Soil	Treatment Class	Treatment designation	Metals									
			Cd[a]		Cr		Cu[b]		Ni		Zn[b]	
			Addition	Actual T	Addition	Actual T	Addition	Actual T	Addition	Actual T	Addition	Actual T
			(mg kg^{-1})		(mg kg^{-1})		(mg kg^{-1})		(mg kg^{-1})		(mg kg^{-1})	
Sandy	Control	CK	0 (0.05)[c]	0T	0 (14.75)	0T	0 (10.2)	0T	0 (6.9)	0T	0 (31.0)	0T
	Low	"0.5T"	0.56		5.0	0.5T	1.6	0.125T	3.0	0.5T	6.4	0.25T
	Intermediate	"1T"	1.12		10.0	1T	3.2	0.25T	6.0	1T	12.8	0.5T
	Intermediate	"2T"	2.24		20.0	2T	6.4	0.5T	12.0	2T	25.6	1T
	High	"3T"	3.36		40.0	3T	12.7	1T	24.0	3T	51.2	2T
Loessial	Control	CK	0 (0.18)	0T	[d]		0 (13.9)	0T	0 (17.9)	0T	0 (39.7)	0T
	Low	"0.5T"	0.05	0.5T			7.5	0.5T	10.8	0.5T	20.9	0.5T
	High	"3T"	0.30	3T			45.0	3T	64.6	3T	125.5	3T

[a] Due to the very low native concentration of Cd in the sandy soil (analytical problems), additions of Cd are not exactly proportional to the total content in the soil.
[b] Additions of Cu and Zn are at different proportion to the total contents than that designated in the treatment. All metals added were nitrate except Zn as sulfate. The actual loading level is used for discussions in this chapter.
[c] Numbers in parentheses are the content of the metal in the native soil (mg kg^{-1}). These values are defined as "T".
[d] Cr was not added to the loessial soil.

Subsamples were taken for selective sequential dissolution analyses (SSD, Table 6.4) after 1hour, 1 day, 3 days, 6 days, 12 days, 18 days, 24 days, days, 30 days, 2 months, 5 months, 10 months and 1 year for the saturation study. Also, subsamples were taken after 1 hour, 1 day, 1 week, 3 weeks, 6 weeks, 12 weeks, 24 weeks, and 48 weeks for the field capacity regime study. For the wetting-drying cycle study, soils were sampled for analyses prior to wetting (time zero) and in the 1st cycle, 2nd cycle, 4th cycle, 8th cycle, 16th cycle, and 24th cycle. Two or three replicates were conducted for SSD analysis.

Table 6.3. Treatment levels of trace metals in the two Israeli arid soils incubated under the field capacity and wetting-drying cycle moisture regimes

Soil	Treatment Class	Treatment designation	Metal added^a (mg kg^{-1})					
			Cd^b	Cr	Cu	Ni	Pb	Zn
Sandy	Control	CK	0.00	0.00	0.00	0.00	0	0.00
	Low	0.5T	0.05	7.4	5.1	3.4	6.2	15.5
	Intermediate	2T	0.17	23.8	16.5	11.1	20.0	50.0
	High	3T	0.31	44.3	30.6	20.6	37.3	93.0
		5T	0.52	73.8	51.0	34.3	62.1	155
Loessial	Control	CK	0.00	0.00	0.00	0.00	0.00	0.00
	Low	0.5T	0.09	14.9	7.5	10.8	8.4	20.9
	High	3T	0.54	89.3	45.0	64.6	50.6	125
		5T	0.90	149	75.0	108	84.4	209.2

^{a}Metals added were nitrates
^{b}Analytical problems related to the very low native concentration of Cd in the sandy soil, lead to the fact that additions of Cd are not at the exact proportionality to the total content in the native soil (T), as are the other metals in the same treatment class.

For the field study, sewage sludge from a local municipal water treatment plant was added to a field (sandy soil) at the rates of 0, 23, 46, 92, and 184 tons/ha. Lettuce was planted after six months of sludge application. Soils were taken at the end of the field experiment (about 8 months of sludge application) (Banin et al., 1990).

Table 6.4. Protocol for the selective sequential dissolution procedure

Step No.	Fraction	Solid-phase desired	Reagents	pH	Ratio of solution/soil	Temperature (^0C)	Time	Note
1	EXC	Exchangeable	1 M HN_4NO_3	7	25:1	25	30 min	
2	CARB	Carbonate	1 M NaOAc-HOAc	5	25:1	25	6 hours	
3	ERO	Easily reducible oxides	0.04 M $NH_2OH.HCl$ + 25% HOAc	2	25:1	25	30 min	
4	OM	Organic matter	30% H_2O_2 + 0.01 M HNO_3	2	25:1	80	3 hours	Water bath
5	RO	Reducible oxides	0.04 M $NH_2OH.HCl$ + 25% HOAc	2	25:1	90	3 hours	Water bath
6	RES	Residual	4 M HNO_3		25:1	180	30 min	Microwave

6.1 PROCESSES OF LONG-TERM TRANSFER OF TRACE ELEMENTS IN ARID SOILS CONTAMINATED WITH METAL SALTS

The addition of trace elements as either salts or sewage sludge and other waste forms changed the original distribution pattern of the metals in arid soils (Figs. 6.1–6.4). The transformation of added soluble metals in arid soils was characterized by an initial (< 1 hour) fast reduction of lability and subsequent slow reduction. Both processes were the result of the transfer of added metals from the labile exchangeable (EXC) fraction to the carbonate bound (CARB), easily reducible oxide (Mn oxide) bound (ERO), organically bound (OM), reducible oxide (Fe oxide) bound (RO), and residual (RES) fractions. At low levels of addition, the trend was towards a distribution more similar to that of the native soil, but at higher levels, the soils remain removed from the quasi-equilibrium in the native soil.

6.1.1 The Initial Fast Processes

Soluble heavy metals initially added to arid soils were quickly transferred from the soluble and exchangeable form into the carbonate bound fraction. Within the first hour from the start of the saturation incubation, and/or during the first step of the SSD procedure, the metals in the freshly salt contaminated arid soils were predominately transferred from the EXC fraction into the CARB fraction (82% for Cd, > 50% for Ni and Cr, 40–85% for Zn, and 30% for Cu) (Han and Banin, 1997). Similarly, under the field capacity regime, added soluble Cd, Pb, Ni, Zn and to some degree, Cu, were also initially transferred from the EXC fraction into the CARB fraction (Figs. 6.1–6.4) (Han and Banin, 1999). The initial rapid redistribution process was affected by the nature of the metals, loading levels, and soil properties (Figs 6.1–6.4). Cadmium was almost exclusively transferred from the soluble fraction to the CARB fraction. Under the saturated paste regime, chromium in the sandy soil primarily moved into the CARB (50%) and the OM fractions (36%). But under the field capacity regime, added Cr in both sandy and loessial soils, and to some extent Cu and Ni in the sandy soil, were mainly incorporated into the OM fraction (40–70% Cr, and 20–30% for both Ni and Cu in the sandy soil). Copper was transferred into the CARB (30%) fraction, followed by the RO, OM and ERO fractions. Still, after one day of saturation incubation, significant amounts of Ni, Cd and Cu remained in the EXC fraction.

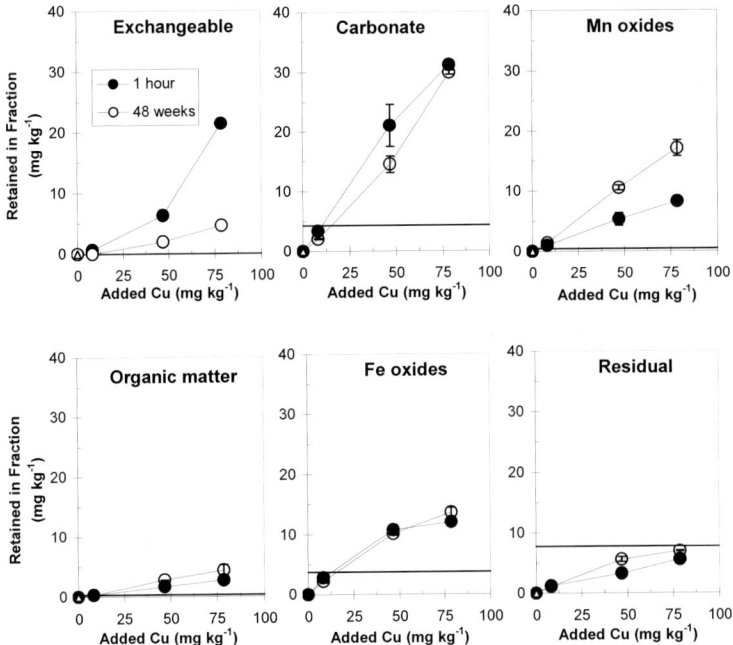

Figure 6.1. The fractional loading isotherms of Cu in a contaminated Israeli loessial soil at an initial (one hour) period and after 48 weeks. The soil was treated with increasing levels of metal nitrates and was incubated under the field capacity regime. Horizontal solid line represents the native content of Cu in the nonamended soil (*Figure 6.1 – Figure 6.4,* after Han and Banin, 2001. Reprinted from Commun Soil Sci Plant Anal, 32, Han F.X and Banin A.,The fractional loading isotherm of heavy metals in an arid-zone soil, pp 2700–2703, Copyright (2001), with permission from Taylor & Francis)

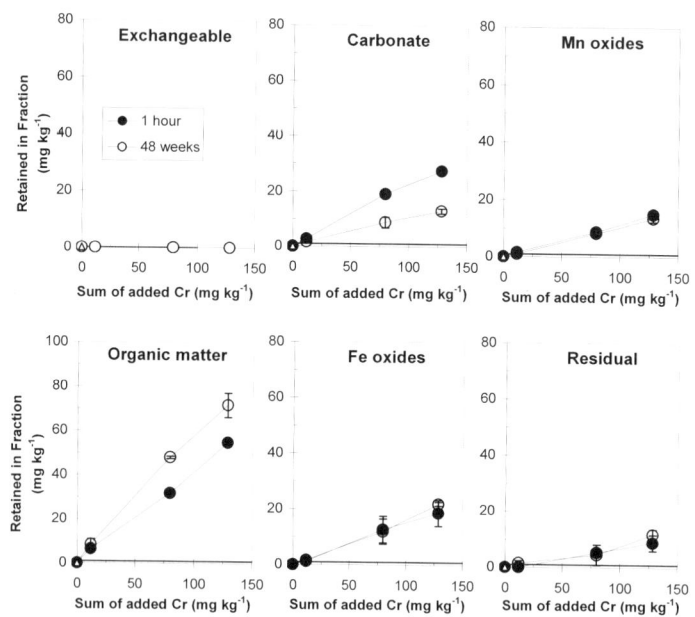

Figure 6.2. The fractional loading isotherms of Cr in a contaminated Israeli loessial soil at an initial (one hour) period and after 48 weeks

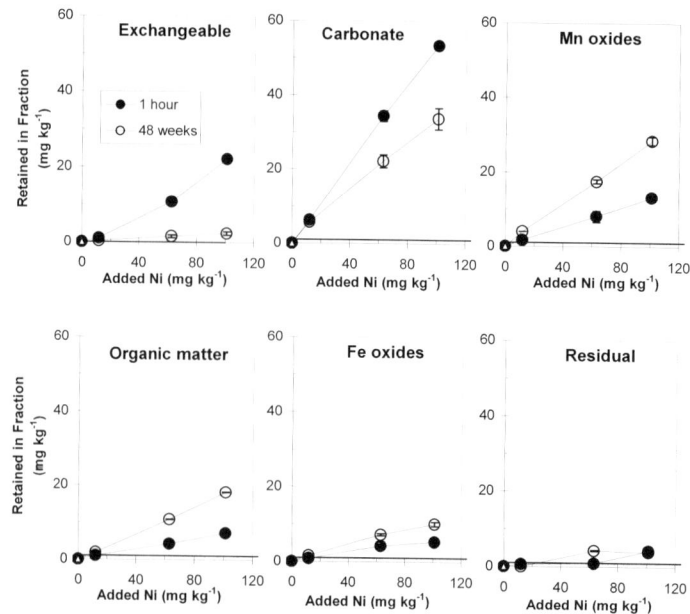

Figure 6.3. The fractional loading isotherms of Ni in a contaminated Israeli loessial soil at an initial (one hour) period and after 48 weeks

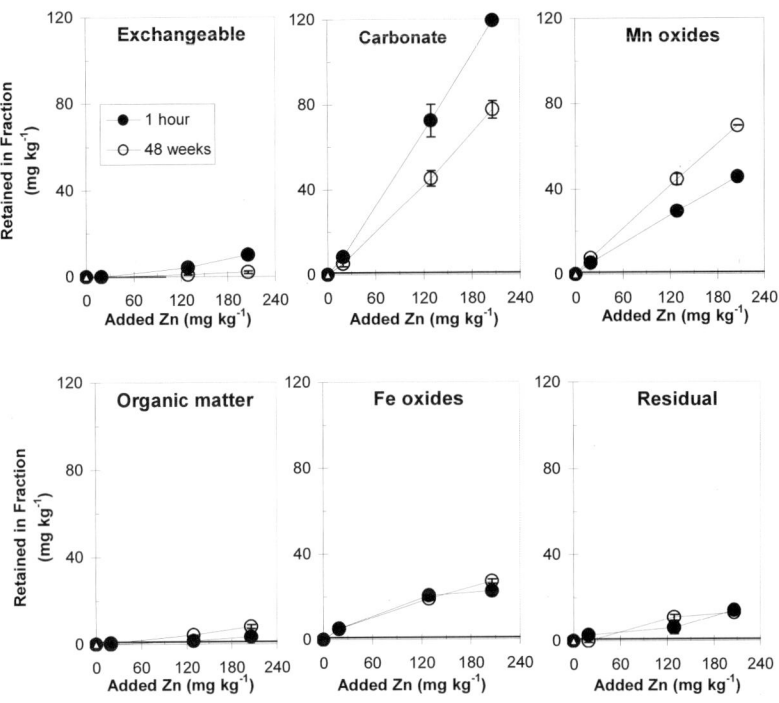

Figure 6.4. The fractional loading isotherms of Zn in a contaminated Israeli loessial soil at an initial (one hour) period and after 48 weeks

The partition index (I_R, which will be discussed in details below) of Cd, Cu, Cr, Ni and Zn in both soils rapidly increased from time zero (calculated value) to one day and further to one year. This was especially true for Cr, Cu and to some extent Ni and Cd (Table 6.5). This result indicates that added trace metals are initially and rapidly transferred from the labile EXC fraction into the more stable fractions. Furthermore, I_R of trace metals in native arid soils incubated under the saturated paste regime decreased at the end of year. This indicates mobilization of trace elements in these soils as saturation (Table 6.5). Also, it can be seen that I_R decreased, for any given time, with an increase of the loading level (Table 6.5, Fig. 6.5). This means that higher additions of soluble metals result in higher metal content in the labile fractions and lower metal binding intensity in soils.

The initial retention of metals in all fractions was linearly dependent upon the loading levels (Figs. 6.1–6.4). This "constant partition" behavior was also maintained after one year of incubation, but the distribution among fractions had shifted. This behavior indicates that there is no saturation metals in any of the solid-phase components of arid soils at the loading levels of metals employed in the present experiments.

Table 6.5. Reduced partition index (I_R) of trace metals in arid-zone soils incubated under saturated paste regime (after Han and Banin, 1997. Reprinted from Water Air Soil Pollut, 95, Han F.X., Banin A., Long-term transformations and redistribution of potentially toxic heavy metals in arid-zone soils. I: Incubation under saturated conditions, p 411, Copyright (1997), with permission from Springer Science and Business Media)

Time	Loading Level (T)	Metals									
		Cd		Cr		Cu		Ni		Zn	
		Sandy	Loessial	Sandy	Loessial	Sandy	Loessial	Sandy	Loessial	Sandy	Loessial
	Native	0.444	0.249	0.830	0.908	0.711	0.669	0.758	0.680	0.552	0.735
0 hour[a]	Non-Amended	0.444	0.249	0.830	0.908	0.711	0.669	0.760	0.680	0.552	0.735
	Low (0.5T)	0.063	0.195	0.491		0.437	0.469	0.400	0.522	0.394	0.574
	Intermediate (1T)	0.046		0.354		0.320		0.280		0.309	
	Intermediate (2T)	0.037								0.220	
	High (3T)	0.034	0.110	0.232			0.206	0.180	0.205		0.216
1 day	Non-Amended	0.444	0.142	0.833	0.916	0.705	0.695	0.750	0.682	0.550	0.836
	Low (0.5T)	0.126	0.178	0.610		0.626	0.631	0.468	0.595	0.429	0.724
	Intermediate (1T)	0.117		0.510		0.537		0.409		0.366	
	Intermediate (2T)	0.129								0.288	
	High (3T)	0.125	0.114	0.420			0.432	0.297	0.334		0.422
1 Year	Non-Amended	0.213	0.124	0.819	0.941	0.661	0.594	0.626	0.700	0.536	0.759
	Low (0.5T)	0.325	0.310	0.640		0.623	0.591	0.475	0.649	0.461	0.666
	Intermediate (1T)	0.231		0.570		0.547		0.408		0.440	
	Intermediate (2T)	0.199								0.394	
	High (3T)	0.199	0.174	0.500			0.513	0.338	0.550		0.554

[a] Added metals were assigned to the EXC fraction at 0 hour to calculate I_R.

6.1.2 The Slow Long-term Processes

Following the initial fast retention, the slow redistribution of the added metals occurred over time. During one year of incubation under the field capacity regime, heavy metals were slowly transferred among solid-phase components as shown in Figs. 6.1–6.4. Added Cu and Ni were transferred from the EXC and CARB fractions into the ERO and OM fractions and Zn mainly into the ERO fraction. Chromium and Pb moved from the CARB fraction into the OM and ERO fractions, respectively. Cadmium redistributed from the EXC fraction into the CARB fraction. After one year of incubation under the field capacity regime, 65–100% of the added Cd was transferred to the CARB fraction. About 50% and 20% of the added Pb was redistributed to the CARB and ERO fractions, respectively.

For Cr, 55–80% moved to the OM fraction, while 30–40% and 15–25% of the added Cu moved to the CARB and ERO fractions, respectively, with 35% of Cu moved to the OM fraction in the sandy soil. About 35–40%, 30%, and 20–25% of added Ni were transferred to the CARB, ERO, and OM fractions, respectively, and 40–60% and 15–35% of added Zn were transferred in the CARB and ERO fractions, respectively.

Two parameters, the redistribution index (U_{ts}) and the reduced partitioning parameter (I_R), are used to describe the redistribution processes of trace elements in contaminated arid soils (Figs. 6.5–6.6) (Han et al., 2003a). The redistribution index depicts the removal or attainment of element-contaminated soils from or to the fractional distribution pattern characteristic of non-amended soils. However, the reduced partitioning parameter quantifies the relative binding intensity of trace elements in soils.

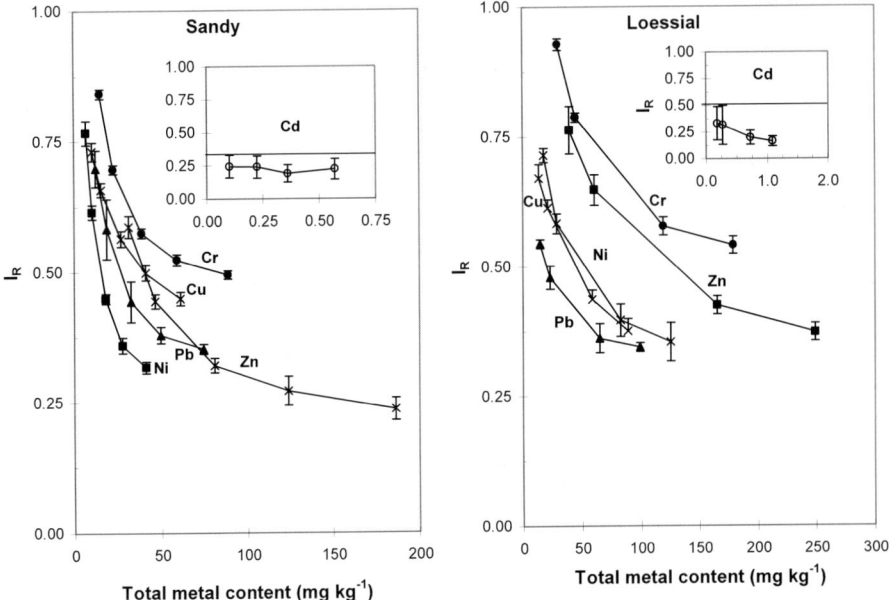

Figure 6.5. The initial reduced partition index, I_R, of six metals in two Israeli arid soils. Two soils were treated with metal nitrates at various loading levels. Soils were incubated under the field capacity moisture regime (modified after Han and Banin, 1999, with permission from Springer Science and Business Media)

The fractional redistribution index (U_{tf}) is defined as follows (Han and Banin, 1996, 1997; Han et al., 2003a):

$$U_{tf} = Fa/Fc \qquad (6.1)$$

where Fa is the percentage of the total amount of the element that is bound to a given component in a metal-contaminated soil, and Fc is the percentage of the total amount of the element that is bound to the same component in the control soil over the same period. U_{tf} describes the approach of trace elements in each solid-phase component in contaminated soils to the same solid-phase component in control soils.

The whole-soil redistribution index for a given metal (Ut-soil) is defined as the sum, for $i = 1, \ldots, k$ components, of the products of the percentage of the total amount of metal bound to a given component by its U_{tf} (Han and Banin, 1996, 1999, Han et al., 2003a):

$$U_{ts} = \sum_{i=1}^{k} U_{tfi} \times F_i \qquad (6.2)$$

where Fi is the percentage of the total amount of the element that is bound to component i in a contaminated soil, and U is the U_{tf} function value of a given solid-phase component. This enables the description of the removal or attainment of the element in metal-contaminated soils from or to the fractional distribution pattern of the element in non-amended soils using a single parameter. By definition, Uts of each element in non-amended soils equals one. However, in contaminated soils it is initially farther from one, and it slowly converges towards one over time. Whenever the element concentration in a given fraction (individual solid-phase component) is below the analytical detection limit (e.g., for trace elements in non-contaminated soils), the detection limit may be substituted for the actual concentrations of the metal. In such cases, it is difficult to apply this parameter to quantify the processes due to the U_{tf} and U_{ts} fluctuation. In addition, sequential selective dissolution procedures affect the values of U_{ts} because different protocols use different extractants with varying extractability and specificity.

Over time, the salt-spiked and sludge-amended soils approached the fractional distribution pattern of non-amended soils. Element species, loading levels and soil properties affect the rates of redistribution of elements in soils. Metals in the contaminated soils approached the fractional distribution pattern of non-amended soil at low loading levels more rapidly than those at high loading levels. In fact, most added metals at the 0.5T loading level (T is the concentration of heavy metals in native soils) were redistributed within one year to a similar relative pattern as observed in the control (CK) soil. Thus, the U_{ts} of the soil approached the value of one (Fig. 6.6). At higher levels of addition (3T), however, the amended soils did not return to the steady-state distribution characterizing the incubated non-amended soil, but the slow return towards equilibrium was clearly observed. The sequence order of approach by metals to the fractional distribution pattern of non-amended soil was: Cd > Cu > Ni > Zn > Cr (Han et al., 2003a). The outcome of this complex process is a return to the steady state of fractional relative distribution which was typical of the incubated native soil, and it was apparently independent of the loading level. However, in the amended soils the absolute concentration of metals (mg kg^{-1} soil) in any given phase was higher than in the non-amended soils, and it was practically linearly proportional to the loading level.

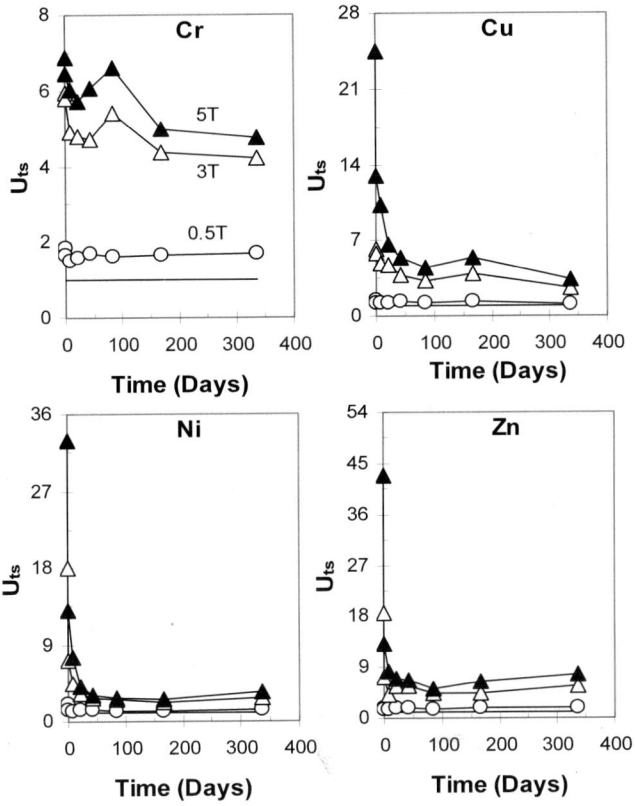

Figure 6.6. Long-term changes of U_{ts} of metals in a loessial soil from Israel. The soil received metal nitrates under field the capacity moisture regime (after Han and Banin, 1999. Reprinted from Water Air Soil Pollut, 114, Han F.X and Banin A., Long-term transformations and redistribution of potentially toxic heavy metals in arid-zone soils. II: Incubation under field capacity conditions, p 245, Copyright (1999), with permission from Springer Science and Business Media)

The partition index of an element, I, and the reduced partition index, I_R, are defined as follows (Han and Banin, 1997; Han et al., 2003a):

$$I = \sum_{i=1}^{k} (Fi \times (i)^n) \qquad (6.3)$$

$$I_R = I / (k^n) \qquad (6.4)$$

where i is the extraction step number (1, least aggressive; k, most aggressive) with the following values assigned to i in Eq. (3) representing various steps and components:
 1: the exchangeable fraction
 2: the carbonate bound fraction
 3: the easily reducible oxide bound fraction
 4: the organically bound fraction
 5: the reducible oxide bound fraction
 6: the residual fraction

Fi is the percentage (fractional content) of the element in the solid-phase component i out of the total amount extracted, and n is an integer (usually 1 or 2). In this case, k equals 6 (Table 6.4) and n equals 2, resulting in minimal I_R equaling 0.03 and maximal I_R equaling 1. The choice of n is arbitrary. The square relationship ($n = 2$) clearly expresses the increasing binding strengths of metals with increasing i in the sequential selective dissolution process, and it keeps calculations simple and convenient. This parameter is introduced to semi-quantitatively describe the relative binding intensity of elements in soils. It also enables comparison of the binding intensity of a given element among soils and of different elements in the same soil. Thus, a low value of I (i.e., close to the minimum) represents a distribution pattern in which much of the element resides in the soluble and exchangeable fractions, whereas a high value (close to 1) results from a high proportion of the metal strongly bound in the residual fraction. Intermediate values represent patterns involving metal partitioning among all solid-phase components. The values of I_R are affected by sequential selective dissolution procedures. If the protocol uses aqua regia instead of 4 M HNO_3 for the residual fraction, the I_R values will increase. This occurs because more metal is extracted in the residual fraction.

The reduced partitioning index (mean I_R values) of Cd, Cr, Cu, Ni and Zn in 45 Israeli arid-zone soils is presented in Fig. 6.7. The trends for mean I_R values for five heavy metals in these soils were as follows: Cr > Zn = Cu > Ni > Cd (Fig. 6.7) (I_R is calculated from Banin et al., 1997a). This indicates that Cr is strongly bound to solid-phase components, possessing lower bioavailability. However, Cd is weakly bound to soils, resulting in higher bioavailability. In both native and contaminated soils (Figs. 6.5, 6.7), Cr had the highest binding intensity, Cd had the lowest, and Cu, Ni and Zn had intermediate values. In addition, these two indices have been used to evaluate the redistribution processes of trace elements in worldwide contaminated soils including sewage sludge and salt-amended soils (Han et al., 2003a).

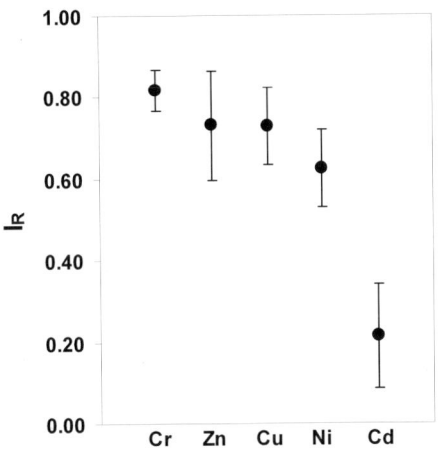

Figure 6.7. Ranges (arithmetic mean and standard deviation) of the reduced partition index (I_R) of Cd, Cu, Cr, Ni and Zn in 45 Israeli arid-zone soils (after Han et al., 2003a. Reprinted from Adv Environ Res, 8, Han F.X., Banin A., Kingery W.L., Triplett G.B., Zhou L.X., Zheng S.J., Ding W.X., New approach to studies of redistribution of heavy metals in soils, p 118, Copyright (2003), with permission from Elsevier)

6.1.2.1 Cadmium

Cadmium in Israeli arid soils under the saturated paste, field capacity, and wetting-drying cycle moisture regimes was predominately present in the CARB fraction, followed by the EXC fraction (Fig. 6.8) (Han and Banin, 1997, 1999; Han et al., 2001a). During one year of incubation under the three moisture regimes, Cd was transferred from the EXC fraction into the CARB fraction. In the loessial soil under the saturated paste regime, Cd was primarily in the CARB, EXC and partly OM fractions after one week of incubation. Afterwards Cd was mainly in the CARB fraction. In the sandy soil Cd was mainly in the CARB fraction, followed by the EXC, ERO, and OM fractions at higher loading levels. In the amended soils, Cd rapidly redistributed at higher levels of addition. Its $U_{tf-CARB}$ function was never far from 1.0. Because Cd was mainly bound in the CARB fraction, it had high lability. Cadmium remained more labile than the other metals as depicted by its reduced partition index, I_R, which was among the lowest of the studied metals (Fig. 6.5, Table 6.5). Han and Banin (1997, 1999) found that different water regimes did not change the directions and pathways of transformation of Cd in arid soils.

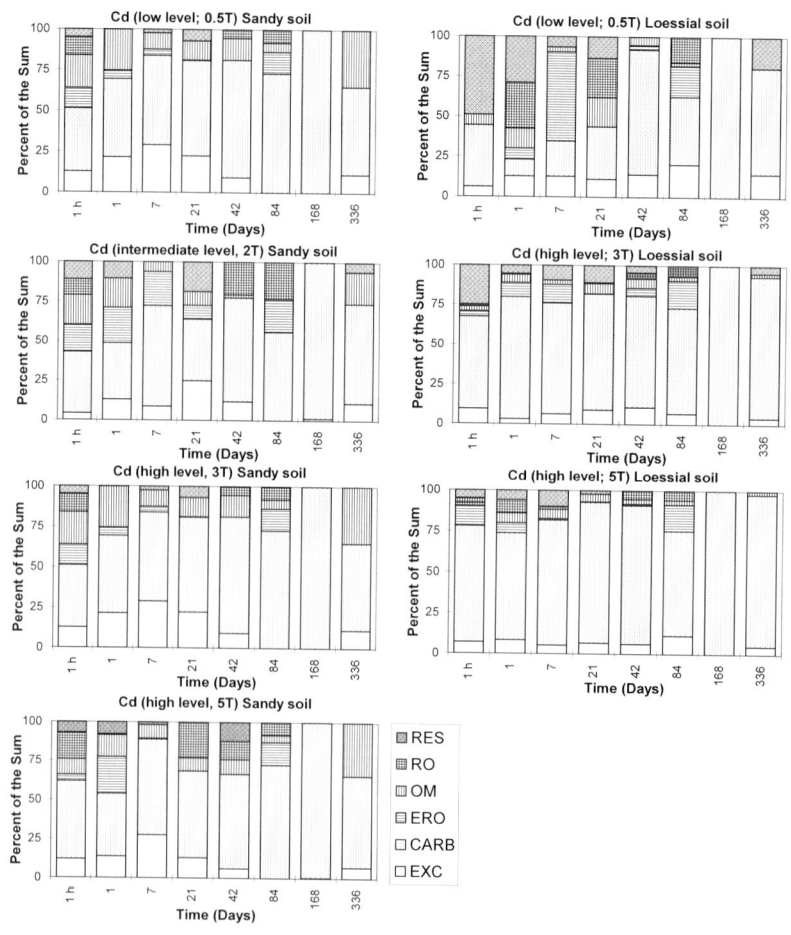

Figure 6.8. Changes of Cd fractions in two Israeli soils during 336 days of incubation at field capacity moisture regime (after Han and Banin, 1999. Reprinted from Water Air Soil Pollut, 114, Han F.X and Banin A., Long-term transformations and redistribution of potentially toxic heavy metals in arid-zone soils. II: Incubation under field capacity conditions, p 238, Copyright (1999), with permission from Springer Science and Business Media)

6.1.2.2 Chromium

Chromium in the native Israeli arid soils was mostly bound to the residual phase, possessing the highest partition index among the metals (Figs. 6.5, 6.7, Table 6.5). Amendment with soluble Cr in arid sandy soil under the saturated paste, field capacity, and wetting-drying cycle regimes initially resulted in the binding of Cr to the carbonate and organic matter fractions, but in the loessial soil a higher proportion of Cr was found in the RO and ERO fractions (Fig. 6.9). During incubation at the three moisture regimes, Cr was slowly transferred from the CARB and the ERO fractions to the OM fraction at all levels of addition in both soils, but the soil still remained removed from soil equilibrium (U_{ts} = 0.42, Table 6.5) due to limited conversion to the residual fraction. This indicates that Cr is more preferentially bound to the organic matter component than other metals. However, compared to the wetting-drying cycle and field capacity regimes, more Cr was transferred into the carbonate fraction under the saturated paste regime. This resulted in less Cr in the organic matter fraction (Han and Banin, 1997) (Table 6.6). At the end of the year, the quasi-equilibrium state present in control soils was not attained in the two soils because of the limited transfer of Cr into the residual fraction.

Drying and remoistening air-dry soils greatly lowers their ability to oxidize Cr (Bartlett and James, 1980). Since Cr^{3+} has a similar ionic radius (0.64 x 10^{-10} m) to Mg (0.65 x 10^{-10} m) and trivalent Fe (0.65 x 10^{-10} m), it is possible that Cr^{3+} could readily substitute for Mg in silicates and for Fe^{3+} in iron oxides. This explains the high proportion of Cr found in the residual fraction in the native arid soil. On the other hand, humic acids have a high affinity for Cr (III) (Adriano, 1986). Thus, present results show that when soluble Cr was added to soils, Cr^{3+} was initially and immediately bound to the organic matter fraction. Due to its slow conversion into the reducible oxide and residual fractions, Cr in the amended soils departed and remained removed from the quasi-equilibrium. However, Cr approached the quasi-equilibrium with time.

6.1.2.3 Copper

Amendment with soluble copper in arid Israeli soils, particularly at high loading levels, resulted in initial perturbation and increased the amounts of copper in the exchangeable and the carbonate fractions in the loessial soil. This also occurred with the carbonate and the organic bound fractions (Fig. 6.10, Fig. 6.1). Even so, the soils converged quite rapidly

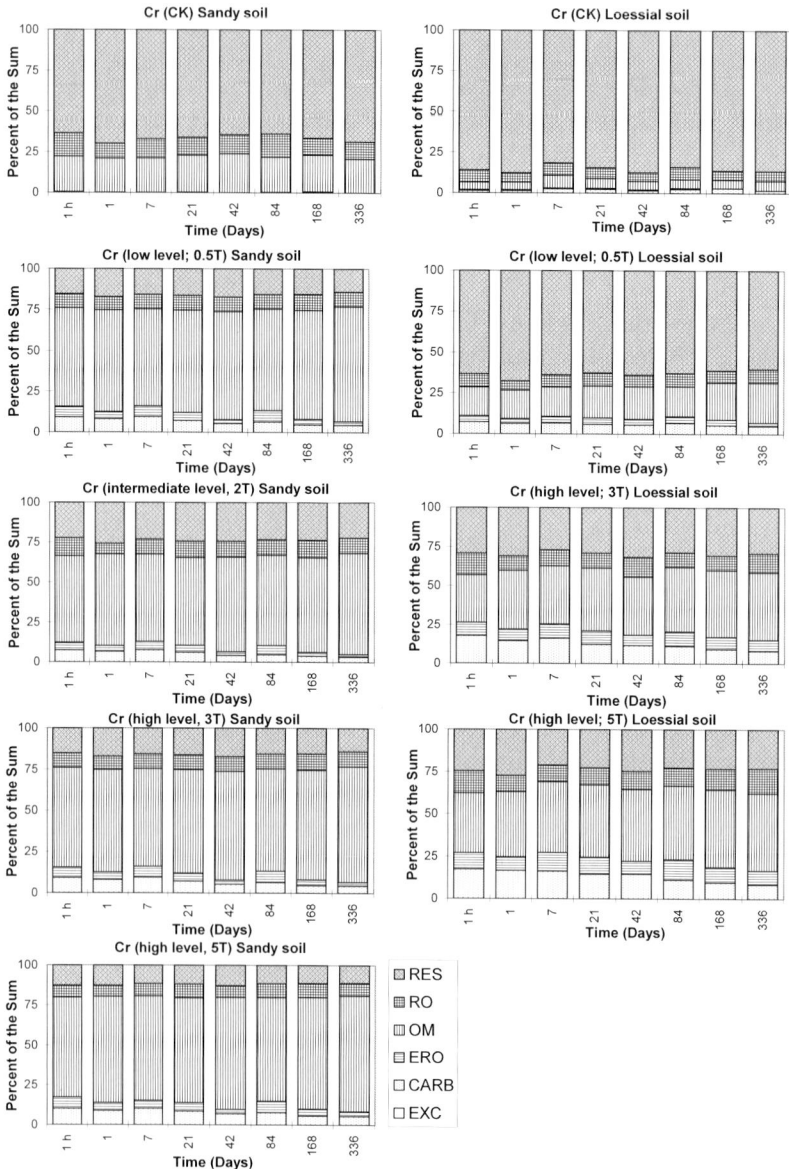

Figure 6.9. Changes of Cr fractions in two Israeli soils during 336 days of incubation at field capacity moisture regime (after Han and Banin, 1999. Reprinted from Water Air Soil Pollut, 114, Han F.X and Banin A., Long-term transformations and redistribution of potentially toxic heavy metals in arid-zone soils. II: Incubation under field capacity conditions, p 239, Copyright (1999), with permission from Springer Science and Business Media)

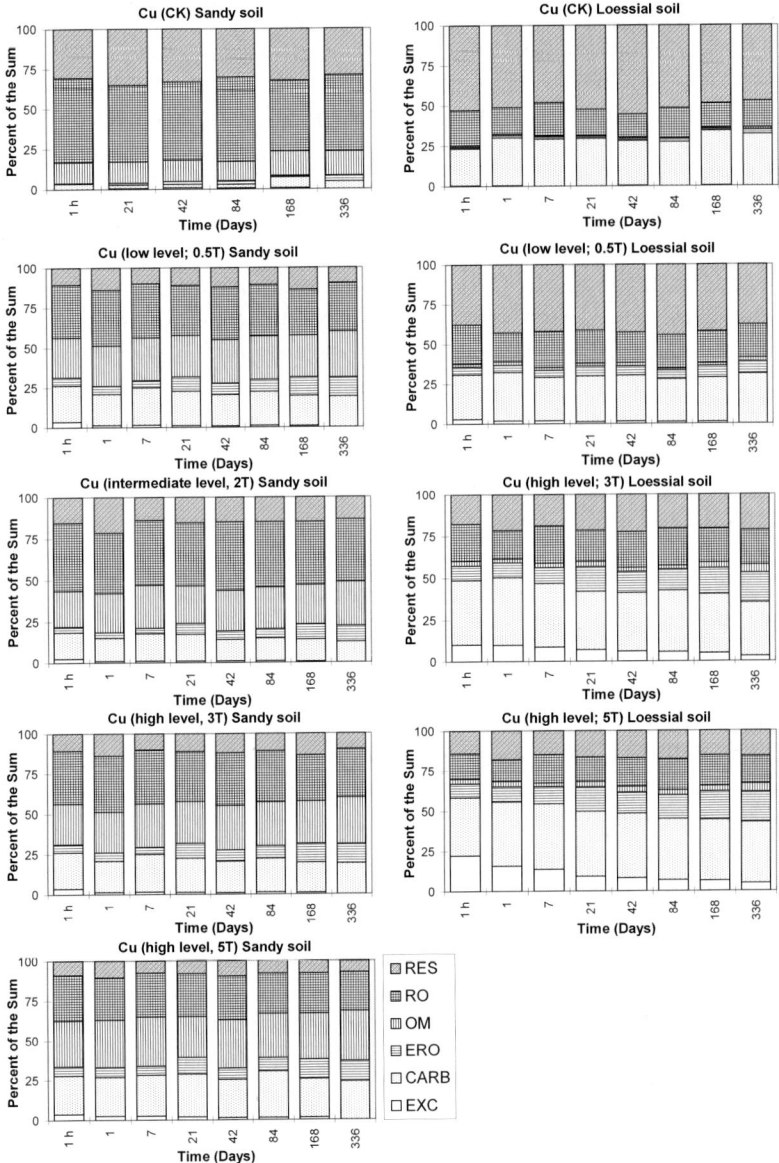

Figure 6.10. Changes of Cu fractions in two Israeli soils during 336 days of incubation at field capacity moisture regime (after Han and Banin, 1999. Reprinted from Water Air Soil Pollut, 114, Han F.X and Banin A., Long-term transformations and redistribution of potentially toxic heavy metals in arid-zone soils. II: Incubation under field capacity conditions, p 240, Copyright (1999), with permission from Springer Science and Business Media)

towards the typical distribution acquired by the incubated non-amended soil. The U_{ts} function, therefore, decreased from high initial values after one year of incubation (Fig. 6.6). However, under the field capacity regime, amendment with soluble Cu, particularly at high loading levels, increased the presence of copper in the EXC and CARB fractions in the loessial soil. This also occurred in the OM and CARB fractions in the sandy soil (Fig. 6.10). Copper was found to be mainly transferred from the EXC fraction into the RO and CARB fractions at the saturated paste regime due to the higher pCO_2 after saturation, while Cu was transferred from the EXC and CARB fractions mainly into the ERO fraction in arid soils at the field capacity regime, and into the organic matter fraction under the wetting-drying cycle regime (Table 6.6). In the loessial soil, Cu concentrations in the EXC and CARB fractions at the highest loading level (5T) decreased more rapidly than those in the sandy soil. The moisture regimes strongly affected the pathways of the slow redistribution processes of Cu in arid soils.

6.1.2.4 Nickel

Nickel distribution in arid soils was affected by the moisture regime (Han and Banin, 1997, 1999; Han et al., 2001a). In arid soils, added soluble Ni was initially bound to the carbonate, the exchangeable, and the easily reducible oxide fractions (Figs. 6.3, 6.11). After one year of incubation, the carbonate fraction remained an abundant fraction (Fig. 6.11). However, during prolonged incubation at the field capacity moisture regime, Ni was slowly transferred from the exchangeable and the carbonate fractions into the easily reducible oxide and the organically bound fractions in soils (Fig. 6.11). Under saturated conditions, added Ni was predominately transferred from the exchangeable and the carbonate fractions into the reducible oxide fraction, but high proportions of Ni remained in the carbonate fraction (Table 6.6). In the wetting-drying cycle regime, added Ni in soil was mainly transferred into the organically bound fraction, and Ni, in all stable fractions, significantly increased after a year of incubation (Table 6.6). Incubation at the wetting-drying cycle regime resulted in more transfer of added Ni into the stable fractions than at the field capacity and saturated paste regimes. This is indicated by the lowest concentration of Ni in the exchangeable fraction both initially and after one year of incubation at the wetting/drying cycling regime.

Table 6.6. Comparisons of the redistribution of metals in two Israeli soils at 3T treatment. Soils were incubated for one year under the saturated paste (SP), field capacity (FC) and wetting-drying cycle (Cycle) moisture regimes (% of the sum of fractions, as means of two replicates) (from Han et al., 2001a, with permission from Lippincott Williams & Wilkins)

Soil	Metal	Moisture	Fractions					
			EXC	CARB	ERO	OM	RO	RES
Sandy	Cr	Cycle	BD[b]	8.7a[c]	3.8a	64.3a	8.9a	14.2a
		FC	BD	4.4b	2.3b	70.1b	9.0a	14.1a
		SP	BD	13.5a	6.4c	56.3a	7.9a	15.9a
	Cu	Cycle	1.6	19.4a	7.7a	32.7a	29.7a	8.9a
		FC	BD	19.3a	11.8b	28.8b	30.4a	9.6a
		SP	NA[d]					
	Ni	Cycle	1.9a	35.6a	15.1a	27.8a	4.8a	14.9a
		FC	2.8ab	29.9b	24.4b	24.2b	4.8a	13.9a
		SP	4.2b	41.1c	13.6c	24.6b	5.7a	10.8b
	Zn	Cycle	2.0ab	50.5a	12.7a	16.7b	9.5a	8.6a
		FC	3.0a	48.3b	15.3b	13.0a	8.9b	11.6ab
		SP	0.8b	39.5c	14.2b	16.4b	15.9c	13.1b
Loessial	Cr	Cycle	0.1	4.8a	6.0a	51.1a	11.4a	26.6a
		FC	BD	8.1b	7.2a	43.3a	12.1a	29.3a
		SP	NA					
	Cu	Cycle	3.6a	31.7a	13.0a	6.7a	21.1a	23.8a
		FC	3.1b	32.0a	17.8b	4.9b	20.7a	21.5b
		SP	2.2ab	35.4a	7.3c	1.2c	28.9b	24.9b
	Ni	Cycle	1.6a	17.6a	15.0a	23.6a	15.6a	26.6a
		FC	1.7a	31.5b	19.9b	13.5b	11.5b	21.9b
		SP	2.4a	27.1b	10.7c	5.7c	23.9c	30.1a
	Zn	Cycle	0.2a	34.2a	17.6a	4.5a	17.6a	25.9a
		FC	0.5b	34.5a	25.3b	2.4b	13.3b	24.0b
		SP	0.4c	31.6b	11.5c	0.3c	24.2c	32.1c

[a]T is the total metal concentration (mg kg^{-1}) in native soils and given in parentheses.
[b]BD : below detection limit of ICP-AES.
[c]Means for a metal within a column followed by the same letter not significantly different (p <0.05).
[d]NA: There was no metal treatment at this loading level in the soil.

6.1.2.5 Zinc

Under the saturated paste regime, Zn in the non-amended arid soils generally tended to be transferred to the CARB phase during incubation. Added Zn in the sandy soil, at an intermediate loading level, was initially and predominately attached mostly to the CARB fraction, but during incubation, redistribution primarily happened to the OM and RO fractions. As a result, initial U_{ts} of Zn in both soils decreased with time at the beginning of the incubation, and only slight change occurred beyond 100

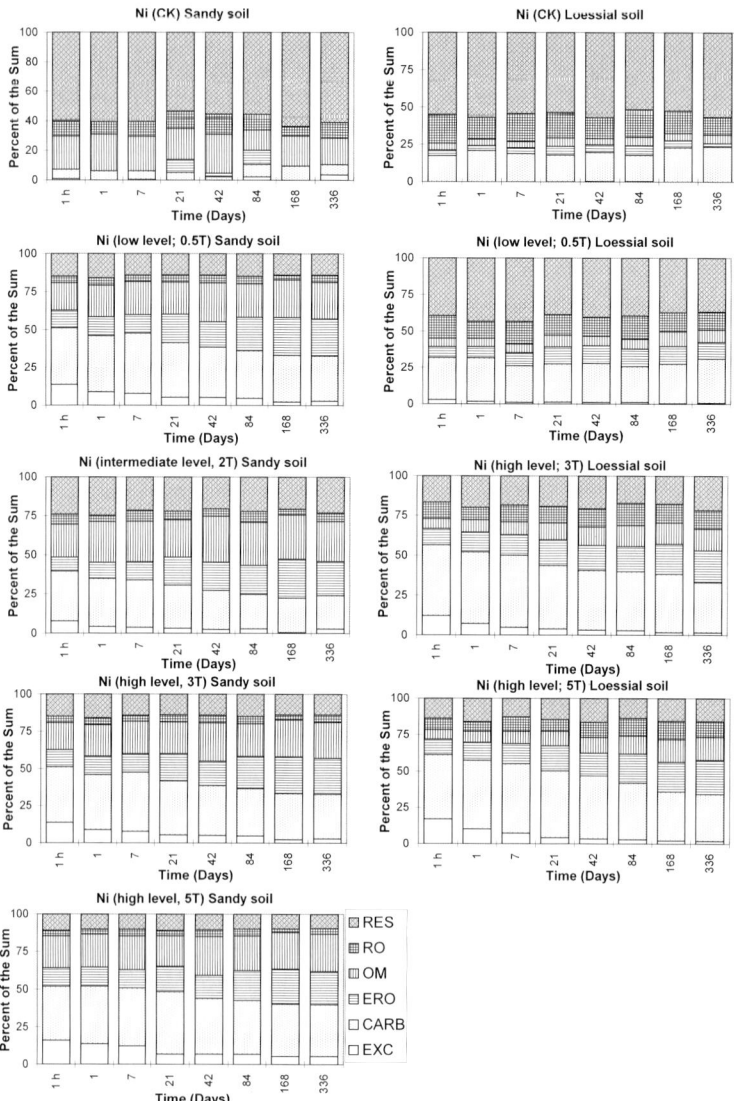

Figure 6.11. Changes of Ni fractions in two Israeli soils during 336 days of incubation at field capacity moisture regime (after Han and Banin, 1999. Reprinted from Water Air Soil Pollut, 114, Han F.X and Banin A., Long-term transformations and redistribution of potentially toxic heavy metals in arid-zone soils. II: Incubation under field capacity conditions, p 241, Copyright (1999), with permission from Springer Science and Business Media)

Transfer Fluxes

days of incubation. After one year of incubation the partition index differences (Table 6.5) and the U_{ts} values (Fig. 6.6) indicated that the soils approached a similar distribution pattern. Added Zn in both soils at the field capacity regime was initially and mostly attached to the CARB and EXC fractions. Zinc also transferred from both of these fractions into the ERO fraction, as well as a small portion into the OM and RO fractions in the sandy soil (Fig. 6.12). Zinc, like Ni, had a faster transfer into more stable fractions under the saturated paste condition.

6.1.2.6 Lead

Native Pb in Israeli arid soils mainly resided in the CARB and the RES fractions in the loessial soil, and it resided in the RES and the RO fractions in the sandy soil. During incubation under the field capacity regime, added Pb was slowly transferred from the CARB fraction into the ERO fraction (Fig. 6.13). However, the majority of Pb still remained in the CARB fraction.

In summary, the moisture regime does not significantly affect pathways of redistribution and transformation of added Cd and Cr in two Israeli arid soils. However, at the field capacity regime, added soluble Cu, Ni, Pb, and Zn are slowly and preferentially transferred into the ERO (Mn oxide bound) fraction during incubation. In contrast, with the saturated paste regime, the added soluble Cu, Ni, Pb and Zn are transferred mostly into the RO fraction and to some extent to the OM fraction, especially at high loading levels. Soils incubated with the saturated paste regime are reduced, and Mn oxides in the ERO fraction are reduced and transferred into the EXC and the CARB fractions in the initial stage of incubation (Han and Banin, 1996, discussed below). Thus heavy metals added to the soil are further transferred into the RO and OM fractions. However, under the field capacity incubation, with enough time, metals from the ERO fraction will be further transferred into the more stable fractions, such as RO and RES fractions.

Zinc adsorption can occur via exchange of Zn^{2+} and $Zn(OH)^+$ with surface-bound Ca^{2+} on calcite (Zachara et al., 1988). Zinc and Ni form surface complexes on calcite as hydrate until they are incorporated into the structure via recrystallization (Zachara et al., 1991). The selectivity of metal sorption on calcite is as follows: Cd > Zn > Ni (Zachara et al., 1991). The easily reducible oxide bound metals are primarily from Mn oxides (Chao, 1972; Shuman, 1982 and 1985a). At pH > 6, Zn sorption on Mn oxide abruptly increases because of hydroxylation of the ions (Loganathan et al., 1977), and a high soil pH in arid soil may favor Zn sorption on Mn oxides due to a great

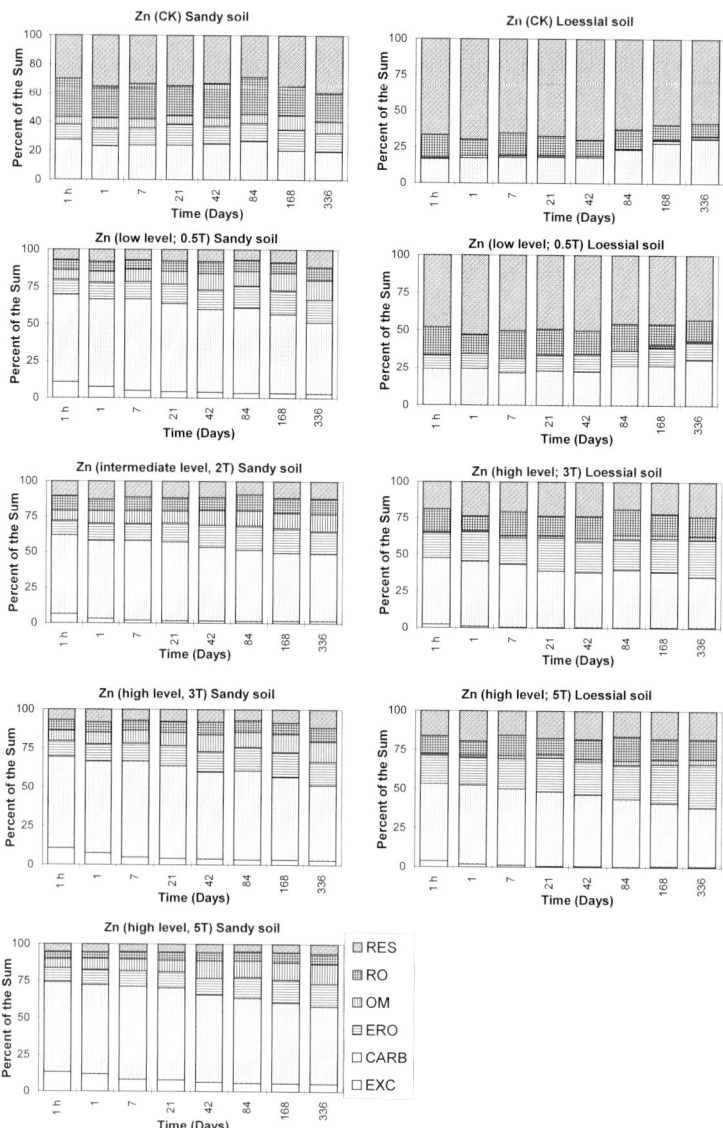

Figure 6.12. Changes of Zn fractions in two Israeli soils during 336 days of incubation at the field capacity moisture regime (after Han and Banin, 1999. Reprinted from Water Air Soil Pollut, 114, Han F.X and Banin A., Long-term transformations and redistribution of potentially toxic heavy metals in arid-zone soils. II: Incubation under field capacity conditions, p 243, Copyright (1999), with permission from Springer Science and Business Media)

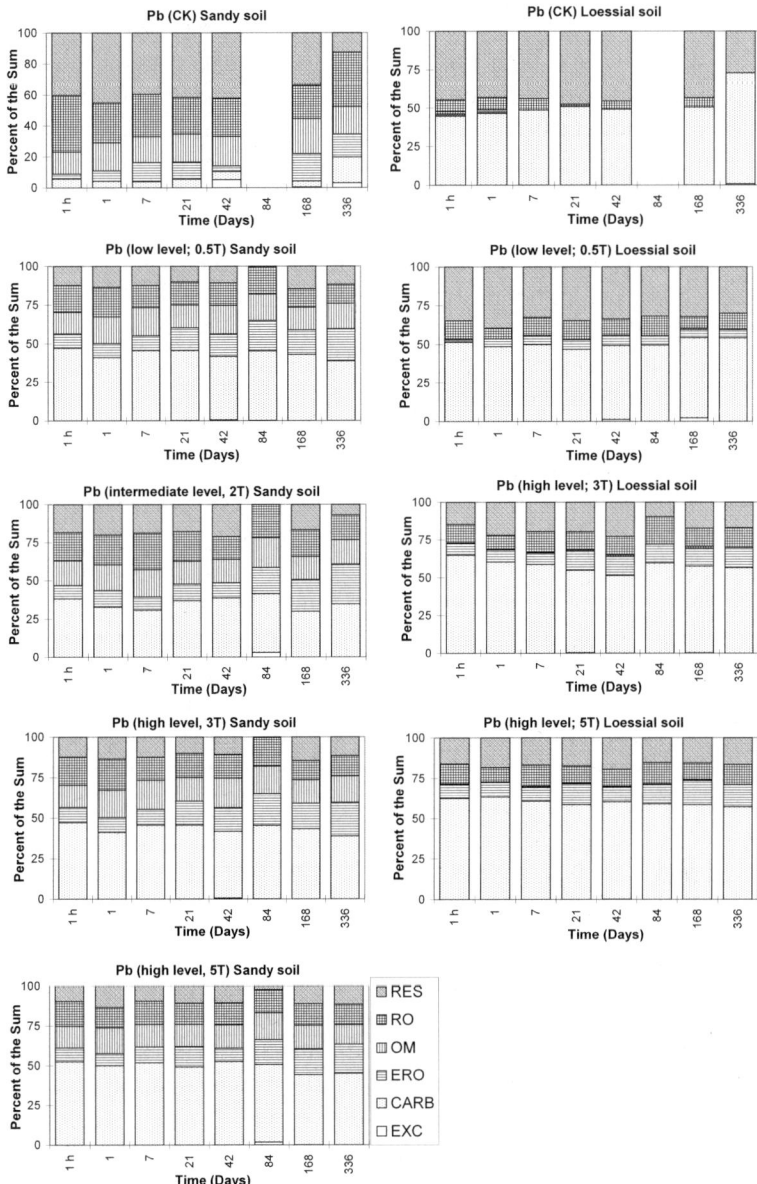

Figure 6.13. Changes of Pb fractions in two Israeli soils during 336 days of incubation at the field capacity moisture regime (after Han and Banin, 1999. Reprinted from Water Air Soil Pollut, 114, Han F.X and Banin A., Long-term transformations and redistribution of potentially toxic heavy metals in arid-zone soils. II: Incubation under field capacity conditions, p 242, Copyright (1999), with permission from Springer Science and Business Media)

proportion of Zn hydroxyl species (Zasoski and Burau, 1988). On the other hand, Cr and Cu have higher stability constants with organic matter. These facts may, in part, explain the observation that a higher concentration of Zn (20–25% of the total) existed in the easily reducible oxide fraction than Cu, Ni and Cr in the loessial soil at 3T level after one year of incubation. Soil properties play an important role in determining sinks for redistribution of trace elements in arid soils. Organic matter is important in binding added trace elements in sandy soil because the soil is relatively poor in oxides, carbonate and clay minerals. Due to a higher affinity of Cu for organic matter compared to Ni and Zn, Cu is preferentially bound to the organic matter fraction. This helps explain the higher amounts of Ni and Zn and the lower Cu amount in the exchangeable fraction in the sandy soil. The 5T treatment in the sandy soil under the wetting-drying cycle moisture is as an example. About 28% of the added Cu (14 mg kg^{-1}) was in the organic matter fraction, while 21% of Ni (7 mg kg^{-1}) and 10% of Zn (14 mg kg^{-1}) were present in the organically bound fraction, respectively (note: the total of added Zn was much higher than Cu). On the other hand, the loessial soil is rich in carbonate, Mn and Fe oxides and clay minerals. Zinc and Ni successfully compete with Cu, thus, both of them are preferentially bound to carbonate and Mn oxides, resulting in lowering Ni and Zn concentrations in the exchangeable fraction (compared to a higher amount of Cu in the soluble and exchangeable fraction) in the loessial soil. The deep discussion can be found in the Chapter 5.

6.2 TRANSFER PROCESSES OF TRACE ELEMENTS IN SLUDGE-AMENDED ARID SOILS

The transfer of trace elements among solid-phase fractions in arid soils amended with sludge depends upon soil and environmental parameters, such as soil type, pH and Eh, agricultural management, as well as the metal and metal-complex type. Banin et al. (1990) investigated the kinetics of the transformation of metals (Cu, Zn, Cd, Pb, Fe and Mn) in Israeli soils eight months after the application of sewage sludge. Following the application of sludge to soils, metal transfer is governed by a combination of two processes. One process is the decomposition of the organic matter of the sludge in soils. The other process is the release of trace elements into the soil solution and is incorporated into solid phases through adsorption, precipitation, co-precipitation, complexation, etc. Therefore, the transfer of trace elements with time depends upon the rate of decomposition of the sludge, the relative affinities of the elements to each of the solid phases, and kinetic factors. The hypothesis also has been proposed that with increasing time after sludge application, the metal distribution between the different phases in the soil is expected to evolve towards the steady-state or quasi-equilibrium situation (Banin et al., 1990). This steady state exists in the native nontreated soil, provided no phase in the soil is becoming saturated with the added metal. The redistribution of the elements defined below is assessed by using the function U_t, the "removal from soil quasi-equilibrium". With time, as equilibrium is re-approached due to element redistribution, U_t will approach the value of 1.

$$U_t = F_a/F_c \qquad (6.5)$$

$$F_a = (Q_s^f + Q_{sl}^f) \times 100 / (Q_s^t + Q_{sl}^t) \qquad (6.6)$$
$$= [C_s^f(1-p) + C_1^f \times p] \times 100 / [C_s^t(1-p) + C_{sl}^t \times p] \qquad (6.7)$$

where F_a is the fraction of the element in a given phase (f) in a treated soil at a given time (% of total extracted), and F_c is the fraction of the element in the same phase for the nontreated control soil at the same time. Q is the quantity of the element, C its concentration, and p the weight fraction of the sludge added. The subscript s and sl represent the soil and sludge, and the superscript f and t a given phase and total in soil or sludge, respectively.

It was assumed by Banin et al. (1990) that at t = 0, the distribution of an element will be an additive property of the distributions in the nontreated soil and the added sludge. The results showed that initial U_t increased with the increase in the sludge application rate (i.e., far from the quasi-equilibrium state) (Fig. 6.14). According to the hypothesis and assessing function above, they summarized that the sludge-amended arid soils tended to return to their quasi-equilibrium pattern. This was typical of the native soils during eight months. Almost complete return to the quasi-equilibrium state was found for Cu except for the RES fraction, which was slow. Also, a slower rate of quasi-equilibrium was observed for Zn. The Zn bound to the RO and the RES phases did not return to the initial quasi-equilibrium state. A complete return to the quasi-equilibrium state was observed for Cd, due to the similar partition pattern of the metal in the sludge and in the non-treated soil. The results also showed that the return to soil quasi-equilibrium at low sludge additions was more complete than at high sludge additions for Cu. Returning to equilibrium in the clayey soils is more complete than in sandy soils for Zn. Copper was transferred from the OM fraction of the sludge into the CARB and the ERO fractions of the soil, Zn was transferred from the OM and the ERO phases of the sludge into the CARB fraction, and Cd slightly moved from the OM and the ERO fractions into the RO and the RES phases with a higher amount in the EXC and the CARB phases.

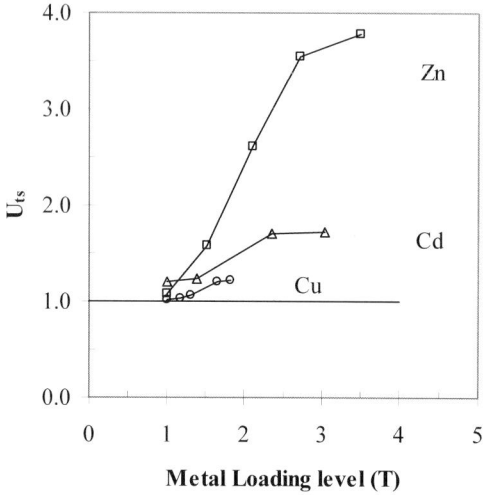

Figure 6.14. Changes of U_{ts} of Cd, Cu and Zn in an Israeli sandy soil after 8 months of sludge application (T is the total metal content in the non-amended soil) (after Han et al., 2003a. Reprinted from Adv Environ Res, 8, Han F.X., Banin A., Kingery W.L., Triplett G.B., Zhou L.X., Zheng S.J., Ding W.X., New approach to studies of redistribution of heavy metals in soils, p 117, Copyright (2003), with permission from Elsevier)

The distribution of trace elements in arid soils amended with sludge is affected by many factors. These factors include soil texture, organic acid, phosphorus application, liming, tillage and crop systems. Sposito et al. (1983) found that the proportion of Ni, Cd, Pb and Zn in the carbonate fraction increased with time in Californian soils amended with sewage sludge. This is especially pronounced in coarse textured soils. They suggested that this shift may be the result of the relatively rapid decomposition of the composted sludge in the coarse textured soil. However, Cd and Pb in finer-textured soil seem to shift into the residual fraction (extracted with 4 M HNO_3).

6.3 TRANSFER FLUXES OF Cu, Cr, Ni, AND Zn AMONG SOLID-PHASE COMPONENTS IN SALT-CONTAMINATED ISRAELI ARID SOILS

The flux of a metal in and out of a given soil fraction, during a certain time period of incubation, is calculated as follows (Han, 1998):

$$Si_{t2-t1} = (Ci_{t2} - Ci_{t1}) / (t_2 - t_1) \qquad (6.8)$$

where Si_{t2-t1} is the flux of a metal in the ith fraction between time t_2 and t_1, mg/kg/hour; Ci_{t2} and Ci_{t1} are the concentrations (mg/kg) of the metal in the ith fraction at the time t_2 and t_1, respectively; t is the time measured in hours. The sum of the positive flux (ΣS^+) or the sum of the negative flux (ΣS^-) of the metal during the transfers among various fractions is defined as the sum of the flux of the fractions with a positive flux value and the sum of the flux of the fractions with a negative flux value during a certain period of time, respectively.

The average sum of the positive fluxes and the average sum of the negative fluxes of Cu, Cr, Ni and Zn at 3T loading level under the field capacity moisture were presented in Figs. 6.15–6.16. The average negative fluxes of the metals in a given treatment are mostly equivalent to the average positive fluxes (standard deviation, indicated by bar, is relatively low). In general, the transfer flux balance in metal salt-spiked soils was better than in

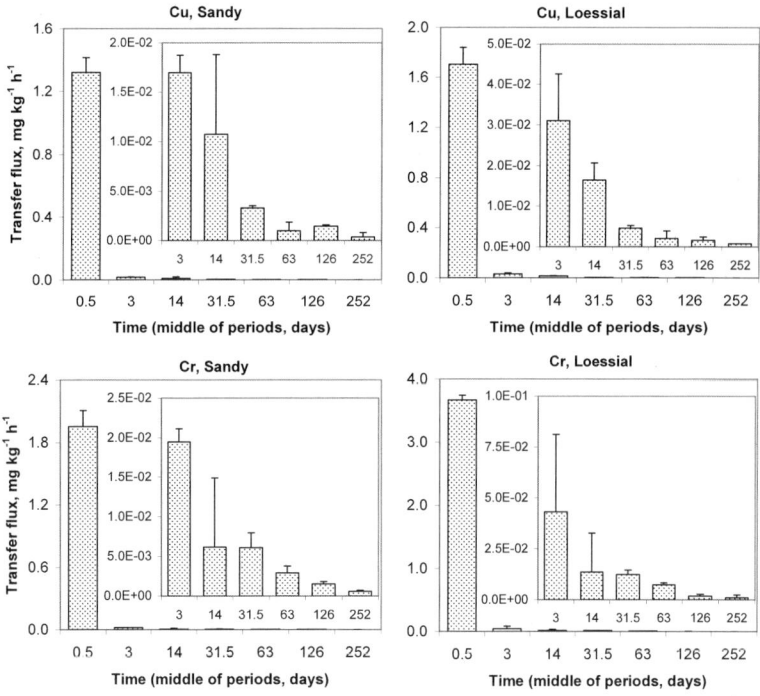

Figure 6.15. The average transfer fluxes of Cr and Cu among solid-phase components in Israeli soils. The soils received metal nitrates at the 3T level under the field capacity moisture regime (T: total metal content in non-amended soils)

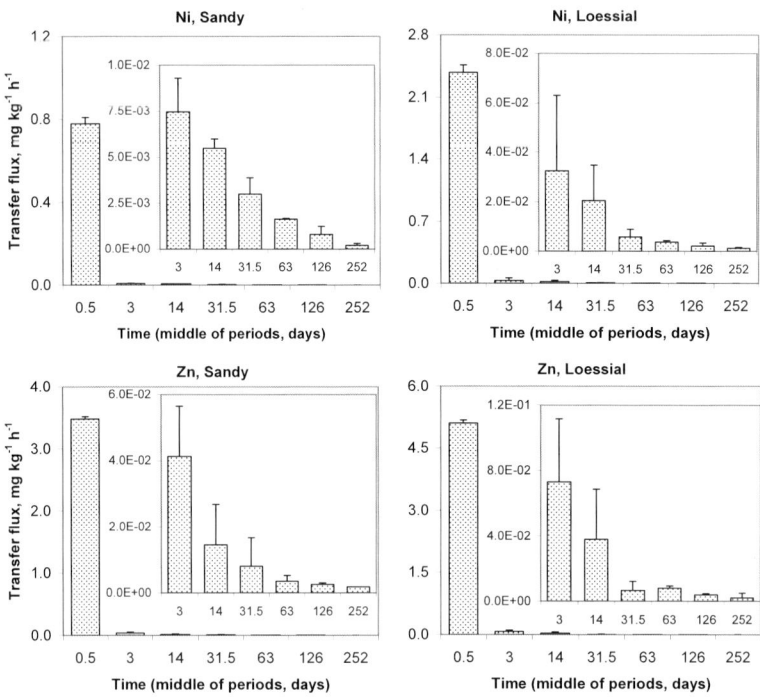

Figure 6.16. The average transfer fluxes of Ni and Zn among solid-phase components in Israeli soils. The soils received metal nitrates at the 3T level under the field capacity moisture regime (T: total metal content in non-amended soils)

the non-amended soil, and it is improved at elevated loading levels. It was worth pointing out that the net fluxes of the metal observed in the present experiments by Han (1998) are the combination of real changes and experimental errors.

The transfer fluxes of metals (Cu, Cr, Ni and Zn) were initially high and quickly decreased with time in both of the metal salt-spiked soils (Figs. 6.15–16). The initially high flux implies the fast reactions of heavy metals with soil matrix as observed by many studies (Brümmer et al., 1988; Xu et al., 1994; Han 1998). The high initial transfer flux was mainly composed of high flux rates out of the soluble plus exchangeable fractions. The initial fluxes of metals were characterized by the nature of the metal (Cr > Zn > Cu and Ni, as indicated by coefficients in the linear relationships between fluxes and loading levels in Fig. 6.17) and the loading levels of the metal. The initial transfer fluxes out of the soluble plus exchangeable fraction increased linearly with loading levels. This indicates that during the initial fast processes of immobilization, the metal is controlled by the nature of the metal and its concentrations. The initial transfer fluxes of the metal were similar between the two soils and two moisture regimes (Fig. 6.17). Copper had the similar first-order coefficients of the transfer flux out of the soluble plus exchangeable fraction with Ni, which averages around -0.034 h^{-1}. Zinc had -0.040 h^{-1} first-order coefficients, while Cr had the highest first-order coefficients (-0.049 h^{-1}) of the transfer flux out of the soluble plus exchangeable fraction (Fig. 6.17).

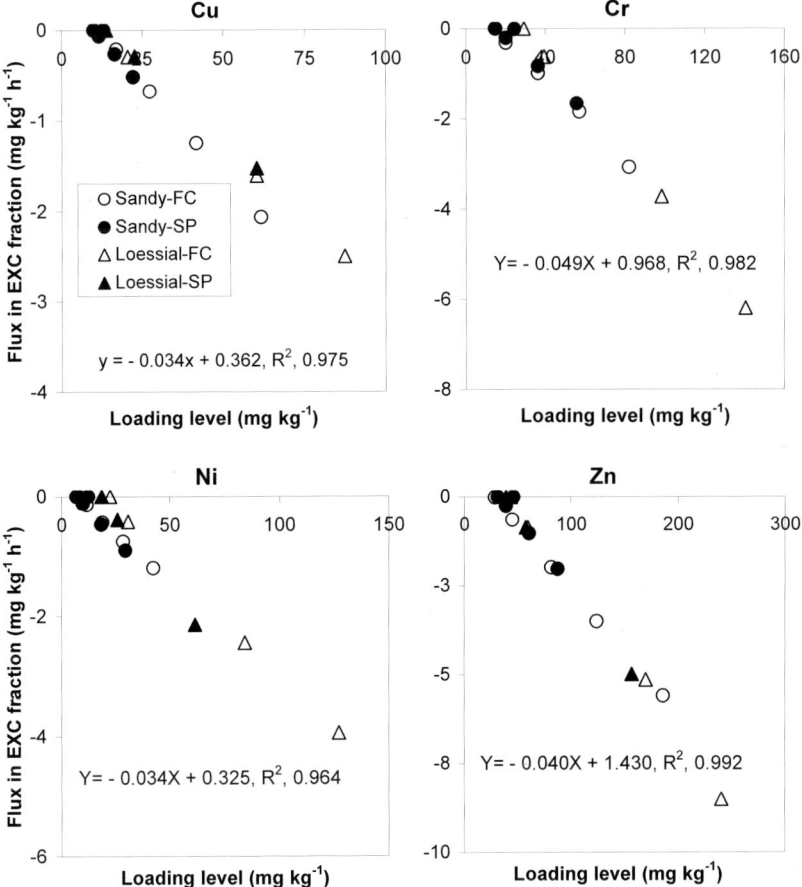

Figure 6.17. Relationships between the initial (the first day) fluxes of metals in the soluble plus exchangeable fraction and metal loading levels in Israeli soils. Soils received metal nitrates and were incubated in the saturated-paste (SP) and field capacity (FC) moisture regimes

Within the first week, the fluxes of these four metals during transformations among their solid-phase components were very high (1.0 x 10^{-1}–1.0 x 10^{-2} mg kg^{-1} $hour^{-1}$ and even higher); then they quickly decreased. After 1–2 weeks, the fluxes of these metals slowly continued to decrease with time (Figs. 6.15–6.16). Bidwell and Dowdy (1987) reported that Cd and Zn uptake by plants decreased with time, while DTPA-extractable Cd and Zn changed slightly in the sludge-treated soils after terminal application. This may be explained by Han's (1998) results that even though a chemical extractable metal such as a DTPA extractable is constant, metal redistribution takes place among various solid phase components. The fluxes of the metal transfer in soils might control the bioavailability of metals and plant uptake.

After about 100 days, the fluxes of Cu, Cr, Ni and Zn in the sandy and loessial soil changed slowly and almost constantly over time (Figs. 6.15–6.16). The slow transfer fluxes of heavy metals in soils represent slow redistribution processes among various solid-phase components. It may also imply possible slow diffusion of surface sorbed or precipitated metals into particles via meso- and micro-pores (Fischer et al., 1996; Almas et al., 2000b). Ionic diffusion processes in soils include macropore diffusion, microscopic diffusion in water films and diffusion inside particles. Proper initiation of the experiment and sampling of soils for determination of metal concentrations during incubation initially required thorough mixing of the added salts and soil matrix, and mixing soil when the subsamples were taken for analyses also was necessary. To some extent this procedure may reduce the importance of the macropore diffusion processes. However, this mixing technique does not affect microscopic diffusion processes in water films and inside particles. Moreover, the soils were only mixed infrequently when the samples were taken for analyses (Han, 1998), and the soils were left for long periods of incubation without mixing, especially after the first month. It should be pointed out that the overall transformation flux is calculated by averaging the sums of the "positive" and "negative" (in absolute values) fluxes of the individual fractions. Thus, the estimated fluxes indicate the net change or net transformation. The fluxes are composed of different sets of fluxes between individual solid-phase components. The overall fluxes are similar but the detailed pathways of heavy metals out of or into the individual fractions may be different as discussed above.

The annual transfer fluxes of the four metals among solid-phase components between one day and one year of incubation at the two moisture regimes are presented in Fig. 6.18. The overall transfer fluxes of the slow processes were differentiated between the two soils and the two moisture regimes. In general, the transfer fluxes of the metal in the loessial soil was higher than that in the sandy soil because of its abundant solid-phase components such as clays, organic matter and carbonate, and those in the saturated paste regime were higher than those in the field capacity regime. Compared to the initial transfer fluxes, the annual transfer fluxes of the four metals were much lower. This implies that heavy metals become more strongly bound to soil matrix during the first year of interactions between the added metals and soil solid-phase components.

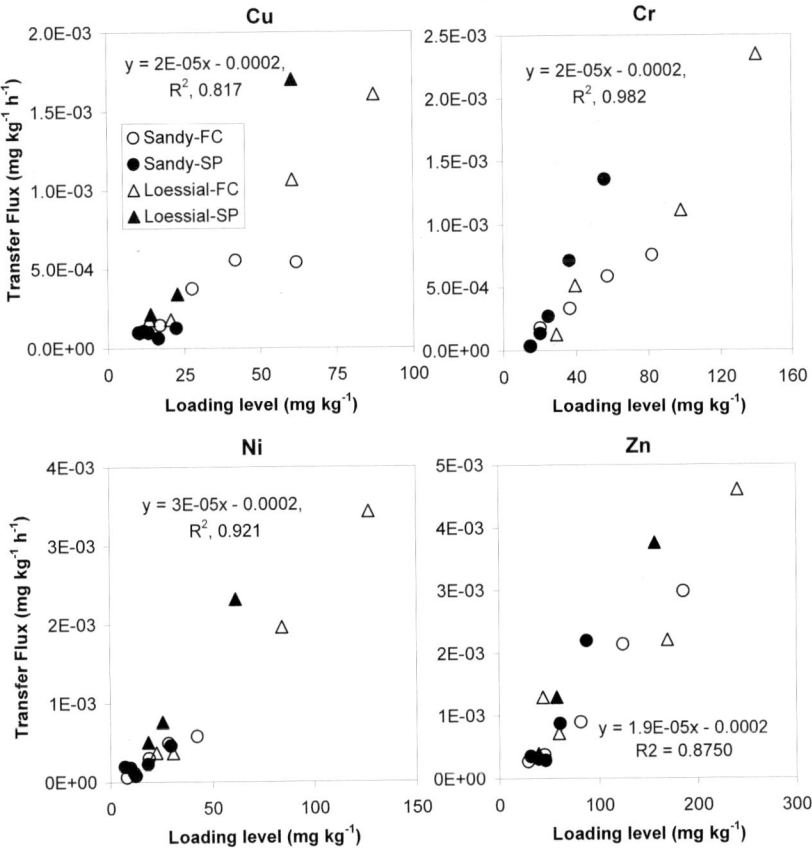

Figure 6.18. Changes of the annual transfer fluxes (from one day to one year) among solid-phase components with metal loading levels in Israeli soils. Soils received metal nitrates and were incubated in the saturated-paste (SP) and field capacity (FC) moisture regimes

As reported earlier by Han (1998) and Han et al. (2001c), heavy metals were transferred and redistributed from the soluble and labile forms into the stable forms with time. The transformation pathways of Cu, Ni and Zn in soils were strongly affected by the soil moisture regime, while Cd and

Cr showed little differences in the major pathways of redistribution as affected by the moisture regimes.

Information on the fluxes of heavy metals during redistribution and transformations in soils is very limited. Some data reported here are recalculated from literature reports on the uptake of Zn by plants as proxy information to compare to the present data. The fluxes of Zn into navy bean plants grown in the Haplargids from Wyoming (pH 7.7–8.1) during the period from planting to harvest (about 110–144 days) were recalculated from the uptake of Zn by the plants. The estimated fluxes decreased with time and increased with the loading levels of addition (fluxes: 1.0×10^{-5} and 1.27×10^{-5} mg kg^{-1} h^{-1} in non-amended soil and metal salt-amended soils, respectively, during the first year; and 3.33×10^{-6} and 3.81×10^{-6} mg kg^{-1} h^{-1} for non-amended soil and metal salt-amended soils, respectively, during the third year) (Blaylock, 1995). The fluxes of Zn into plants during navy bean growth have been found to be very similar to those measured in an Oxisol (pH 6.1) from Puerto Rico (flux, 3.72×10^{-5} mg kg^{-1} h^{-1}) (Blaylock, 1995, Goenaga and Chardon, 1995). However, the fluxes of Zn calculated from plant uptake under field conditions (Blaylock, 1995, Goenaga and Chardon, 1995) were much lower, and were usually three magnitudes of order less than those found in the study by Han (1998) under laboratory controlled conditions. This may be partly related to the much higher soil amount under field conditions compared to the soil amount in the pots under controlled conditions when the fluxes are calculated. If the volume of plant root over bulk field soil (20 cm plowing layer) is assumed to be around 0.2% (v/v) for plants such as wheat (recalculated from Barber, 1984), the fluxes calculated above from field conditions are actually as a measured flux of Zn from root-contact soil to plant. Accordingly, the fluxes of Zn from soil to plant, after correction of the soil volume displaced by the root, equals the measured fluxes calculated above divided by 0.2%, which are similar to the range of fluxes obtained in the Han's study (1998). The fluxes of Zn in the Haplargids (corrected to the soil volume displaced by the root), range from 1.76×10^{-3} to 5.88×10^{-3} mg kg^{-1} h^{-1}, and the fluxes of Zn in the Oxisol are 1.76×10^{-2} mg kg^{-1} h^{-1}.

6.4 RESIDENCE TIME OF METALS IN SOIL SOLUTION OF METAL-CONTAMINATED ISRAELI ARID SOILS

The residence time is calculated by dividing the concentration of the metal in the soil solution (mg metal/liter soil solution) by the sum of positive fluxes or the sum of the negative fluxes of the metal among various solid-phase fractions:

$$\Gamma = C_s / (\sum_{i=1}^{k} S^+), \quad \text{or} = C_s / (\sum_{i=1}^{k} S^-), \quad (6.9)$$

where Γ is the residence time (hour); S^+ or S^- is the positive or negative fluxes; ΣS^+ or ΣS^- is the sum of the positive or the sum of the negative fluxes of the metal to/from various solid phases, respectively; and C_s is the concentration of the metal in solution (mg/l). In theory, the sum of the positive fluxes should equal the sum of the negative fluxes of the metal in the soil. However, due to experimental variation, these two fluxes may not be exactly equivalent to each other.

The residence time of metals in the Israeli arid soil solution was calculated according to the concentration of the metal in the soil solution at a given time. Since the measured concentrations of Cr in soil solution were very low, usually below or close to detection limits of ICP-AES, the residence time of Cr could not be estimated. The residence time of Zn, Ni and Cu was calculated from the field capacity and saturation paste regimes. Results showed that the residence time of the metals was similar in soils incubated at both the saturated paste and field capacity regimes (Table 6.7). In general, the residence time of the three metals was in the order: Ni (7 hours, as an average) > Cu (2 hours, as average) > Zn (1 hour, as average); and sandy > loessial soil. This indicates that the reaction of Zn is faster than Ni, and the loessial soil has faster reaction rates than the sandy soil due to the abundance of oxides and clay minerals in the loessial soil. This observation fits well with the metal transfer fluxes as discussed above. The literature data are very limited on the residence time of heavy metals in arid soil solution.

Table 6.7. The average residence time (hour) of heavy metals in Israeli soil solutions

Metal	Sandy soil								Average
	Saturation		Field capacity		Saturation		Field capacity		
	Range	Average	Range	Average	Range	Average	Range	Average	
Cu	<1-30	6.5	1-16	3	<1-30	2	<1-8	1	2
Ni	3-80	15	3-65	10	<1-7	3.3	<1-40	3.8	7
Zn	<1	0.4	<1-2	0.2	<1-9	1.6	<1-7	1.7	1

Ahnstrom and Parker (2001) observed the fast migration of ^{111}Cd into the exchangeable and the sorbed or carbonate bound fractions in uncontaminated arid soil and contaminated soils with sewage sludge, smelter emission and Pb-Zn mine soils. A Cd isotopic exchange reaction half-time is in minutes for the labile portion of the Cd. However, a slower migration of ^{111}Cd into the organically bound fraction is noted with a reaction half-time on the order of hours and a few days, and the slowest migration is found in the reducible oxides bound and the residual fractions, with a reaction half-time of weeks to years. Theoretically, the reducible oxides bound and the residual fractions are comprised of very refractory Cd bound in the lattices of reducible, secondary oxides (e.g., Fe_2O_3) and aluminosilicate minerals. Lattice diffusion, a process necessary for isotopic exchange, can require years.

6.5 TRANSFER OF OTHER TRACE ELEMENTS IN UN-CONTAMINATED ISRAELI ARID SOILS

6.5.1 Changes of Manganese Partitioning Among Solid-phase Components in Arid-zone Soils: Pathways and Short- and Intermediate-term Kinetics

Manganese supply may be a limiting factor for plant growth in arid-zone regions. Calcareous soils with a high pH and a high content of $CaCO_3$ may affect Mn availability. It was found that in some cases excess Mn was taken up by plants from calcareous soils undergoing intermittent reduction periods (e.g., Moraghan and Freeman, 1978; Moraghan, 1985a). In other cases, Mn deficiency developed in plants grown on calcareous soils (e.g., Moraghan, 1985b; Schaffer et al., 1988). Changes in the soil moisture regime shift the partition of Mn among the components of the solid-phase and its transformations.

Submergence increases the soluble plus exchangeable Mn fraction at the expense of the reducible Mn fraction (Swarup and Anand, 1989; Mandal and Mitra, 1982). This explains why DTPA-extractable Fe and Mn increase upon the submergence of soils (Sajwan and Lindsay, 1988). The transformations of Mn and Fe during submergence are related to changes in pH, pCO_2 and Eh (Swarup and Anand, 1989). Pasricha and Ponnamperuma (1976) showed that $MnCO_3$ may have controlled Mn^{2+} levels in flooded rice soils. This is established by Schwab and Lindsay (1983) in a highly calcareous, Mn-rich soil. They found, however, that at (pe + pH) > 16, Mn-oxides may determine Mn concentration in solution.

The kinetics of transformation of Mn and Fe among soil fractions is related to the redox potential (Eh changes) as well as the content of reductants or oxidants in the soil. Generally, a drop in Eh is observed within a few days of the waterlogging of a soil. In a coastal saline silty-clay soil, Eh was reduced to 210 mV after four days of submergence (Bandyopadhyay and Bandyopadhyay, 1984). The decrease in Eh mobilized Mn, and the maximal soluble Mn concentration was found after 14 days of submergence in a sandy-loam soil (Sadana and Takkar, 1988).

The transformation pathway(s) and kinetics of Mn in the solid-phase of two Israeli arid-zone soils incubated under saturated paste and field capacity conditions for a prolonged period of time is discussed below.

6.5.1.1 Transformations of Mn Among Solid-phase Fractions

Changes in the Mn percentages in the various soil fractions during one year of incubation under the saturated paste and field capacity regimes are presented in Figs. 6.19 and 6.20. Details of the rapid initial changes in Mn content in its three more labile fractions (EXC, CARB and ERO) and long-term variation (one year of incubation) in the major fractions are shown in Figs. 6.21 and 6.22.

During the saturation regime, a rapid initial stage of transformation of Mn from the ERO fraction to the exchangeable and carbonate fractions was observed. This stage was finished within three days in the sandy soil, but the stage lasted 12 days in the loessial soil (Figs 6.19, 6.21). This stage was then followed by more subtle and slow changes among fractions during the rest of the year. The slow transformations involve the OM and RO fractions from which some Mn was transferred to the more mobile fractions (Fig. 6.21). On the other hand, when the soils were incubated in the field capacity regime, a rapid initial stage of transformation of Mn from the CARB, the RO and, to some extent, the EXC fractions to the ERO fraction was found. This stage was finished within the first day in both soils (Figs. 6.20 and 6.22). However, the changes of Mn in the absolute content (mg/kg soil) were much smaller during the field capacity regime compared to the saturation regime (Figs. 6.19–6.22).

The initial transformations of Mn among soil fractions during incubation in both moisture regimes were concomitant with changes in pH and Eh. After one hour of saturation incubation, the overall redox potential (pe + pH) was 12.0 and 13.6 in the sandy and loessial soils, respectively (Fig. 6.23).

In both soils (pe + pH) decreased to a value of about four after 7–9 days of the saturated-paste incubation, and remained stable thereafter. Most of the change resulted from pe changes: Eh decreased from 290 mV and 332 mV after one hour of incubation to –200 mV and –190 mV after 7–9 days of incubation in the sandy and loessial soil, respectively, and changed very little thereafter. During the same period, pH in the sandy soil was slightly increased from 7.2 to 7.5–7.7, and in the loessial soil it was decreased slightly from 8.0 to 7.0–7.4. With the decrease of Eh and (pe + pH) of the soils as a result of incubation in the saturated paste condition, Mn was chemically reduced and transformed between soil fractions. The changes in the (pe + pH) of the two soils in the field capacity condition are not available.

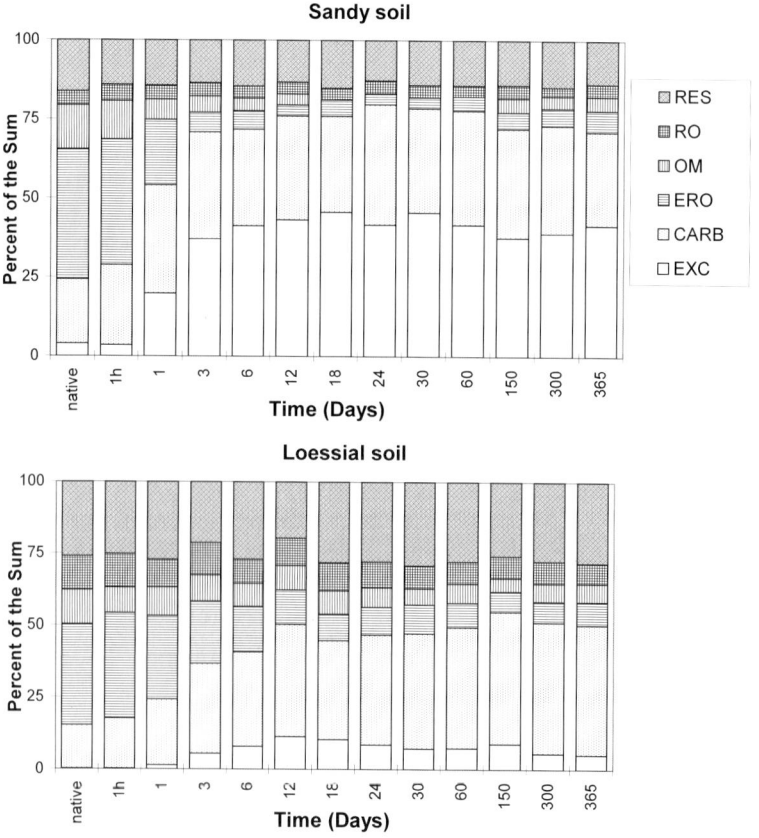

Figure 6.19. Distribution of Mn in solid-phase fractions in Israeli soils. Soils were incubated at the saturated paste regime (after Han and Banin, 1996. Reprinted from Soil Sci Soc Am J, 60, Han F.X., Banin A., Solid-phase manganese fractionation changes in saturated arid-zone soils: Pathways and kinetics, p 1075, Copyright (1996), with permission from Soil Sci Soc Am)

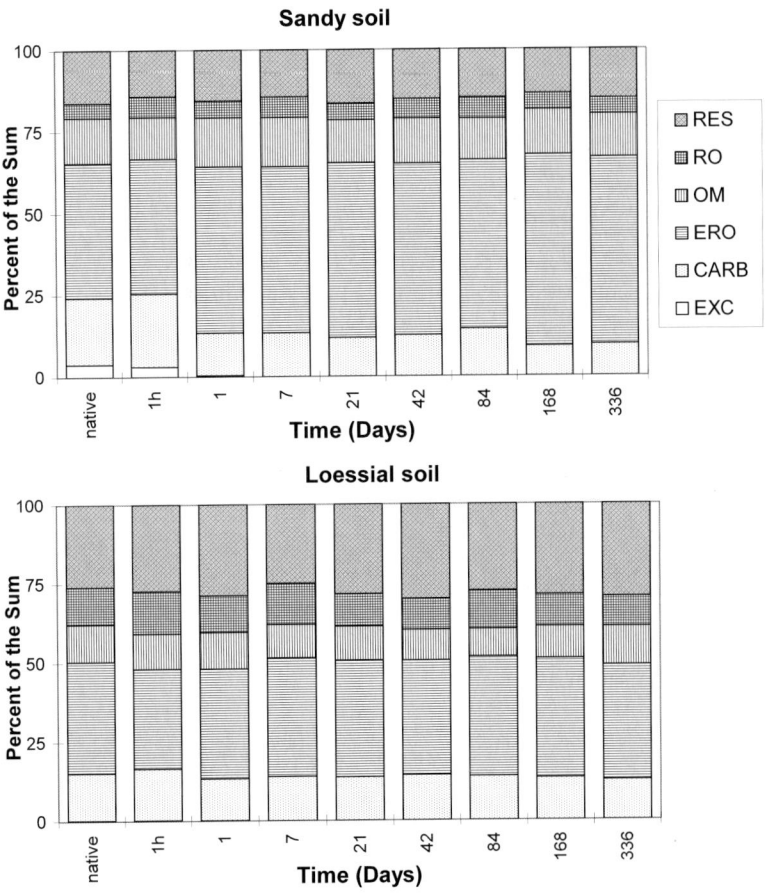

Figure 6.20. Distribution of Mn in solid-phase fractions in Israeli soils. Soils were incubated at the field capacity regime

It is generally known that ionic and mineral forms of Mn in soils are dependent upon (pe + pH) of the soil (e.g., Bartlett, 1986; Banin et al., 1990; McBride, 1979; Sajwan and Lindsay, 1988; Warden and Reisenauer, 1991). Soil pe (Eh) is controlled by the soil-water regime through its effect on soil water and soil air ratio, gas-exchange and biological activity in the soil. The present observation is a possible indication that during incubation, as anaerobic activity predominates, Mn was reduced from minerals such as pyrolusite (β-MnO_2; Mn^{4+}) and possibly, manganite (γ-MnOOH; Mn^{3+}), originally present in the ERO fraction. The reduced Mn^{2+} ion was present in the EXC fraction or as rhodochrosite ($MnCO_3$; Mn^{2+}) present in the CARB fraction (Han and Banin, 1996). In contrast, the soils incubated under the field capacity regime underwent the opposite changes as the soils under the saturated condition.

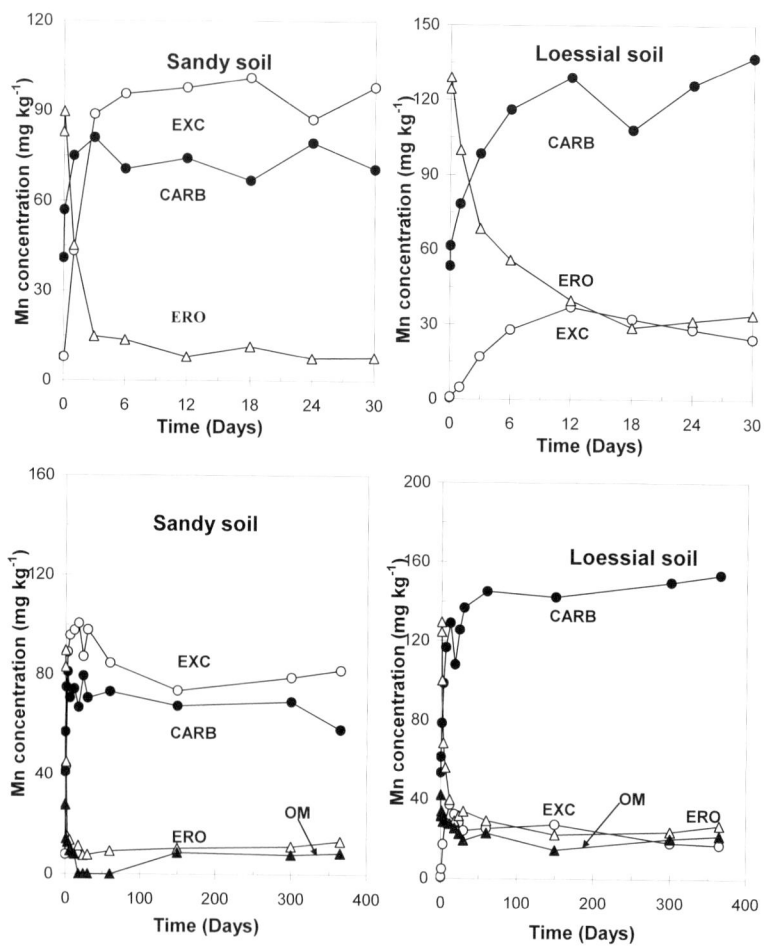

Figure 6.21. Initial and annual distribution and transformations of Mn in the major solid-phase fractions in two Israeli soils. Soils were incubated at the saturated-paste regime (after Han and Banin, 1996. Reprinted from Soil Sci Soc Am J, 60, Han F.X., Banin A., Solid-phase manganese fractionation changes in saturated arid-zone soils: Pathways and kinetics, p 1076, Copyright (1996), with permission from Soil Sci Soc Am)

However, the two soils differed significantly in their detailed transformation pathways. During saturation in the sandy soil, Mn was transformed from the ERO and the OM fractions predominately into the EXC fraction (Figs. 6.19 and 6.21). In the loessial soil, Mn was transformed from the ERO fraction into the CARB fraction, and only a small portion initially moved into the EXC fraction. This small portion then continued the transformation slowly into the carbonate fraction (Figs. 6.19–6.20). The Mn distribution in the reduced sandy soils stabilized at 40% and 35% in the EXC and the CARB fractions, respectively. However, in the loessial soil, these two fractions contained 10% and 45% of Mn, respectively. This difference between the two soils was attributed to their capacity to accommodate the reduced Mn in the carbonate fraction. Due to its low carbonate content, the sandy soil had a much more limited carbonate-capacity for Mn than the loessial soil, and therefore, the excess Mn^{2+} was forced to remain in the exchangeable fraction. The transformation to the carbonate phase in the loessial soil was also kinetically limited either by the rate of neoformation of carbonate and/or by the rate of diffusion of Mn into existing carbonate in the soil.

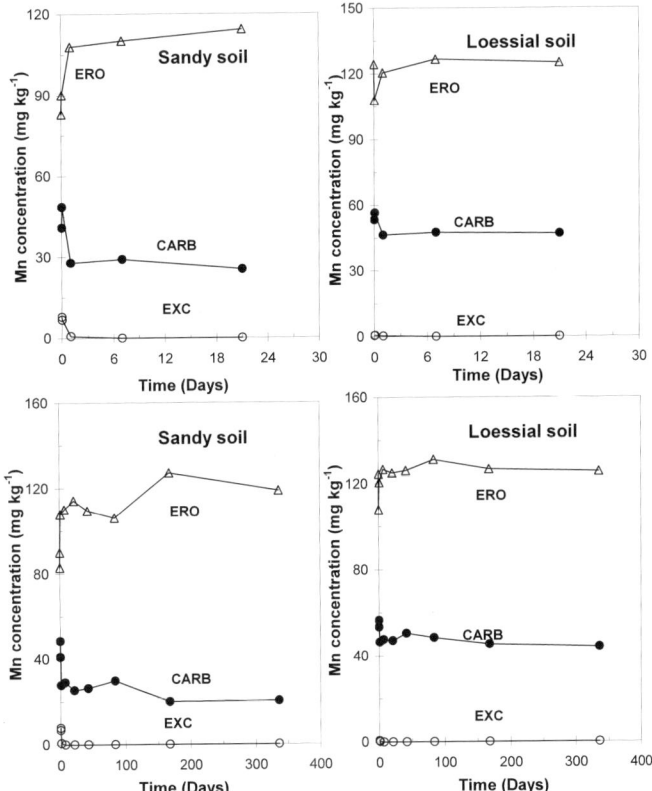

Figure 6.22. Distribution and transformations of Mn in the major solid-phase fractions in two Israeli soils. Soils were incubated at the field capacity regime

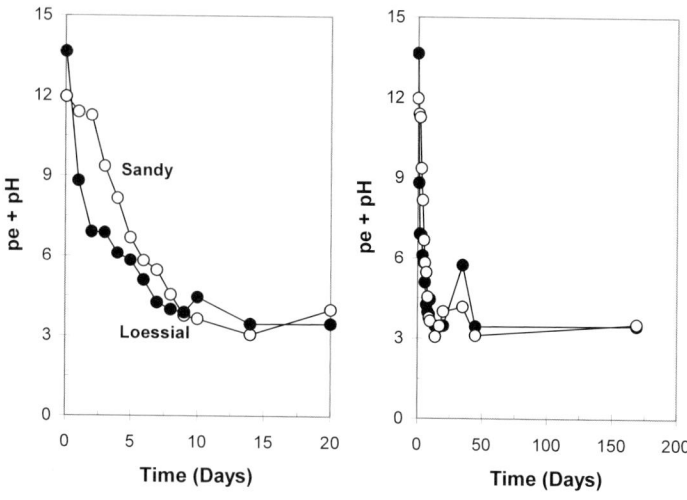

Figure 6.23. Changes of the redox parameter (pe + pH) in two Israeli arid soils during saturated paste incubation (after Han and Banin, 1996. Reprinted from Soil Sci Soc Am J, 60, Han F.X., Banin A., Solid-phase manganese fractionation changes in saturated arid-zone soils: Pathways and kinetics, p 1076, Copyright (1996), with permission from Soil Sci Soc Am)

During incubation at the field capacity regime in the sandy soil, Mn was transformed mainly from the CARB, and to a small extent the EXC and the RO fractions into the ERO fraction (Figs. 6.20 and 6.22). It was noted that the Mn changes in the CARB fraction constituted 75% of the transfer to the ERO fraction. In the loessial soil, equal amounts of Mn were transferred from the CARB and the RO fractions into the ERO fraction. The difference in Mn transformation in the two soils may be related to the initial Eh status at the field capacity regime. Under the field capacity regime, Eh in the sandy soil after one hour was 420 mV, but it was 250 mV in the loessial soil. Perhaps this was due to the higher aeration rate in the sandy soil. The lower Eh status in the loessial soil resulted in greater Mn transformation from the RO fraction into the ERO fraction. The difference in Eh between the two soils, however, became smaller with time.

When the saturated arid soil (the sandy soil) was dried, the (pe + pH) increased rapidly. Manganese was transferred back from the EXC (the initial flux: -3.60×10^{-5} mg/kg/sec) and the CARB (-4.16×10^{-6} mg/kg/sec) fractions primarily into the ERO (3.89×10^{-5} mg/kg/sec) and partly into the OM (-4.16×10^{-6} mg/kg/sec) fractions (Han and Banin, 1997) (Fig. 6.24). Most of the Mn reoxidation was complete within 15 days after the beginning of the drying process (Banin et al., 1997b). Actually it took a similar period of time (about 2 weeks) for Mn reduction in the soil (Han and Banin, 1997).

The rate of Mn transfer between the EXC and the ERO fractions decreased after 15 days, with only partial recovery of ERO-Mn to its initial and pre-saturation value. The missing Mn resided in the OM and the RO fractions, and apparently it was held in non-accessible sites, which were slow to equilibrate with the solution. This required a much longer period of time to be reoxidized.

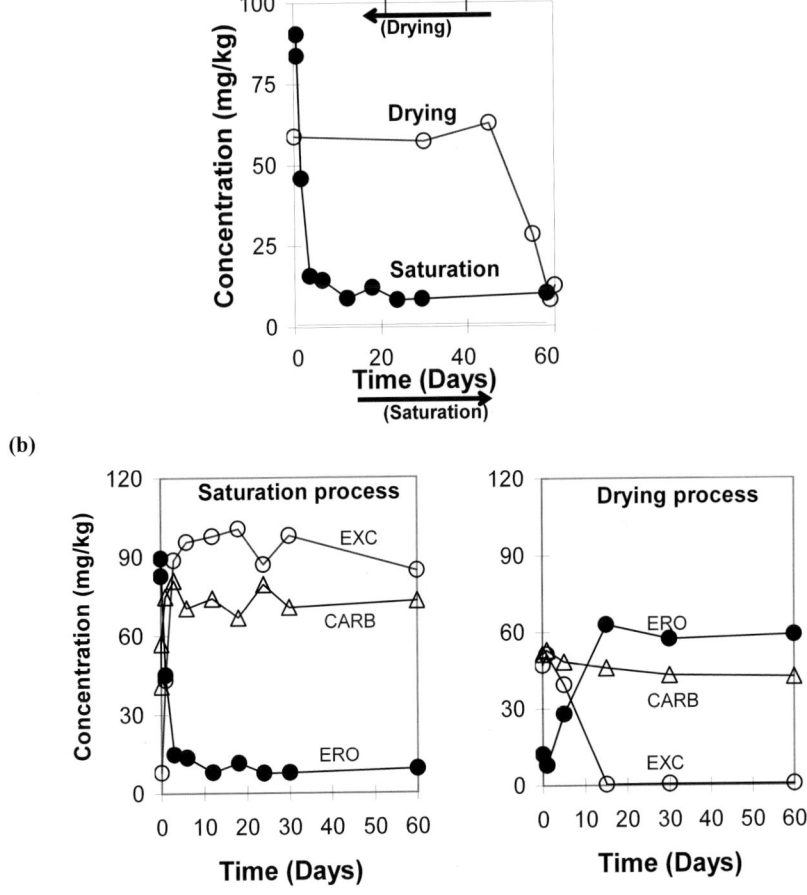

Figure 6.24. Comparisons of Mn changes in the ERO (a) and EXC and CARB fractions (b) in an Israeli sandy soil during one year of saturation incubation and subsequent drying processes

6.5.1.2 "Equilibrium" Parameters

Manganese transformations can be seen from changes in the "equilibrium" parameters of the various fractions (U_{tf},) and the whole soil (U_{ts}) with time of incubation under saturation (Fig. 6.25). In both soils incubated under the saturation condition, the U_{tf} parameters of the EXC and the CARB fractions increased and departed from the native state ($U_{tf\text{-}EXC}$ and

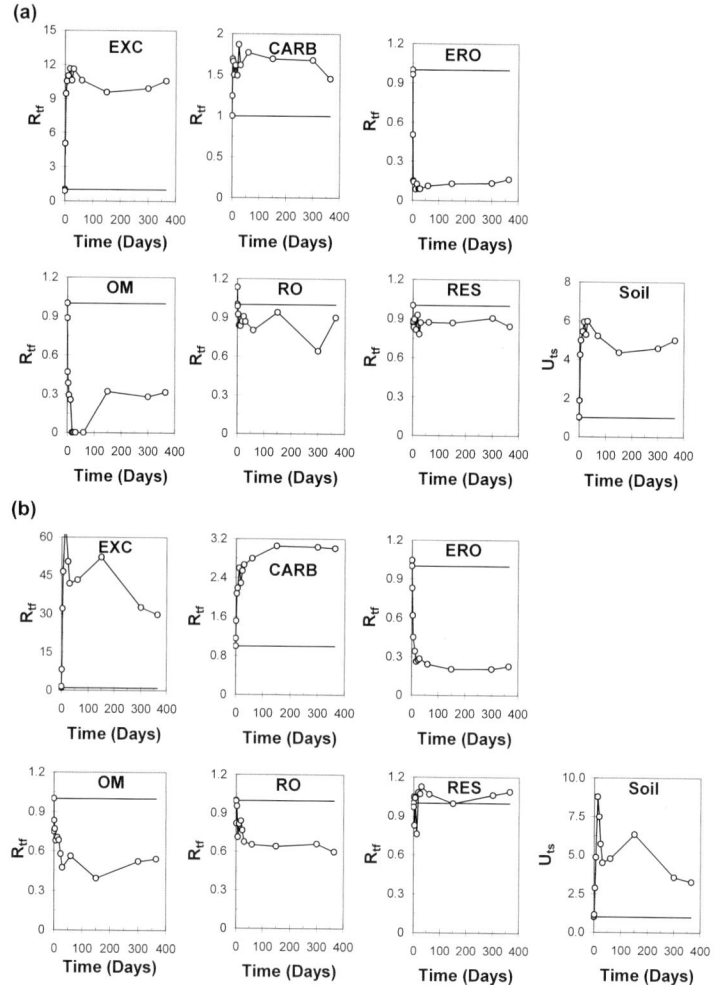

Figure 6.25. Changes of R_{tf} of Mn fractions and U_{ts} in Israeli sandy (a) and loessial (b) soils during one year of incubation at the saturated paste regime (after Han and Banin, 1996. Reprinted from Soil Sci Soc Am J, 60, Han F.X., Banin A., Solid-phase manganese fractionation changes in saturated arid-zone soils: Pathways and kinetics, pp 1077–1078, Copyright (1996), with permission from Soil Sci Soc Am)

$U_{tf-CARB} > 1$). In parallel, Mn in the ERO and the OM fractions decreased, and their U_{tf} also departed from the native state (U_{tf-ERO} and $U_{tf-OM} < 1$). The U_{tf} of the RO fraction was also slightly reduced. This reduction especially occurs during the first stage of the incubation, whereas the RES phase was relatively stable ($U_{tf-RES} \cong 1$).

The soil "equilibrium" parameter, U_{ts}, measures the departure from the Mn fractionation in the oxic native soil, which is caused by the saturation and incubation of the soil. This parameter can acquire values equal to or larger than one. Compared to the native state, an increase in the U_{ts} indicates that some fraction(s) have been grossly enriched in Mn. Note the fact that a fraction that has lost some Mn will only reduce its own contribution to the U_{ts} value, but its contribution will not be subtracted from that of the other fraction(s) since the product $F_i \times U_{tfi}$ can not acquire negative values.

It is observed that U_{ts} had been changed considerably by the saturated incubation in both soils. However, after the initial stage where U_{ts} peaked, a decrease in its value indicated that the fraction(s) that were enriched in Mn were slowly losing some of their Mn to the other fractions. During the saturated paste incubation, both soils acquired fractionation patterns differing from their native pattern. These were depicted by U_{ts} of 4–4.5 and 3–3.5 for the sandy and loessial soils, respectively. In contrast, U_{ts} of Mn in the fractions and for the whole soil during incubation at the field capacity regime changed little and the data are not shown here.

6.5.1.3 Kinetics of Transformation of Mn

The zero-order, first-order and second-order reactions, the parabolic diffusion equation, the two-constant rate equation and the simple Elovich equation were employed attempting to describe the transformations of Mn in the two arid soils during the initial period of incubation. This lasted six and twelve days in the sandy and loessial soils under the saturated paste regime, respectively (Table 6.8 and Fig. 6.26). The changes of Mn in fractions in the two arid soils during incubation under the field capacity regime were relatively small compared to changes of Mn in the soils under the saturation regime.

During the saturation regime, fluxes of reduced-species of Mn, as measured by the decrease in its content in the ERO fraction (mg kg^{-1} sec^{-1}), were somewhat faster (by a factor of 1.43/0.828 = 1.7) in the sandy soil than in the loessial soil. The reduced Mn species was split in both of the soils between the exchangeable and the carbonate fractions. The rate of increase of Mn in the carbonate fraction was somewhat faster (6.9/4.14 = 1.7) in the loessial soil than in the sandy soil, whereas the attachment of Mn^{2+} to the exchangeable phase was considerably faster in the sandy soil than that in the loessial soil (1.77/0.37 = 4.8). These differences in rates of attachment primarily seemed to be related to the different sorption capacities of the carbonate phase in the two soils, as discussed above. The overall reaction appeared to be controlled by the reduction of Mn from the ERO fraction.

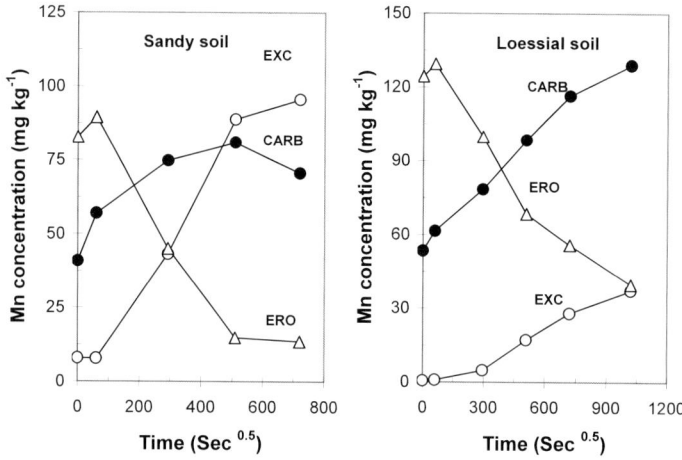

Figure 6.26. Kinetics of Mn transformations (described by the parabolic diffusion equation) among three main labile solid-phase fractions of Israeli arid soils. Soils were incubated at the saturated paste regime (after Han and Banin, 1996. Reprinted from Soil Sci Soc Am J, 60, Han F.X., Banin A., Solid-phase manganese fractionation changes in saturated arid-zone soils: Pathways and kinetics, p 1079, Copyright (1996), with permission from Soil Sci Soc Am)

Generally, kinetic fitting to the various mechanisms does not give highly significant linear regression coefficients. However, the parabolic diffusion equation, the two-constant equation and the simple Elovich equation all described the initial changes of the three main Mn fractions in both soils incubated under the saturation regime better than the zero-order, first-order and second-order reactions (Table 6.8). In both soils, the parabolic diffusion equation produced the highest coefficients for all three fractions, and this equation also seemed to provide the best kinetic description of the processes (Fig. 6.26). This suggests that the diffusion processes controlled the rate of the Mn transformations. The rate-limiting step for the processes was possibly the transport of Mn^{2+} after the reduction of Mn-oxides. However, it is also possible that the transformation processes were highly controlled by the rate of change in the soils' redox potential. The present data are too limited to enable detailed analysis of the factors controlling the reaction kinetics, and thus this further clarifies their mechanisms.

Table 6.8. The kinetics of the initial changes of Mn concentration of three fractions in incubated Israeli soils under saturated regime (C: mg kg^{-1}; t: sec) (Han and Banin, 1996, with permission from Soil Sci Soc Am)

Model	Fraction		R^2	N
Sandy soil (6 days)				
Zero-order	EXC	$C_t = 1.77 \times 10^{-4} t + 17.9$	0.841	5
($C_t = -k_o t + C_o$)	CARB	$C_t = 4.14 \times 10^{-5} t + 57.7$	0.321	5
	ERO	$C_t = -1.43 \times 10^{-4} t + 74.0$	0.754	5
First-order	EXC	$\ln C_t = 4.77 \times 10^{-6} t + \ln 12.9$	0.705	5
($\ln C_t = k_1 t + \ln C_o$)	CARB	$\ln C_t = 7.15 \times 10^{-7} t + \ln 55.7$	0.325	5
	ERO	$\ln C_t = -3.79 \times 10^{-6} t + \ln 71.1$	0.838	5
Second-order	EXC	$1/C_t = -2.13 \times 10^{-7} t + 1/10.4$	0.577	5
($1/C_t = k_2 t + 1/C_o$)	CARB	$1/C_t = -1.16 \times 10^{-8} t + 1/53.8$	0.320	5
	ERO	$1/C_t = -1.31 \times 10^{-7} t + 1/68.5$	0.875	5
Parabolic diffusion	EXC	$C_t = 0.14 t^{0.5} + 5.3$	0.958	5
($C_t = k_p t^{0.5} + C_o$)	CARB	$C_t = 0.04 t^{0.5} + 52.1$	0.581	5
	ERO	$C_t = -0.11 t^{0.5} + 85.5$	0.917	5
Two-constant	EXC	$C_t = 0.11 t^{0.53}$	0.989	4
($C_t = at^b$)	CARB	$C_t = 36.7 t^{0.06}$	0.697	4
	ERO	$C_t = 2484 t^{-0.39}$	0.892	4
Simple Elovich	EXC	$C_t = 1/0.06 \ln (0.006 \times 0.06) + 1/0.06 \ln t$	0.921	4
($C_t = 1/\beta \ln(\alpha\beta) + 1/\beta \ln t$)	CARB	$C_t = 1/0.26 \ln (6321 \times 0.26) + 1/0.26 \ln t$	0.663	4
	ERO	$C_t = -1/0.06 \ln [(-1.57 \times 10^{-5}) \times (-0.06)] - 1/0.06 \ln t$	0.977	4
Loessial soil (12 days)				
Zero-order	EXC	$C_t = 3.65 \times 10^{-5} t + 3.1$	0.928	6
($C_t = -k_o t + C_o$)	CARB	$C_t = 6.90 \times 10^{-5} t + 67.6$	0.846	6
	ERO	$C_t = -8.28 \times 10^{-5} t + 122$	0.806	6
First-order	EXC	$\ln C_t = 3.64 \times 10^{-6} t + \ln 1.9$	0.673	6
($\ln C_t = k_1 t + \ln C_o$)	CARB	$\ln C_t = 7.64 \times 10^{-7} t + \ln 66.7$	0.766	6
	ERO	$\ln C_t = -1.12 \times 10^{-6} t + \ln 112$	0.908	6
Second-order	EXC	$1/C_t = -1.13 \times 10^{-6} t + 1/1.1$	0.414	6
($1/C_t = k_2 t + 1/C_o$)	CARB		small	6
	ERO		small	6
Parabolic diffusion	EXC	$C_t = 0.04 t^{0.5} - 1.9$	0.972	6
($C_t = k_p t^{0.5} + C_o$)	CARB	$C_t = 0.08 t^{0.5} + 56.4$	0.984	6
	ERO	$C_t = -0.09 t^{0.5} + 126$	0.961	6
Two-constant	EXC	$C_t = 0.003 t^{0.69}$	0.979	5
($C_t = at^b$)	CARB	$C_t = 19.80 t^{0.13}$	0.937	5
	ERO	$C_t = 728.16 t^{-0.19}$	0.866	5
Simple Elovich	EXC	$C_t = 1/0.16 \ln (0.0008 \times 0.16) + 1/0.16 \ln t$	0.779	5
($C_t = 1/\beta \ln(\alpha\beta) + 1/\beta \ln t$)	CARB	$C_t = 1/0.08 \ln (0.35 \times 0.08) + 1/0.08 \ln t$	0.886	5
	ERO	$C_t = -1/0.06 \ln[(-7.94 \times 10^{-7}) \times (-0.06)] - 1/0.06 \ln t$	0.944	5

6.5.2 Pathways and Kinetics of Transformation of Cobalt Among Solid-phase Components at Saturated Arid-zone Soils

An adequate supply of Co in a pasture is important to the health of grazing animals. Ruminants suffering from Co deficiency are locally recognized under different names such as Bush Sickness, Denmark Wasting Disease, Coasty Disease and Pining (Mckenzie, 1975). In plants, Co is required by legumes for the optimal functioning of symbiotic nitrogen fixation. Soil is the main source of Co to the grazing grass and crops.

The bulk of Co residing in the residual fraction is probably from isomorphous substitution in the primary and secondary minerals (Mclaren et al., 1986a). There is a close relationship between Co and the easily reducible fraction of Mn (Mn oxides), and Co is accumulated in Mn nodules in soils (Jarvis, 1984; Mckenzie, 1975). Submergence affects the redox potential, pH and the breakdown of organic matter with the production of CO_2. These processes influence the multi-equilibria of Fe and Mn in soils. Therefore, other trace metals, including Co, may be transformed from one component to another in the soil as the redox potential changes. Extractable Co and plant uptake of Co increase significantly in waterlogged soils (Beckwith et al., 1975; Berrow et al., 1983).

The section below will briefly address the distribution of Co, its transformation pathway(s), and kinetics among solid-phases in two selected Israeli arid-zone soils incubated under saturated paste conditions for prolonged periods of time (Han et al., 2002b). The parallel relationships of Co and Mn transformations among their solid-phases in soils are demonstrated.

6.5.2.1 Transformations of Co Among Solid-phase Fractions

Changes in the percentages of Co in the various soil fractions in the two selected soils during one year of incubation are presented in Fig. 6.27. The details of the rapid initial changes in Co content in its three major active fractions (CARB, ERO, and RO fractions) are presented in Fig. 6.28.

The initial transformations of Co among solid-phase fractions in two arid soils during incubation were concomitant with changes in Eh and pH. A rapid initial stage of transformation of Co was observed, mainly from the ERO fraction, and to some extent, from the RO and the OM fractions to the CARB fraction. Transformation pathways of Co followed those of Mn in the soils during incubation (Fig. 6.29). Most of the changes were completed within three days in the sandy soil, but these changes lasted 18 days in the loessial soil (Fig. 6.28). It was followed by the much slower changes among fractions during the rest of the year. The slow transformations involved the RO and the OM fractions from which some Co was transported to the more soluble fractions.

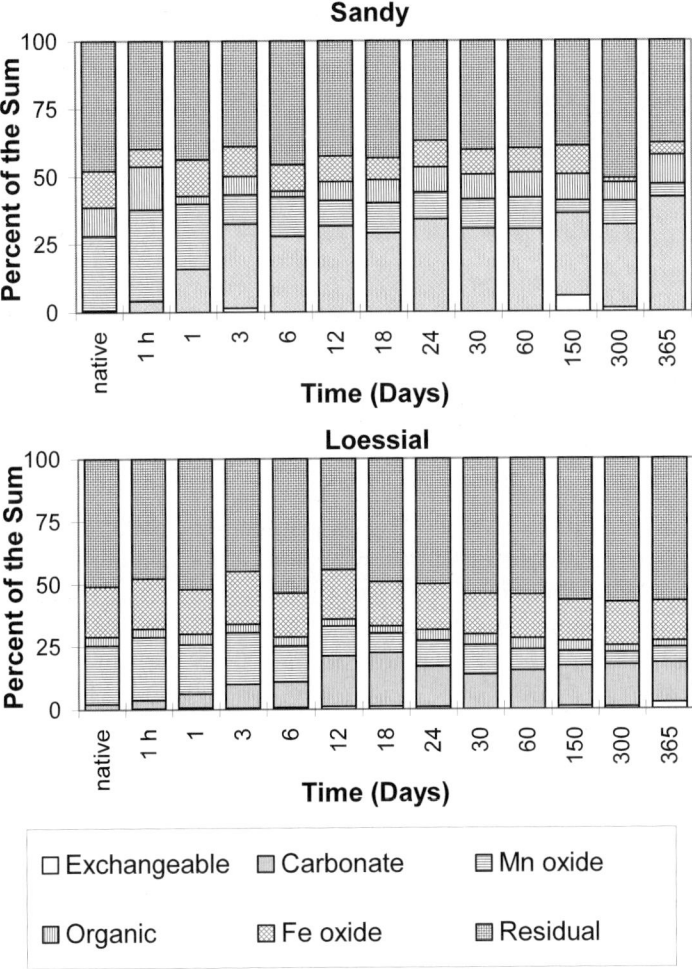

Figure 6.27. Distribution of Co in the solid-phase components in two Israeli arid soils incubated at the saturated paste regime (after Han et al., 2002b. Reprinted from J Environ Sci Health, Part A, 137, Han F.X., Banin A., Kingery W.L., Li Z.P., Pathways and kinetics of transformation of cobalt among solid-phase components in arid-zone soils, p 187, Copyright (2003), with permission from Taylor & Francis)

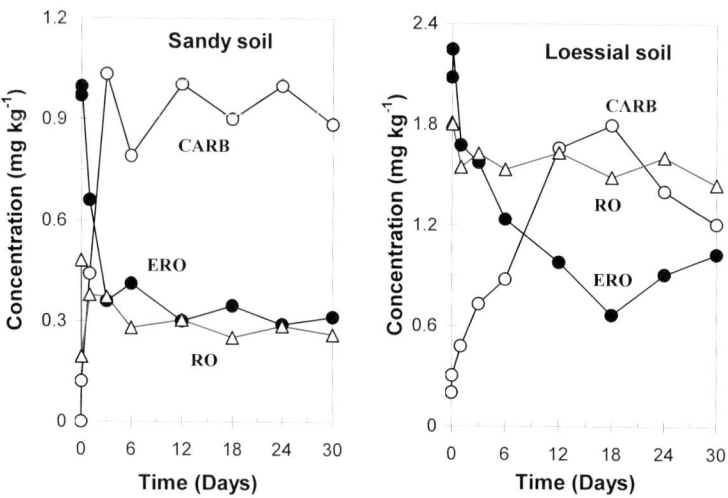

Figure 6.28. Initial changes of Co concentrations in the main solid-phase fractions of two Israeli arid soils. Soils were incubated under the saturated paste regime (after Han et al., 2002b. Reprinted from J Environ Sci Health, Part A, 137, Han F.X., Banin A., Kingery W.L., Li Z.P., Pathways and kinetics of transformation of cobalt among solid-phase components in arid-zone soils, p 188, Copyright (2003), with permission from Taylor & Francis)

It was reported that the HOAc-extractable Co in soils and the Co content in herbage increased in poorly drained soils, and approximate linear relationships between herbage and HOAc-extractable soil-Co were obtained (Berrow et al., 1983). The above finding may be explained by the present observation that Co in soils was transformed from the Mn oxides fraction (ERO) into the carbonate fraction (CARB) upon saturation. This transformation increased the availability of Co, and the HOAc solution mainly extracted Co from the carbonate fraction. Mclaren et al. (1985) reported no relationship between Co content in herbage and acetic acid extractable Co from a wide range of soils with different drainage conditions. They found, however, good correlation between herbage and HOAc-extractable Co when different drainage classes of soils were separately considered.

6.5.2.2 Relationship of Co and Mn Transformation Under the Saturated Regime

During incubation, anaerobic activity predominated as discussed previously, and Mn was reduced. Cobalt transformation closely followed the pathways of Mn transformation in the two soils throughout the whole period of the saturated paste incubation. The transformation of Co by itself may

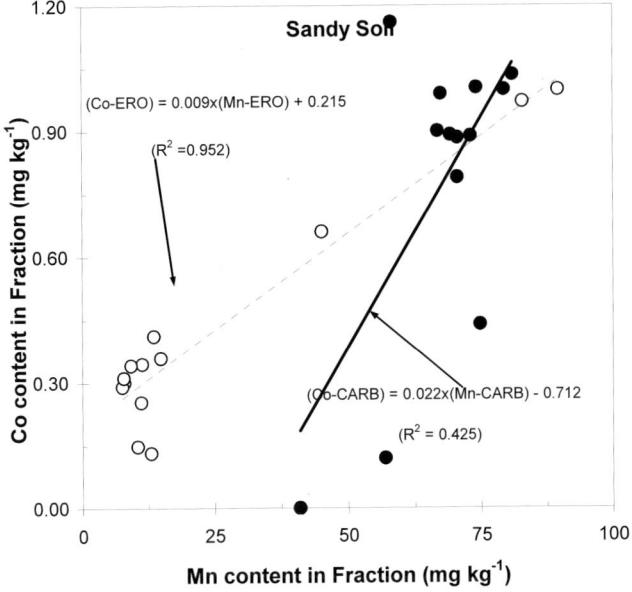
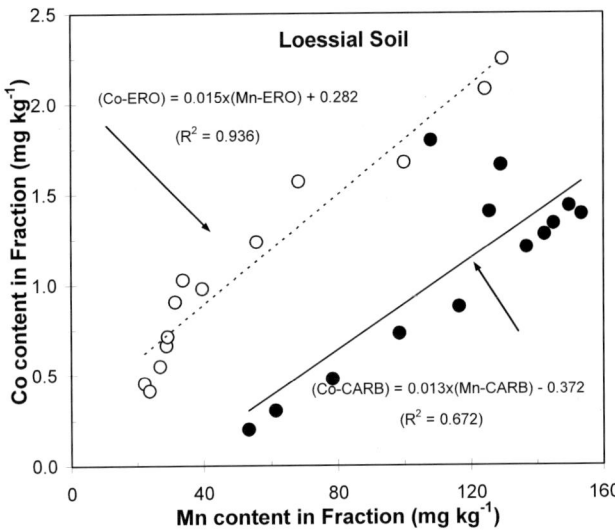

Figure 6.29. Relationships between the Co and Mn contents in the CARB and ERO fractions in Israeli soils during the saturated paste incubation

involve a reduction process. Changes of Co content (in mg kg^{-1}) in the ERO, the CARB and the RO fractions in the two soils were significantly correlated with changes of Mn content in the same fraction during a one year period of incubation (data on the ERO and the CARB fractions shown in Fig. 6.29). This shows that Co chemistry in reduced soils is similar to that of Mn, especially concerning transformation pathways and kinetics.

It was reported that cobalt (Co^{2+}) was specifically adsorbed by Mn and Fe oxides, and the concentration of Co sorbed by Mn oxides was much greater than those by Fe oxides (Backes et al., 1995). Cobalt can replace the Mn ion in Mn oxides. Furthermore, it was suggested by Mckenzie (1975) that Co only replaces the low-valence Mn (Mn^{3+}) ion in Mn oxides. Traina and Donor (1985) suggested that the Mn release during Co^{2+} sorption resulted not only from the oxidation of Co^{2+} to Co^{3+}, but also from a direct exchange of Co^{2+} for Mn^{2+} produced during the redox reaction. Cobalt in the ERO fraction may be, therefore, in part, as Co^{3+} ion, while in the CARB fraction it is present as Co^{2+}.

6.5.2.3 Kinetics of Co Transformation Among its Major Solid-phase Fractions

The initial fluxes of the transformation of Co, as measured by the initial decrease in its content in the main source, the ERO fraction (mg kg^{-1} sec^{-1}), were faster (by a factor of 2.43/0.90 = 2.70) in the sandy soil than in the loessial soil (data not shown here). The rate of increase in the Co content in the carbonate fraction was also faster (3.79/1.04 = 3.64) in the sandy soil than in the loessial soil. The fluxes of Co in and out of the two main fractions in the sandy soil during the first three days of incubation were as follows: CARB: 3.79×10^{-6} (mg kg^{-1} sec^{-1}) and ERO: -2.43×10^{-6} (mg kg^{-1} sec^{-1}). Part of the Co in the CARB fraction was transferred from the RO and the OM fractions. In the loessial soil, Co content in the EXC fraction also increased during incubation. Also in the loessial soil, the average fluxes of Co during the first 18 days of incubation were 1.04×10^{-6}, 0.045×10^{-6}, -0.901×10^{-6}, and -0.137×10^{-6} (mg kg^{-1} sec^{-1}) for the CARB, EXC, ERO and RO fractions, respectively. The overall reaction was controlled by the transfer of Co from the ERO (and RO) fractions.

The two-constant equation and the simple Elovich equation effectively described the change of Co in the main source, the ERO fraction in both soils (Fig. 6.30). This suggests that the diffusion processes controlled the rate of Co transformation. It is noted above that the kinetics of Mn transformation in the three main fractions (EXC, CARB, and ERO), in the

Transfer Fluxes

same incubated soils, was also well depicted by the parabolic diffusion and the simple Elovich equations. This indicates that Co and Mn transformations were controlled by the diffusion processes. However, the present data are too limited to enable detailed analysis of the factors controlling the reaction kinetics.

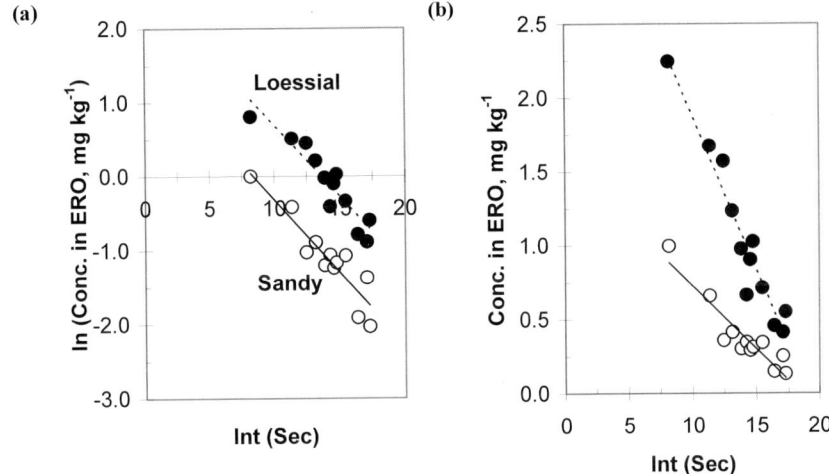

Figure 6.30. Kinetics of transformation of Co in the ERO fraction in two Israeli soils according to the two-constant rate model (a) and the simple Elovich model (b), respectively. Soils were incubated under the saturation paste regime (modified after Han et al., 2002b. Reprinted from J Environ Sci Health, Part A, 137, Han F.X., Banin A., Kingery W.L., Li Z.P., Pathways and kinetics of transformation of cobalt among solid-phase components in arid-zone soils, p 192, Copyright (2003), with permission from Taylor & Francis)

6.6 SUMMARY

The transfer fluxes of Cu, Cr, Ni and Zn increased with the metal loadings, and they were initially high and quickly decreased with time. After 1–2 weeks, the fluxes of these metals continued to decrease with time. The saturated paste moisture regime had higher overall fluxes of metal transfer among solid-phase components of arid-zone soils than the field capacity regime. This implies that the overall metal-redistribution rate would be controlled by both the actual reactions of metals with the solid-phase compounds in soils and the transport processes in the soil solution. Soil management strongly affected the metal redistribution and immobilization processes. The residence time of the metals was similar in soils incubated at both the saturated paste and field capacity regimes. In general, the residence time of the three metals was in the order: Ni > Cu > Zn; and sandy soil > loessial soil. This indicates that the reaction of Zn in soils was faster than Ni, and the loessial soil had faster reaction rates than sandy soil. These findings may be of high significance in managing metal-contaminated soils, and controlling and minimizing heavy metals pollution in agricultural land and in crops irrigated with reclaimed sewage effluents and amended with sewage sludge.

Chapter 7

BIOAVAILABILITY OF TRACE ELEMENTS IN ARID ZONE SOILS

Bioavailability of trace elements refers to the availability of trace elements to biological organisms such as plants, animals and human beings. An available or bioavailable trace element to plants is one that is present in a pool of ions in the soil and can move to the plant roots during plant growth. An available form must be absorbed by the roots. This chapter addresses the uptake, storage, translocation and contents of trace elements in plants, rhizosphere soil chemistry, chemical extractants, bioavailability of trace elements in the soil solution/solid phase and soil factors controlling their bioavailability. In addition, the distribution of bioavailable trace elements in global arid and semi-arid soils, micronutrient deficiency, and toxicity will be discussed.

7.1 PLANT UPTAKE, STORAGE, TRANSLOCATION, AND CONTENTS OF TRACE ELEMENTS

The uptake of trace elements by plants is regulated by active (metabolic) and passive (nonmetabolic) mechanisms at the soil-root interface. The former processes are mainly dependent on ion exchange reactions between the soil matrix, solution phase, and plant roots (Dufey et al., 1999). There exists a very large electrical potential gradient across the root plasma membrane, which drives trace elements/heavy metals into the root cells. The root exchange sites are essentially located in the apoplasm. Uronic acid from pectin and hemicellulose in the cell walls and proteins (mainly carboxylic groups) are responsible for the exchange properties of roots (Dufey et al., 1999). Kochian (1993) found that Zn, Cu and Ni were transported into the root cell via a common transport system - voltage-gated cation channels. In general, essential trace elements, such as Zn and Cu, are taken up into the plant cells by metabolic mechanisms, while non-essential trace elements may be taken up by plants by both metabolic and nonmetabolic mechanisms (Hardiman et al., 1984a). Nickel and Cd can be readily taken up by plants, while Pb tends not to be taken up by most plants.

Most Cd, Pb and Cu absorbed by plants from soils are retained in the roots either precipitated in the cells or chelated with an organic compound, lowering the trace element translocation to the shoots. Furthermore, increases in Cd in the roots of Indian mustard (*Brassica juncea*) have been associated with a rapid accumulation of phytochelatins (Salt et al., 1995). It was reported that an appreciable quantity of Zn could be bound as Zn phytate (*myo*-inositol *kis*-hexaphosphate) within small vacuoles of cells in the root elongation zone of lucerne (*Medicago sativa* cv WL514), soybeans (*Glycine max* cv Riddley), lupins (*Lupinus angustifolius* cv Gungurre), tomatoes (*Lycopersicon esculentum* cv Rouge de Marmand), rape seed (*Brassica napus*), cabbage (*Brassica oleracea* cv Savoy King), radishes (*Raphanuc sativus* cv Mars), maize (*Zea mays* cv SR103), and wheat exposed to high levels of Zn in a hydroponic study (Steveninck et al., 1994). Globular deposits of Zn phytate were found in the endodermis of the dicotyledonous species and in the pericycle of the monocotyledonous species, but also may occur in the stele and inner cortex after long-term exposure to toxic levels of Zn (Steveninck et al., 1994). This deposit could not be found in Zn-treated sunflowers (*Helianthus annuus* cv Hysun), field peas (*Pisum arvense* cv Dunn) and Italian rye grass (*Lolium multiflorum*). In lucerne, soybeans, and maize, Zn-induced phytate globules were present, while a high level of Cd did not induce the formation of Cd-containing phytate globules in these plants (Steveninck et al., 1994). Simultaneous Zn and Cd treatments induced the formation of Zn phytate globules as effectively as Zn alone, and Cd was not detected in the deposits (Steveninck et al., 1994).

Trace elements are translocated from roots to shoots via a number of physiological processes, including metal unloading into root xylem cells, long distance transport within the xylem to the shoots and metal re-absorption from the xylem stream by leaf mesophyll cells (Blaylock and Huang, 2000). After the trace metals are unloaded into the xylem vessels, the metals are transported to the shoots by the transpiration stream. Salt et al. (1995) reported that Cd accumulation in the leaves of Indian mustard was driven mainly by mass flow due to transpiration. This type of transpiration-driven transport also has been observed for other ions, including Cd, B, Si and Cl (Marschner, 1986). Moreover, the root uptake of Cd and the shoot accumulation of Cd appear to be independent processes (Salt et al., 1995). The amount of trace elements translocated from roots to shoots is generally lower than those absorbed by roots. Huang and Cunningham (1996) found that Pb translocated to the shoots was less than 30% of the Pb absorbed by roots for any fixed time period. For long-distance transport, toxic trace elements/ heavy metals may be chelated with organic compounds inside the cell. Experimental data showed increases in Pb translocation from roots into shoots by adding the synthetic chelates EDTA (Blaylock and Huang, 2000). Cd is transported to shoots of rye grass when Cd is taken up from soils with the majority of Cd retained in the roots. However, Cd accumulated in roots is not readily further translocated into the shoots during the subsequent periods even without Cd supply from media (Jarvis et al., 1976).

Hardiman et al. (1984a) suggested that Cu transport to the stele from the roots was metabolically controlled, whereas Cd and Pb reached the stem by leakage across non-suberised areas of the endodermis. The transport of nonessential elements to the tops of plants is often restricted to entering the symplast and being secreted into the stele by metabolic processes in order to bypass the casparian strip in the root endodermis. The accumulation of Pb and Cu in the endodermal and pericycle cells was reported earlier (Hardiman et al., 1984a). Moreover, the transport in stems to leaves may be governed by the interaction of trace elements/heavy metals in the xylem fluid with adsorption sites in the xylem. Salt et al. (1995) found that the majority of the Cd in the roots of Indian mustard was associated with sulfur ligands, probably as a CdS_4 complex (phytochelatins), but Cd in the xylem sap was primarily coordinated with oxygen or nitrogen ligands (Salt et al., 1995). The restricted entry of heavy metals into tops may be due to the precipitation of metal oxalates in the intercellular spaces (Hardiman et al., 1984a). Cd, Pb and Cu are bound to cell walls in the roots and restricted to apoplastic transport by the Casparian strip and little passage via the symplast.

Nickel is localized predominantly in the epidermal and subepidermal of the leaves. However, in leaves of some hyperaccumulator plants such as *T. caerulescens* and *T. goesingense*, Ni and Zn are found mainly in vacuoles (Salt and Kramer, 2000). Trace elements also are inactivated in the vacuoles as high-affinity low-molecular-weight metal chelators (such as Cd-phytochelatin complex), providing plants with trace element tolerance. Some Ni in leaves is found to be associated with cell wall pectates as well.

Plants exhibit different uptakes of trace elements from soils. In general, lettuce, spinach, and radishes have higher Cd, Pb, Zn, Cu and Ni uptake from soils than other plants. Concentrations of selected trace elements in arid and semi-arid soils are presented in Table 7.1. Lettuce takes up more trace elements than do oats from arid soils. The Cd concentration decreases for various vegetable crops: lettuce (*Lactuca sativa* L.) > radish tops (*Raphanus sativus* L.) > celery stalks (*Apium graveolens* L.) > celery leaves >> green peppers (*Capsicum frutescens* L.) > radish roots. The concentrations of Cd in oats grown in an uncontaminated arid soil from Colorado ranged from 0.22–0.30 mg/kg, but was from 1.6–1.8 mg/kg in lettuce (Simeoni et al., 1984). However, both oats and lettuce had similar concentration ranges of Cu and Zn. The Cu concentration in oats and lettuce was 3.8–5.4 and 5.1–6.1, respectively; while the Zn concentration was 14.1–22.8 and 12.5–19.5 mg/kg, respectively (Simeoni et al., 1984). Fenn and Assadian (199) reported bermuda grass accumulated some heavy metals (Pb, Zn, Cu, Cd) from alkaline calcareous Texas soils. On the other hand, fruits and seeds usually accumulate less Cd than do vegetable parts. Higher trace element concentrations are very often found in plant roots than in shoots. Further, the uptake of trace elements by the root systems is more efficient than that of the foliar placement. Haghiri (1973) reported that the uptake of Cd by soybeans via the root system was more efficient than that via the foliage system.

Table 7.1. Concentrations of selected trace elements in plants grown on noncontaminated arid and semi-arid soils[a]

Element	Soil pH	Plants	Shoots/Leaf mg/kg	Seed/bean mg/kg
Cd	7.5	Field bean	0.6	0.05
	7.5	Soybean	0.4	0.6
	7.5	Corn	3.9	0.05
	7.5	Wheat	0.1	0.1
	7.5	Rice	0.1	0.1
	7.5	Lettuce	1.6	
	7.8	Blue grama	0.5	
	7.8	Galleta	0.5	
	7.8	Bottlebush squirreltail	0.5	
Zn	7.5	Field bean	47	23
	7.5	Soybean	57	61
	7.5	Corn	38	32
	7.5	Wheat	44-63	45-73
	7.5	Rice	15	39
	7.5	Lettuce	82	
	7.5	Yarrow	30	
	7.8	Blue grama	24	
	7.8	Galleta	27.5	
	7.8	Bottlebush squirreltail	17.5	
Ni	7.5	Wheat	3.4	<1.0
	7.5	Lettuce	4.5	
Cu	7.5	Wheat	11.5	7.2
	7.5	Lettuce	6.2	
	7.8	Blue grama	4	
	7.8	Galleta	5.5	
	7.8	Bottlebush squirreltail	5	
Pb	7.8	Blue grama	1	
	7.8	Galleta	1	
	7.8	Bottlebush squirreltail	1	

[a]Data from Bingham et al., 1975; Mitchell et al., 1978; Fresquez et al., 1990

Concentrations of Cd, Pb and Cu in the roots of bush beans grown in Israeli arid soils increased in response to the soil concentrations, whereas in stems only, Cd and Pb concentrations increased and Cu concentrations were relatively constant (Hardiman et al., 1984a). In Israeli arid soils with long-term irrigation with reclaimed sewage effluents, the soil solution contained about five times the Pb concentration that was taken up by the bush beans via the transpiration stream, while the amount of Cd in the plant was up to 10 times greater than that apparently available in the soil solution; furthermore, the concentration of Cu in the soil solution also was inadequate to supply the amount actually taken up by the plant (Hardiman et al., 1984b). This indicates that the actual concentrations of Cd and Cu in the Israeli soil solution cannot account for the amounts taken up by the plants during the growing season. This implies that the rate of absorption of Cd and Cu by the roots is in excess of the apparent amount in the transpiration stream. A continual depletion of Cd and Cu has to take place in the soil solution to be replenished by trace elements adsorbed on the solid phase. Therefore, a concentration gradient develops around the roots and in rhizosphere soils. Moreover, the availability of Cd and Cu to the roots would depend upon the binding strength of Cd and Cu to the soil surface. The binding strength of Cd and Cu in these arid soils was found to be controlled by clay contents in highly clayey soils and by calcite contents in highly calcareous soils (Hardiman et al., 1984b). However, uptake of Pb by plant roots is governed by mechanisms not related to adsorption onto the soil surface. Because Pb concentrations in plant roots are poorly correlated with the Pb concentration in the soil solution, and the Pb concentration in the soil solution is higher than the Pb plant uptake, only a fraction of the soluble Pb is available to the plant. This also may be related to the strong complexation of Pb with macromolecules in the soil solution.

However, inputs of sewage sludge or mining activity increase trace element concentrations in plants. For wheat seedlings grown on a calcareous soil from Israel with pH 8.2 and 62% $CaCO_3$, Cd, Ni and Zn in shoots increased with the loading levels as well, ranging from 0.1–1.1, 0.5–2.1 and 20–188 mg/kg, respectively, compared to the ranges of Cd, Ni and Zn in the roots of wheat, which was 0.2–3.8, 2.0–9.3 and 71–325 mg/kg, respectively. The Cd, Cu and Zn concentrations in lettuce grown in sludge-amended soil increased in the ranges of 2.1–3.1, 4.2–12 and 39–75 mg/kg, respectively, and in oats in the ranges of 0.24–0.76, 5.7–9.4 and 44–95 mg/kg, respectively. The concentrations of Zn, Cu, Ni and Cd in plants grown in California Domino soils (pH 7.5) amended with metal sulfate-enriched sludge increased linearly with the metal levels in soils (Fig. 7.1) (Mitchell et al., 1978). Lettuce leaves contained much higher Zn, Ni and Cd than wheat, but Cu in wheat leaves was higher than those in lettuce and wheat grain.

Mahler et al. (1980) reported that Cd concentrations in the shoots of corn (*Zea mays* L.), tomatoes (*Lycopersicon esculentum* Mill.), and Swiss chard (*Beta vulgaris* var.) ranged from 0.11–251 mg/kg, and linearly increased with total Cd inputs as sewage sludge in arid California soils, but negatively correlated with soil pH. Gerard et al. (1999) reported that *Thlaspi caerulescence* took up 8.7–647 mg/kg Cd in the shoots, followed by lettuce (0.4 –8.3 mg/kg), and rye grass (0.1–2.3 mg/kg) from a northern France calcareous soil with pH 8 and contaminated with zinc smelter activities. Plants (such as lettuce, spinach, and turnip greens) grown on sewage sludge-amended Californian soil with pH 7.5 and 1% $CaCO_3$ accumulated 175–354 mg/kg Cd, whereas fruit and seed tissues of field beans, corn, and wheat accumulated no more than 10–15 mg/kg Cd and those of soybeans had up to 30 mg/kg Cd (Bingham et al., 1975, 1976). Rice grains contained 2 mg/kg Cd. Grain Cu concentrations in wheat grown in arid soils amended with 5–6 annual applications of biosolids ranged from 3.2–8.9 mg/kg, indicating that Cu concentrations in plants are not significantly affected by biosolids (Barbarick et al., 1995). Grain Zn concentrations in wheat ranged from 12–60 mg/kg, with the plateau concentration around 52.5 mg/kg, while Ni concentrations in wheat grain ranged from 0.01– 2.4 mg/kg. Both Zn and Ni concentrations in wheat grain linearly increased with cumulative Zn and Ni in arid soils with biosolids.

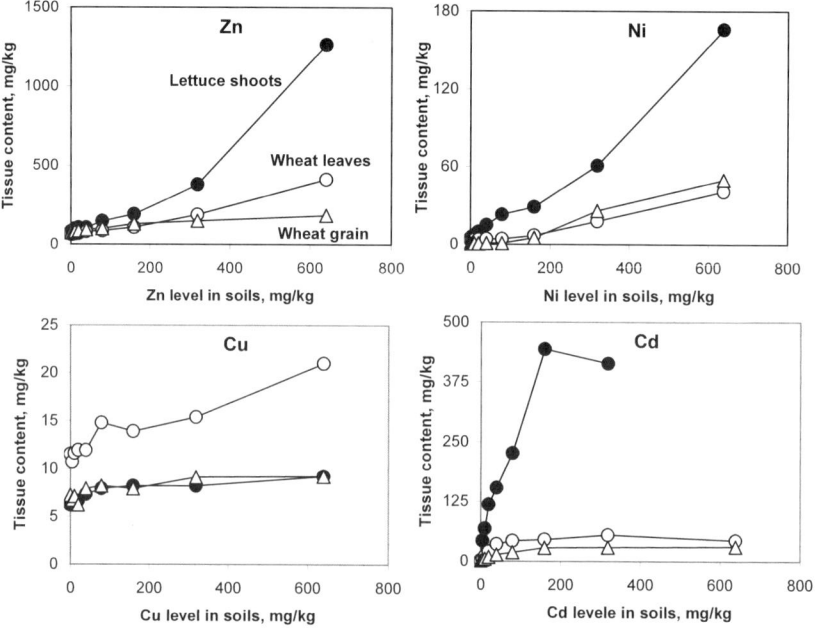

Figure 7.1. Trace element concentrations in plants on California Donimo soil (pH 7.5) amended with metal sulfate-enriched sludge (Data from Mitchell et al., 1978)

However, some plants can accumulate more than 0.1% of Pb, Co, Cr, and more than 1% of Mn, Ni and Zn in the shoots. These accumulator plants are called hyperaccumulators. To date, there are approximately 400 known metal hyperaccumulator plants in the world (Baker and Walker, 1989). *Thlaspi caerulescens*, *Alyssum murale*, *A. lesbiacum*, *A. tenium* are Zn and Cd hyperaccumulators. *Brassica juncea*, a high-biomass plant, can accumulate Pb, Cr(III), Cd, Cu, Ni, Zn, Sr, B and Se. *Thlaspi caerulescens* accumulates Ni. Hybrid poplar trees are reported to phytoremediate Cd and As contaminated soils. A Chinese brake fern, *Pteris vittata*, is an As hyperaccumulator (Ma et al., 2001).

7.2 RHIZOSPHERE CHEMISTRY

The term rhizosphere originates from Greek - *rhiza*, meaning "root" and *shaira* meaning "world" or "globe" - thus, "root world". It was first used by Hiltner to indicate the zone of soil where the root exudates released from plant roots can stimulate, inhibit or have no effects on the activities of soil microorganisms (Pinton et al., 2001). The rhizopshere is generally considered to be a narrow zone of soil subject to the influence of living roots, where the root exudates stimulate or inhibit microbial populations and their activities (Brimecombe et al., 2001). It is the interface between the plant roots and the soil environment surrounding the roots. The rhizosphere controls the plant uptake of nutrients and toxic matters, plant growth, and microbial activity. Rhizosphere soil differs from the bulk soil in physical, chemical, and biological aspects (Dinkelaker et al., 1993; Lombi et al., 1999), which might affect trace element distribution and the overall bioavailability to plants.

In the rhizosphere soil, water and nutrients dynamically move and are concentrated around the roots. The water potential regime in the rhizosphere soil is generally lower than in the bulk soil. The redox potential is more negative in the rhizosphere soil than in the bulk soil, due in part to the higher oxygen consumption of respiration of the roots and microbes (Badalucco and Kuilman, 2001). Soil mineral weathering rates are generally higher in the rhizosphere soils as well. Fenn and Assadian (1999) reported less $CaCO_3$ content in plant rhizospheres than in bulk calcareous soils of Texas, resulting in a significant reduction in soil pH.

Plant roots and root-induced chemical changes in the rhizosphere strongly affect the bioavailability of trace elements (Hinsinger, 1999). First, root-induced changes in the ionic equilibria influence the bioavailability of trace elements. The differential rates of plant uptake of water and ions in the soil solution result in a depletion or an accumulation of the ions in the rhizosphere. In general, trace elements present at low concentrations in the soil solution are most likely to be depleted as a consequence of plant uptake. Secondly, root-induced changes in the pH and redox potential in the rhizosphere affect the dynamics of trace elements, either through changes of speciation or through dissolution of the trace element bearing mineral phases. The root-induced changes in pH is primarily a consequence of both the excretion of protons and the differential rates of uptake of cations and anions by plants, which are compensated for by a release of protons or hydroxyl/bicarbonate ions in the rhizosphere. Changes in redox potentials are especially important for trace elements such as Cr, Mn, Fe or Se. Plant roots are likely to alter the speciation of these elements. Many carbonates and oxides of trace elements can be soluble under acidic or reducing conditions occurring in the rhizosphere. Finally, root exudates as well as microbe-induced organic compounds (phytosiderophores and organic acid) containing complexing substances tend to increase the solubility and thus the bioavailability of trace elements in soils. These released organic molecules are prime sources of energy for rhizosphere microorganisms. Some exudates containing carboxylic anions such as citrate, oxalate or malate and phytosiderophores exhibit strong complexation properties with respect to a whole range of trace elements (Fe, Cd, Cu and Zn). Awad and Romheld (2000a) and Treeby et al. (1989) reported that higher uptakes of Fe, Zn, Ni and Cd from highly calcareous soils were found in a wheat species with a higher release of phytosiderophores in its rhizosphere. In addition, reductive enzymes may also play a role in the solubilization of As, Cu and Mn (Lombi et al., 1999). The main sources of enzymes in the soil are plant roots, animals and microorganisms. Enzymes from microorganisms are the most important in rhizosphere soils.

The bioavailability of Fe, Zn, Ni and Cd increases in the rhizosphere through the release of strong organic chelators by plant roots and microbes (Marschner and Romheld, 1996; Awad and Römheld, 2000a). Motaium and Badawy (1999) reported that in Egyptian arid soils, long-term irrigation with sewage water for 10, 40, and 80 years significantly increased Diethylenetriaminepentaacetic acid (DTPA)-extractable Fe, Mn, Cu, Zn, Cd, Co and Pb in the rhizosphere soils of orange trees and cabbage plants (Fig. 7.2). The DTPA-extractable trace elements in the rhizosphere of both of the plant systems were much higher than bulk soils (Average ratios of the DTPA-extractable trace elements in the rhizosphere over bulk soils were 1.44.) Fenn and Assadian (1999) reported that in alkaline-calcareous Texas soils, more Pb, Cu and Mn were found in the soluble forms in the rhizosphere soil in comparison to the bulk soil. It also has been observed in non-calcareous soils that lowering the pH and increasing the bacteria in the rhizosphere led to increased concentrations of soluble Zn relative to bulk soils (McGrath, 1997; Whiting et al., 2001).

Bioavailability of Trace Elements

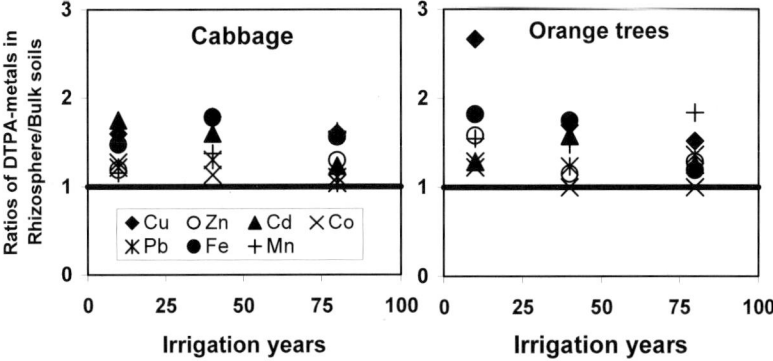

Figure 7.2. Ratios of DPTA-extractable trace elements in the rhizosphere over bulk Egyptian soils irrigated with sewage water for 10, 40, and 80 years (Data from El-Motaium and Badawy, 1999).

The distribution of trace elements among solid-phase components is different between rhizosphere soil and bulk soil. In forest soils, exchangeable Zn and oxalate-extractable Al and Fe are much higher in the rhizosphere than in bulk soils (Courchesne et al., 1999; Courchesne and Gobran, 1997). Shuman and Wang (1997) reported that in rhizosphere soil, Cd increased in both exchangeable and organic matter fractions, and Fe increased in the easily reducible oxide fraction. Fenn and Assadian (1999) reported that in alkaline-calcareous Texas soils, less Pb and Cu were present in insoluble manganese oxide and amorphous iron oxide bound compounds, while more Mn was present in carbonate and easily reduced manganese oxides bound fractions in rhizosphere soils than in bulk arid soils.

7.3 EXTRACTANTS FOR BIOAVAILABLE TRACE ELEMENTS IN ARID SOILS

Bioavailable trace elements in soil correlate with plant uptake and concentrations in plants. Extractants for bioavailable trace elements include chelating agents, diluted inorganic acid, neutral salt solutions, and water (Table 7.2). The most popular extractant for bioavailable trace elements in arid and semi-arid soils is DTPA-TEA (triethanolamine), which was developed by Lindsay and Norvell (1969, 1978) to extract available Cu, Zn, Fe and Mn from neutral and calcareous soils. Use of this chelating agent, DTPA, is based on the fact that it has the most favorable combination of stability constants for simultaneous complexation of Cu, Fe, Mn and Zn

(Lindsay and Norvell, 1978). DTPA-Zn/Cu complexes are water soluble. DTPA also has less dissolution of carbonates. Excessive dissolution of $CaCO_3$ is prevented by the presence of soluble Ca^{2+} as $CaCl_2$ added in the extractant by buffering the solution at pH 7.3 with TEA (Reed and Martens, 1996). This is especially important in arid and semi-arid soils, because trace elements occluded by carbonate are less bioavailable than the soluble and exchangeable forms. Later, DPTA-NH_4HCO_3 (DPTA-AB) was proposed to extract trace elements from arid and semi-arid soils (Soltanpour and Schwab, 1977). The presence of soluble HCO_3^- as NH_4HCO_3 in the extractant solution with pH 7.6 also prevents excessive dissolution of $CaCO_3$ from arid and semi-arid soils, which is similar to the role of Ca^{2+} as $CaCl_2$ in the DPTA-TEA extractant. In addition, since the NH_4^+ component extracts exchangeable cations and the HCO_3^- removes plant available P, DPTA-AB can be used to simultaneously analyze a number of trace elements in a single extract employing ICP-OES.

Table 7.2. Extractants for bioavailable trace elements in arid and semi-arid soils

Extractants	pH	Trace elements
0.005M DTPA+0.1 M TEA+ 0.01M $CaCl_2$	7.3	Cu, Zn, Fe, Mn, Ni, Cd
0.005M DTPA + 1M NH_4NO_3	7.6	Cd, Cu, Ni, Pb, Zn, Fe, Mn, (Se)
0.02 M EDTA		Cu
0.05/0.01 M $CaCl_2$		Cu, Zn, Pb, Ni, Cd
1M NH_4NO_3	6.4	Fe, Zn, Ni, Cd
5 mM Desferal	7.3	Fe, Zn, Ni, Cd
5 mM Phytosiderophores from wheat	7.3	Fe, Zn, Ni, Cd
Oxalate buffer	3.25	Ni, Pb, Mo
Saturation paste (H_2O)		Cd, Cu, Ni, Pb, Zn, Cr
Hot water (H_2O)		B
1mM KNO_3 + 0.5 mM K_2SO_4	6.0	Cr(VI)
1mM HNO_3 + 0.5m MH_2SO_4	2.7	Cr(VI)

Barbarick and Workman (1987) compared the extractability of trace elements from sludge-amended calcareous soils by using DTPA-TEA and DTPA-AB. In general, DTPA-TEA extracts less Cd, Cu, Ni, Pb and Zn than DTPA-AB from sludge-amended soils. The researchers found that extractable Cd, Cu, Ni, Zn and Pb by both extractants were strongly correlated. DTPA-AB-extractable Cd, Cu, Ni and Zn in calcareous soils also significantly correlated with their concentrations in Swiss chard (*Beta vulgaris* L.), with Cd in wheat straw, and with Ni, Pb and Zn in wheat grain. Bingham et al. (1976) reported that DTPA-extractable Cd and saturation extracted Cd from Californian soils had a close correlation with plant Cd uptake and crop yield. DTPA-extractable Zn correlated significantly with the grain Zn concentration. Mitchell et al. (1978) reported that Zn uptake was more closely correlated with DTPA-soluble Zn while water-soluble concentrations of Cd, Cu and Ni better predicted plant uptake (lettuce and wheat) of these metals. Street et al. (1977) reported that DTPA-extractable Cd levels in soils with 0–6% $CaCO_3$ were highly correlated (r = 0.96) with Cd concentrations in corn seedlings grown in soils amended with Cd solutions, sewage sludge or Cd-spiked sewage sludge.

In an arid soil with a pH of 7.3 from Colorado, DTPA-AB extractable Zn, Cu, Pb, Cd and Ni concentrations were 1, 2.9, 2.6, 0.64 and 0.83 mg/kg, respectively (Barbarick et al., 1997). Sewage sludge applications increased DTPA-Zn in arid soils. In arid Colorado soils (0–20 cm) amended with sewage biosolids (5–6 applications over 11 years), the ranges of the concentrations for DTPA-AB extractable Cd, Cr, Cu, Ni, Pb and Zn ranged from 0.09–0.27, 0.12–0.21, 2.0–13.9, 0.64–2.83, 1.5–3.5 and 0.5–13 mg/kg, respectively (Barbarick et al., 1998). DTPA-AB extractable trace elements decreased with soil depth. Because concentrations of biosolid Zn were much greater than other elements, a biosolid application was found to have a greater effect on extractable Zn than on Cd, Cu, Ni and Pb. Concentrations of DTPA-AB-extractable Zn linearly increased with total Zn added as sewage sludge. Further, grain Zn concentration significantly correlated with DTPA-extractable Zn in soil (Barbarick et al., 1997). However, with a single application, DTPA-extractable Cd, Cu, Ni, Pb and Zn concentrations decreased with time from sewage-amended soils (Barbarick et al., 1997). In Colorado garden soils contaminated with silver mine sump, DTPA-AB-extractable Cd, Pb and Zn concentrations ranged from 0.2–14.2, 9.2–808 and 8.4–484 mg/kg, respectively (Boon and Soltanpour, 1992).

Bioavailable trace elements in arid and semi-arid soils are also extracted by EDTA. Chelate complexing agents (EDTA and DTPA) extract higher amounts of trace elements than neutral salts from arid soils. In a calcareous vineyard soil from southern France, Cu extracted by EDTA, DTPA, and $0.05M CaCl_2$ decreased in the order: EDTA > DTPA > $CaCl_2$. Extractable Cu followed the same trend in the sewage sludge-amended vineyard soil (Cherrey et al., 1999). Awad and Romheld (2000b) compared extractable Fe, Zn, Ni and Cd from highly calcareous soils using ammonium nitrate (pH 6.4), DTPA (pH 7.3), DTPA-AB (pH 7.6), desferal (pH 7.3), and phytosiderophores from wheat (pH 7.3). They found that phytosiderophores and desferal were the most efficient and ammonium nitrate the least efficient extractant for these trace elements from highly calcareous soils. Trace elements from contaminated calcareous soils were extracted in the following decreasing order: Phytosiderophoreses > desferal > DTPA-AB > DTPA > NH_4NO_3. Singh et al. (1996) reported that the DTPA- and NH_4NO_3-extractable Cd strongly correlated with Cd concentrations in plants. They pointed out that both extractants were equally effective in relation to the Cd concentration in plants. DTPA-extractable Mn had a much higher correlation between plant Mn and soil Mn than did the EDTA-extractable Mn investigated in global studies involving both arid and semi-arid zone soils and humid and temperate soils (Sillanpaa, 1982).

In a calcareous vineyard soil from southern France, Cu extracted by EDTA, DPTA, and 0.05MCaCl$_2$ accounted for 46%, 35% and 0.4% of the total Cu in soil. The application of sewage sludge increased bioavailable Cu (EDTA-, DTPA- and CaCl$_2$-extractable) (Cherrey et al., 1999). In Chinese calcareous upland and paddy soils, DTPA-extractable Zn ranged from a trace to 3.0 mg/kg with an average of 0.37 mg/kg and from 0.27–1.62 mg/kg with an average of 0.35 mg/kg, respectively (Liu, 1996; Han et al., 1990a). DTPA-extractable Zn and Cu in soils from Syria ranged from 0.40–0.90 and 0.22–2.0 mg/kg, respectively, depending on the soil pH and the contents of carbonates (Ryan et al., 1996).

In addition, oxalate buffer has been reported to be an effective extractant for bioavailable Ni and Pb from arid soils. Oxalate buffer extractable Ni from 32 arid soils of India with a pH of 7.1–9.1 and 1.05–9.20% CaCO$_3$ was strongly correlated with Ni uptake by sorghum (Misra and Pandey, 1974). The extractants with plant uptake decreased in the order of: Oxalate buffer ($r = 0.945$) > 0.1 N HCl (0.742) > 0.5 N CH$_3$COOH (0.705) > 1 N NH$_4$OAc (pH 7.0) (0.945) > 0.1 N HNO$_3$ (0.410) > 0.1 N H$_2$SO$_4$ (0.404). EDTA (0.02 M), K$_4$P$_2$P$_7$ (0.1 M) and citric acid (1%) did not show any significant correlation with plant uptake (Misra and Pandey, 1974). Of four extractants, the oxalate buffer extracted the maximum amount of Pb from 16 Indian alluvial arid soils with pH 6.9–7.2 and CaCO$_3$ 0.1–2.1% and significantly correlated with Pb uptake by wheat plants (Misra and Pandey, 1976). The average extractable Pb from these soils were in the order of: oxalate > 0.1 N HCl > 0.02 M EDTA > 1 N NH$_4$OAc. However, diluted inorganic acid was not suitable for extraction of bioavailable trace elements from arid soils since acids dissolve much of the carbonates in arid soils.

Saturation extraction also has been used to extract available trace elements. Saturation extracted Se was correlated with Se concentration in both four wing salt bush (*Atriplex canescens* (Pursh) Nutt.) and two grooved milk vetch (*Astragalus bisucatus* L.) grown on arid soils from Wyoming with pH 7.4–8.4 as compared to other extracts (DTPA, Na$_2$CO$_3$, and hot water) (Jump and Sabey, 1989). Sodium carbonate extracted the largest fraction of the total soil Se among these extractants. However, DTPA-extractable Se showed the highest correlation with Se concentrations in alfalfa and in wheat grain grown in North Dakota soils (Soltanpour and Workman, 1980; Soltanpour et al., 1982). In alkaline soils enriched with chromite ore processing residue, water, dilute salt solution (1 mM KNO$_3$ + 0.5 mM K$_2$SO$_4$

pH 6), dilute acid (1 mM HNO_3 + 0.5mM H_2SO_4 pH 2.7) and strong acid equally extracted soluble Cr(VI) (James, 1994).

7.4 BIOAVAILABILITY OF TRACE ELEMENTS IN ARID SOILS

Among trace elements of most concern, in general, Cd has the highest bioavailability in arid and semi-arid soils, while Pb and Cr(III) have the lowest bioavailability in arid soils (Table 7.3). Zinc, Cu, and Ni are intermediate. The bioavailability of trace elements in soils depends upon their distribution between the solution and solid-phase and further distribution among solid-phase components. Cadmium is mainly present in the soluble and exchangeable forms, posing higher bioavailability to plants, while Cr and Pb are mostly present in the stable residuals and Fe/Mn oxide bound forms, leading to low bioavailability. The distribution of trace elements in the soil solution and among solid phase components has been discussed in Chapters 3 and 5, respectively.

Table 7.3. Bioavailability of trace elements in arid and semi-arid soils

Category	Order of Bioavailability in Arid/Semi-arid Soils
Trace Elements	Cd > Zn, Cu, Ni > Pb, Cr(III)
Solution/Solid Phase	Solution phase (water soluble, saturation extracts) > Solid phase
Solution Phase	Free trace element ion > Soluble chelated complexes (inorganic / small organic molecular complexes) > Strongly complexed species (insoluble macroorganic molecular complexes)
Solid Phase	Exchangeable > Organically bound > Carbonate bound, Fe/Mn oxide bound > Clay mineral bound, humin bound, and residual fractions

Trace elements in the soil solution are the most bioavailable soil form to plants (Table 7.3). Page et al. (1972) found that the threshold Cd concentration in soil solution for yield reduction of vegetable crops ranged from 0.4–9 mmol/L, depending on the plant species. One great concern is the introduction of Cd in plants into the food chain. Bingham et al. (1976) reported that saturation extracted Cd from Californian soils had a close correlation with plant Cd uptake and crop yield. Mahler et al. (1978) found that saturation extracted Cd was significantly correlated with leaf Cd concentrations in lettuce and Swiss chard grown in sewage sludge-amended Californian soil with 1.2–28.3% $CaCO_3$. Mitchell et al. (1978) observed that water-soluble Cd, Cu and Ni better predicted their plant uptake. Hardiman et al. (1984a, b) reported that Cd and Cu concentrations in roots of bush beans were linearly correlated with their solution concentrations in both non-contaminated Israeli soils and soils with long-term irrigation with reclaimed sewage effluents.

Among the various species of trace elements in soil solution, free ions are the most bioavailable form, followed by chelated soluble species with inorganic and organic ligands, while the strongly complexed trace elements are less bioavailable and toxic (Table 7.3). Free Cd^{2+} is the most bioavailable soil form to plants, and chloride-complexes increase plant uptakes from arid soils. Bingham et al. (1983) reported that leaf Cd concentrations of Swiss chard correlated better with the estimated activity of Cd^{2+} than with the concentration of total Cd in soil solution. Increases in the concentrations of Cd chloride-complexes ($CdCl^+$ or $CdCl_2^0$) in an arid soil solution due to high salinity increased plant uptake of Cd (Weggler-Beaton et al., 2000). Complexation of Cd^{2+} by Cl^- increased soluble Cd in culture and soil solutions, but the calculated activity of Cd^{2+} was not significantly affected by increasing concentrations of Cl^- in solution (Smolders and McLaughlin, 1996, Bingham et al., 1983). As solution Cl^- concentration increased, Cd concentrations in plant shoots increased from 6.5–17.3 mg kg^{-1} and in roots from 47–106 mg kg^{-1}. These complexes also enhanced transport of Cd across the zone encompassing the soil-rhizosphere-apoplast-plasma membrane (Smolders and McLaughlin, 1996). Smolders and McLaughlin (1996) concluded that the enhancement of Cd uptake by Cl in soils was not only related to enhanced diffusion of Cd^{2+} through the soil to roots, but also that: (i) the $CdCl_2$n-n (in addition to Cd^{2+}) species in solution were phytoavailable and/or (ii) Cl enhanced the diffusion of Cd^{2+} through the unstirred liquid layer adjacent to the root surface or through the apoplast to sites of Cd uptake within the root itself. Also, other inorganic complexes became available to plants as well. McLaughlin et al. (1998a) reported that the $CdSO_4^0$ complex was taken up by plants with as equal efficiency as free Cd^{2+} ions. This indicates that $CdSO_4^0$ has similar bioavailability as free Cd^{2+}. Plant growth (Swiss chard, *Beta vulgaris* L. cv. *Fordhook Giant*) was unaffected by increasing SO_4 concentrations in solution. Despite considerable reduction of free Cd^{2+} ion activities in solution by increasing SO_4 concentrations, plant Cd concentrations were unaffected (McLaughlin et al., 1998a). In soils from southern Australia, shoot Cd content in 19-day-old Swiss chard plants marginally but significantly ($p < 0.05$) increased with increasing SO_4^{2+} concentration, but no effect was observed with increasing NO_3^- concentration. These results are comparable with earlier works on the marked effects of Cl^- salinity on Cd availability in Swiss chard. Compared to Cl, the effect of SO_4^{2-} on Cd availability was much smaller.

Because trace element concentrations in the soil solution of arid soils are very low, the exchangeable trace elements on various solid-phase components become important in supplying plants. Among various solid-phase components, the exchangeable trace elements are the most bioavailable to plants, followed by the organically bound fraction, but trace elements bound to the carbonate fraction are potentially bioavailable. As we discussed above, the rhizosphere metabolic products decrease soil pH in the rhizosphere, which increases the bioavailability (plant uptake) of metal carbonates. The bioavailability of Cd and Zn in arid and semi-arid soils can be divided into four categories: (i) Cd and Zn in the exchangeable fraction are rapidly available; (ii) Cd and Zn in the carbonate bound fraction and Cu in the organically bound fraction are easily available, while (iii) Zn in the easily reducible oxide bound and the organically bound fractions is potentially available; and finally, (iv) in contrast, metals bound in the reducible oxides bound and residual fractions are not available (Banin et al., 1990; LeClaire et al., 1984). LeClaire et al. (1984) pointed out that in arid California field soils that received 3–6 years of applications of composted or liquid sewage sludge, exchangeable Zn (as extracted by KNO_3) was the highly labile and soluble pool, dominated by Zn^{2+} and immediately bioavailable. Significantly positive correlations were found between Zn concentrations in barley leaf and Zn extracted by both NaOH and EDTA. Zinc extracted by KNO_3 and NaOH (targeting the organically bound fraction) also had a strong correlation with DPTA-extractable Zn. Zinc in both the exchangeable and organically bound fractions was plant available, while EDTA extractable Zn (mostly as carbonate bound Zn) represented a reservoir of potentially bioavailable Zn. Banin et al. (1990) reported that in sewage sludge-amended Israeli arid soils, heavy metal concentrations in lettuce were best correlated with the concentrations of metal bound to carbonates for Cd and Zn, and to organic matter for Cu. Exchangeable, easily reducible oxide-bound, and organic matter-bound Zn fractions were also significantly correlated to their concentrations in lettuce. This implies that both organically bound and carbonate bound trace elements are bioavailable to plants in arid and semi-arid soils receiving sludge.

In calcareous wetland and paddy soils, in addition to exchangeable and organically bound fractions, the Fe/Mn/Al oxides bound trace elements are also to some extent bioavailable. Murthy (1982) found that the exchangeable and soluble organically complexed Zn (extracted by $Cu(OAc)_2$) were more available to rice in wetland soils and that amorphous iron and aluminum oxides bound Zn (extracted by oxalate reagents) also contributed to bioavailability. The organically complexed Zn played an important role in Zn nutrition of rice in paddy soils (Han and Zhu, 1992). The chelated fraction, organically bound Zn, and oxides occluded fractions were major sources of available Zn to rice in calcareous soils amended with organic material (Prasad et al, 1990). In a paddy soil with pH 10.45 and 3.73% $CaCO_3$ in India, the exchangeable and amorphous sequioxide bound Zn fractions contributed significantly more to Zn uptake by rice than other fractions (Singh and Abrol, 1986). DTPA extracted Zn more readily from the exchangeable and complexed fractions than from the sequioixdes (Singh and Abrol, 1986). In calcareous paddy soils of China, DTPA extractable Zn had significant correlations with the exchangeable, carbonate bound, and chelated organic bound fractions in soils (Han and Zhu, 1992). In arid mined soils from Colorado contaminated with mine tailings, Cd, Zn and Pb concentrations in three plant species (Sedge, *slough grass*, *Carex* spp.; ruch, wire grass, *Juncus* spp.; blue grass, *Poa* ssp.) significantly correlated with exchangeable Cd, soluble and exchangeable Zn, and exchangeable, organic bound, and oxide bound Pb in soils, respectively.

Radioactive tracers have been used to study the bioavailability of trace elements in arid soils. In uncontaminated arid soils with 1.3–6.2% $CaCO_3$ and soils contaminated by smelter emissions and sewage sludge, the total labile pool (%E) of Cd after a 2-week equilibrium was 35–49% of the total soil Cd. About 85–98% of the total labile pool was contributed by NaOAc- and NaOCl-extractable Cd, representing the sorbed/carbonate fraction and the oxidizable (organically bound) fraction, respectively (Ahnstrom and Parker, 2001). In the NaOAc-extractable fraction (sorbed/carbonate bound fraction), 70–75% of the total Cd was isotopically labile, while in the oxidizable (organically bound) fraction, only 35–41% of the Cd was labile within 2 weeks. The oxidizable fraction was mostly organically bound Cd. The oxalate- and HNO_3-extratable fractions, however, were dominated by nonlabile Cd. This indicates that besides soluble and exchangeable Cd, Cd in the sorbed/carbonate bound and organically bound Cd are more labile and bioavailable than Cd in the oxides bound and residual fractions. However, in a Pb-Zn mine contaminated soil, the overall labile pool of Cd was only 13%, 82% of which is present in the oxidizable fraction with 2% being isotopically labile. The nonlability of Cd in the oxidizable fraction was due to the presence of refractory ZnS mineral sphalerite. This implies that not all Cd in the oxidizable fraction is labile and bioavailable.

The bioavailability of trace elements in soil-water-plant ecosystems of arid zones is largely determined by their partitioning between solution and solid-phase components. The redistribution and transformation of trace elements among solid-phase components under various biogeochemical conditions strongly adjust their lability and bioavailability. Soluble, exchangeable and organically bound forms are bioavailable to plants. The carbonate bound fraction represents a reservoir of potentially bio-available trace elements to plants. In paddy soils, the amorphous Fe/Al oxide fraction contributes to trace element availability as well.

7.5 FACTORS AFFECTING BIOAVAILABILITY OF TRACE ELEMENTS IN ARID SOILS

The bioavailability of trace elements in arid soils is strongly affected by soil properties such as soil pH and Eh, texture, $CaCO_3$ content, organic matter, dissolved organic carbon content, total metal inputs, metal speciation and salt concentrations. In general, the bioavailability of trace metals to plants is higher in arid soils with lower pH, lower clay and carbonate contents, higher content of organic matter and dissolved organic carbon and higher total metal inputs.

7.5.1 Soil pH

Soil pH is the most important single soil property controlling trace element bioavailability to plants in soils (Adriano, 2001). Both trace element concentrations in plants and available concentrations in soils decrease with increases in soil pH and carbonate content. At a soil pH above 7, the bioavailability of most trace metals (Cu, Zn, Ni, Cd and Pb) is significantly reduced. At a high pH around 7–8, most trace metals (Cd, Zn, Cr(III), Pb, Ni) are present as the hydrolysis species, which are highly adsorbed by the soil matrix. A high pH decreases the soluble and exchangeable fractions of Zn, Cu, Ni, Cr(III) and Cd and increases adsorption by soils and various solid-phase components such as Fe/Mn oxides and clay minerals. Harter (1983) reported that adsorption of Zn significantly increased above pH 7.0–7.5. Han et al. (1995) reported that Zn bound in both the amorphous and crystalline Fe oxides increased exponentially with soil pH, especially above pH 7–8. A high pH also decreases the solubility of trace element bearing minerals. Therefore, the critical concentrations of trace elements, such as Ni, Cu, Zn, Cr(III), Pb and Hg, are much higher in calcareous soils of arid zones than in acid soils. The critical Ni concentrations in calcareous soils (440–510 mg/kg) were higher than those in acid soils (110–195 mg/kg), which decreased the yield of wheat and romaine lettuce by 50% (Adriano, 2001). Singh et al. (1996) reported that concentrations of Cd, Ni and Zn in plant species (wheat, *Triticum aestivum*; carrots, *Daucuc Carota* L.; and lettuce, *Lactuca sativa*), decreased with increasing soil pH (from pH 5.5–7.5), while the Cu concentration was not consistently affected by soil pH. DTPA- and NH_4NO_3-extractable Cd also decreased with increases in soil pH. Ma and Lindsay (1990) found that Zn, Pb and Cd in the soil solution of ten arid soils from Colorado increased with decreasing soil pH (Fig. 7.3). Zinc uptake by rice plants greatly decreased as soil pH increased from 5.0–8.0 (Jugsujinda and

Patrick, 1977). In addition, the bioavailability of trace elements in arid soils contaminated with mine tailings increased with decreasing soil pH due to oxidation of sulfide minerals.

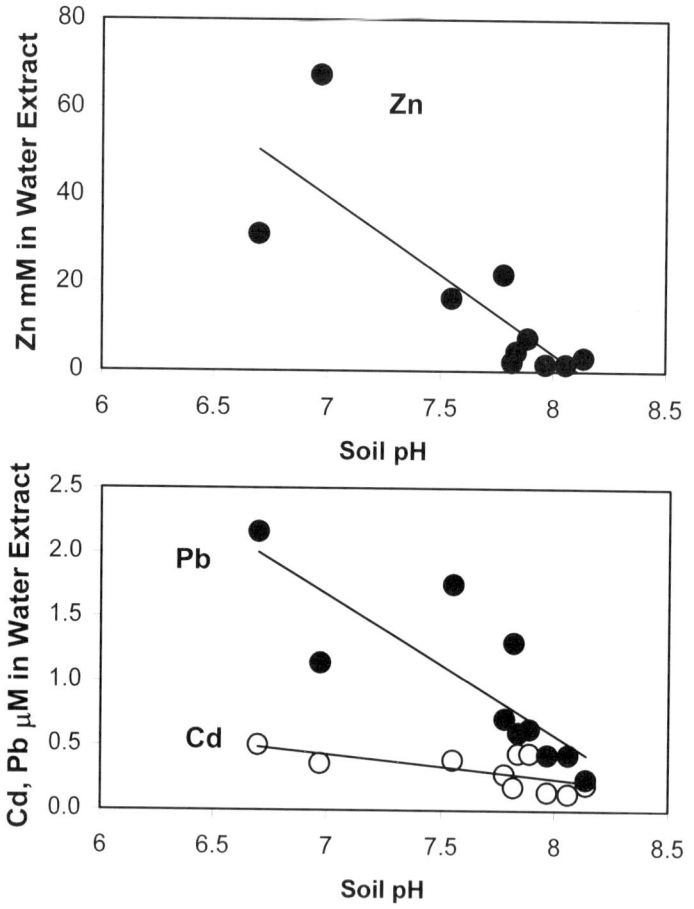

Figure 7.3. Changes of Cd, Pb and Zn concentrations in soil solutions with pH from ten Colorado arid soils (data from Ma and Lindsay, 1990)

Soil pH has been used as a safeguard against excessive uptake of toxic trace elements/heavy metals in sludge-treated soils. In England and Wales, soil pH should be above 6.0 for grasslands and 6.5 for arable lands for sludge application for Cu. Soil pH management also is a mandate for western Europe, Canada, and the United States for sludge application (Adriano, 2001). It is recommended that the soil pH be maintained at pH 6.5 or greater in land receiving biosolids containing Cd and other trace elements.

Soil pH affects the transformation of Cr between Cr(III) and Cr(VI) in soils. Since Cr(VI) has greater bioavailability and mobility in soils than Cr(III), which is strongly bound by soil solid matrix (Han and Banin, 1997). Cr(III) can be oxidized by soil manganese oxides into Cr(VI), while Cr(VI) can be reduced by organic matter, Fe(II) and microorganisms in soils. Reduction of Cr(VI) has been found to occur much slower in alkaline soils compared to acid soils (Cary et al., 1997).

7.5.2 Soil Eh and Soil Moisture

The soil moisture regime strongly affects the bioavailability of trace elements in soils as a result of its influence on trace element redistribution among solid-phase components. Han and Banin (2000) divided the trace element fractions among solid-phase components into the readily labile, potentially labile and less labile groups (Han and Banin, 2000). Trace elements in the exchangeable fraction were readily labile, while those in the carbonate, organic matter and readily reducible oxide bound fractions were potentially labile, which is highly dependent upon soil properties and environmental factors (such as pH and Eh change). Trace elements in the reducible oxide and residual fractions were regarded as less labile fractions. In general, Israeli arid soils contaminated with metal nitrates at the saturated paste regime had the slowest kinetic rate in decreasing bioavailable (NH_4NO_3-extractable) Zn, Cu and Ni, while soils under the wetting-drying cycle had the fastest kinetic rate in lowering the bioavailable trace elements (Fig. 7.4). Soils at the field capacity regime had rates in between (Fig. 7.4). In other words, Israeli arid soils had the highest bioavailable Cu, Zn and Ni under the saturated paste regime and the lowest bioavailable trace elements under the wetting/drying cycle regime, especially within 100 days of reactions for Zn and Ni and 300 days for Cu (Han and Banin, 1997, 1999, Han et al., 2001a) (Fig. 7.4). However, at the end of a year of reactions, the soils in the three moisture regimes approached similar kinetic rates (Fig. 7.4). The moisture effects on metal reactivity were strongly influenced by soil properties and the nature of the metal (Fig. 7.4). Moisture effects in the sandy soils were not significant due to the soils' lower amounts of clay minerals and oxides.

Flooding increased Fe, Mn and Cu concentrations in rice grown on arid soils in Australia with a soil of pH 7.3–7.5 compared to upland conditions, but decreased Cu uptake from flooded soils (Beckwith et al., 1975). Plant uptake of Co was also increased by waterlogging soils since Co was released from the dissolution of Mn oxides. However, the wetting incubation of 24 arid soils from Colorado farms with a wide range in texture and soil pH (7.3–9.2) and relatively low levels of Fe and Zn (5–28 mg/kg), showed significant decreases in DTPA-Fe, Mn, Cu and Zn, while air-drying

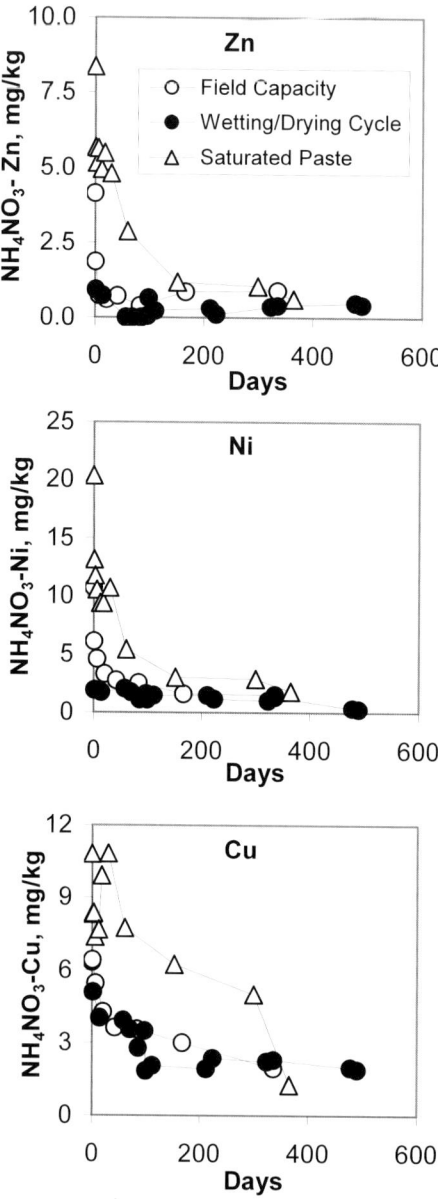

Figure 7.4. Comparisons of decreases in NH_4NO_3-extractable Zn, Ni and Cu in an Israeli loessial soil receiving metal nitrates and incubated under saturated paste, field capacity, and wetting/drying cycle moisture regimes (Han and Banin, 1997, 1999, and Han et al., 2001a)

of these moist-incubated soils increased the levels of Fe, Zn and Cu to values close to their original levels (Khan and Saltanpour, 1978). Reducing conditions enhanced the solubility and mobility of trace metals in soils from France with a soil pH of 8.0 and 0.5–0.8% $CaCO_3$, initially by dissolution of manganese and ferric oxides (Charlatchka and Cambier, 2000). Lead appeared more sensitive to these processes than Zn and Cd since flooding decreased the soil pH and reduced the Mn and Fe oxides, but prolonged flooding may lead to fixing of trace metals, rather by re-adsorption or precipitation than by formation of insoluble sulfides.

In the wetlands of Idaho, the formation of an Fe(III) precipitate (plaque) on the surface of aquatic plant roots (*Typha latifolia*, cat tail and *Phalaris arundinacea*, reed canary grass) may provide a means of attenuation and external exclusion of metals and trace elements (Hansel et al, 2002). Iron oxides were predominantly ferrihydrite with lesser amounts of goethite and minor levels of siderite and lepidocrocite. Both spatial and temporal correlations between As and Fe on the root surfaces were observed and arsenic existed as arsenate-iron hydroxide complexes (82%).

7.5.3 Total Trace Element Inputs into Soils and Aging Effects

The bioavailability of trace elements in arid soils increases with their total inputs in soils as wastes, such as sewage sludge, reclaimed sewage irrigation, and fertilizers. Lerch et al. (1990) reported that a sludge application at all loading rates resulted in significant increases in concentrations of DTPA-AB extractable Zn, Cu, Cd, Ni and Pb compared to the control in an arid soil, and DTPA-AB extractable trace elements linearly increased with increasing sludge rates. Barbarick and Workman (1987) found that DTPA-AB extractable Zn, Cu, Cd and Pb was significantly correlated with the total amount of each element added for 3 years. Mahler et al. (1980) reported that Cd concentrations in shoots of corn (*Zea mays* L.), tomatoes (*Lycopersicon esculentum* Mill.), and Swiss chard (*Beta vulgaris* var.) increased with increases in the total Cd inputs as sewage sludge in arid soils from Californa. DTPA-TEA extractable Zn, Cu, Pb and Ni increased with sludge applications, but four years of sludge applications did not show significant adverse effects on soil microbial populations or activity in an arid soil from Arizona (Brendecke et al., 1993).

The termination of sewage biosolids applications in arid soils often results in decreases in the bioavailability of trace elements. Webber and Beauchamp (1979) found a stronger statistical relationship of Cd uptake in corn to the amount of Cd added in the most recent sludge application than to the cumulative Cd loading in a calcareous soil. But they observed no real change in Cd uptake by corn for three years after the cessation of sludge application. Barbarick and Ippolito (2003) reported that DTPA-AB extractable Zn and to some extent Cu decreased to the levels in non-amended soils after three plantings (six years) following termination in semi-arid Colorado soils, which received continuous alternate year biosolid additions with five to six applications (Fig. 7.5). A decline in plant uptake of Cd and Zn by corn for six years after termination of biosolids application also had been observed in non-arid zones (such as in Minnesota) (Bidwell and Dowdy, 1987). This indicates that dry land in arid zones receiving excessive biosolids can also recover to untreated control levels within a few years. However, in a non-arid field amended with a single heavy application of sewage sludge, the Zn and Cd remaining in the topsoil (pH 6.5–7) after 15 years were still plant-available. Vegetable crops took up excessive amounts and developed severe phytotoxicity. Concentrations of water soluble Cu, Zn and Ni were found to be more than 10 times higher than those in the control soil (McBride et al., 1997).

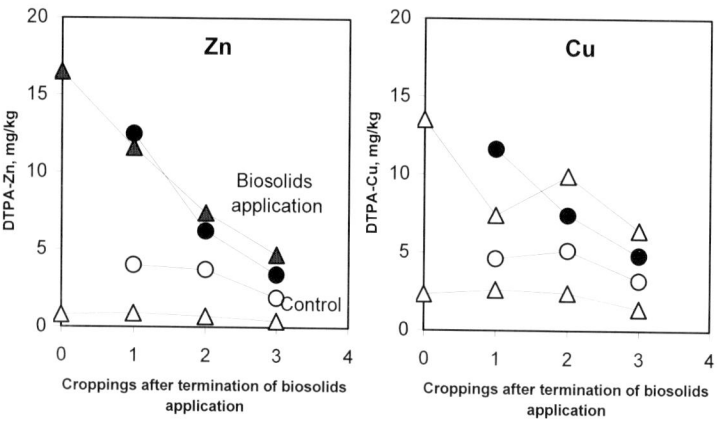

Figure 7.5. The AB-DTPA-extractable Cu and Zn in a semi-arid soil from Colorado after termination of five excessive biosolids applications. Open circle (fine, smectitic, mesic Aridic Argiustolls) and open triangle (fine, smectitic, mesic Aridic Paleustolls) represent control soil without biosolid application, while solid circle and solid triangle indicate biosolid applications (data extracted from Barbarick ad Ippolito, 2003)

Han and Banin (1997, 1999) and Han et al. (2001a) found that concentrations of bioavailable trace elements (Cu, Ni and Zn, as extracted by NH_4NO_3 at pH 7) in general, decreased with time in two Israeli arid soils contaminated with metal salts under the saturated paste, field capacity, and wetting-drying cycle moisture regimes for one year (Fig. 7.6). However, the bioavailability of trace elements was strongly dependent upon the nature of

Bioavailability of Trace Elements

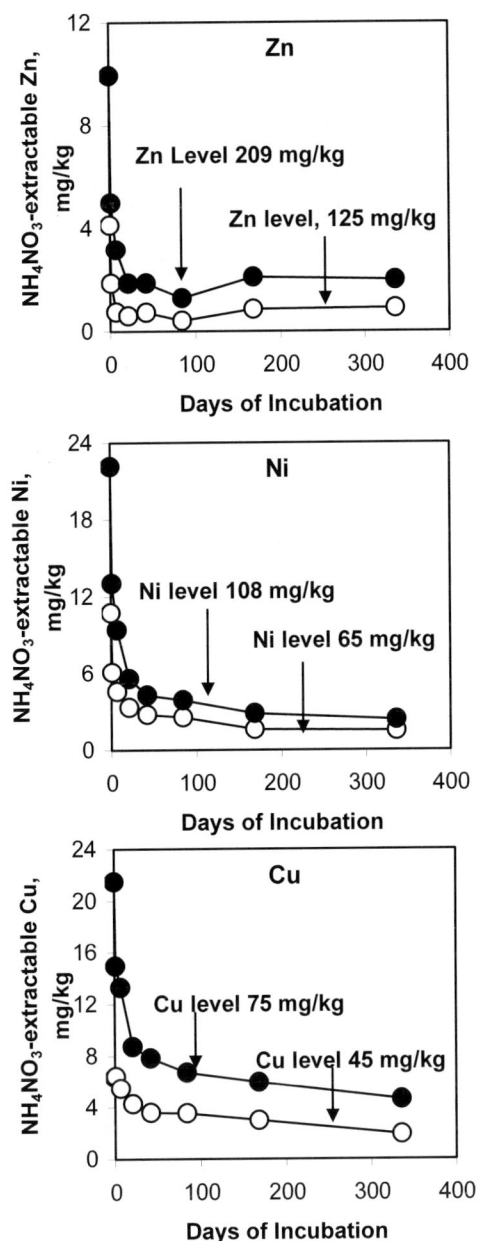

Figure 7.6. Decreases in bioavailability (as NH$_4$NO$_3$-extractable) of Cu, Ni and Zn in an Israeli loess soil receiving metal salts and incubated under field capacity regime for one year (data from Han and Banin, 1999)

the metal and soil properties. In the sandy soil, lability of the metal decreased as metals were redistributed from the readily labile fractions into the potentially labile fractions. In the loessial soil, however, metal bioavailability decreased as metal redistribution took place from both readily labile and potentially labile fractions into the less labile fractions due to the higher amounts of clay minerals and oxides than were found in sandy soil. Overall changes of the bioavailability of metals in loessial soil were larger than those in the sandy soil. This implies that the reactivity of heavy metals in the loessial soil is greater than those in the sandy soil. Furthermore, changes of overall lability of Ni, Zn and Cu are larger than Cr.

7.5.4 Contents of Organic Matter and Dissolved Organic Carbon

Organic matter plays a key role in the bioavailability of trace elements in soils. Organic material degradation and incorporation control the reactivity of soil organic matter with trace elements in soils. In mineral soils with low content of organic matter, organic matter, in general, increases the bioavailability of trace elements in soils. In southern Australia, concentrations of Cu and Zn in water extracts from the soils with pH > 7 increased with those of DOC (Fig. 7.7). However, in the organic soils with a high content of organic matter, organic matter decreased the bioavailability of trace elements in the soils. For example, EDTA-extractable Cu increased with increasing organic matter content of soils up to organic carbon contents of about 1–2% (Sillanpaa, 1982). With further increases in the organic carbon content, the Cu bioavailability began to decrease. Manganese had a similar trend with organic matter. The Mn content of plants and the DTPA-extractable soil Mn concentration tended to increase as organic carbon content of the soil increased, but decreased at higher organic matter levels (Sillanpaa, 1982).

The role of organic matter in controlling the bioavailability of trace elements was found to depend upon the nature of organic matter and trace elements, soil pH, redox potential and other competing ions and ligands (Adriano, 2001; McBride, 1995). Insoluble organic matter very effectively inhibited uptake of metal cations such as Cu^{2+}, which bound strongly with organic matter and were prevented from diffusing to roots (McBride, 1995). Insoluble Cu-humic acid complexes may decrease concentrations of Cu in soil solution (Adriano, 2001). In arid soils, added Cu and Pb were mainly bound to organic matter (Han and Banin, 1997). Actually, organic matter was more important in the immobilization of Pb than carbonates and hydrous oxides in soils, where insoluble Pb-organic complexes were formed. Solid organic matter has been used to reclaim high-Ni soils. On the other hand, soluble organic and dissolved organic carbon and organic ligands increased the bioavailability of trace elements in soils. Soluble organics increased the carrying capacity of soil solutions for Cu^{2+} and other metal cations at any pH by forming soluble metal-organic complexes (McBride, 1995). Copper formed more stable complexes with humic and fulvic acids than Pb, Fe and Zn (Adriano, 2001). The stability constants with humic acids were in the order of: Cu > Pb > Cd (Stevenson, 1976). Further, the presence of dissolved and colloidal organic matter in the soil solution affected the ionic strength and charge balance, as well as the total concentrations of the trace elements.

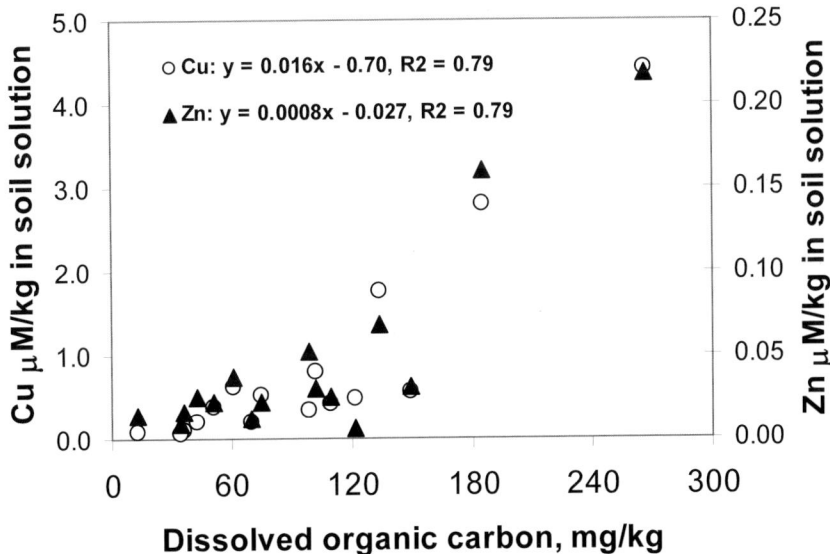

Figure 7.7. Changes of Cu and Zn concentrations in soil solution with DOC from South Australia (Data from Fotovat et al., 1997)

A large amount of dissolved organic matter in the soil solution occurs in forest landscape and agroecosystem with the amendment of biosolids, compost, animal wastes, and sludge. Dissolved organic ligands increase the concentrations of trace elements in soil solutions. The effects of dissolved organic matter on concentrations and speciation of trace elements in arid soil solutions have been discussed in Chapter 3. A high pH in arid soils increases the solubility of organic matter, thus increasing dissolved organic matter in soil solution.

For some trace elements (e.g., Zn and Cu), high organic matter leads to Zn/Cu deficiency. Zinc deficiency has been observed in muck or peat soils due to the formation of insoluble Zn/Cu-organic complexes or strong adsorption by insoluble soil organic matter (Adriano, 2001). In paddy soils receiving organic materials as crop residuals, Zn deficiency in rice occurred due to the formation of bicarbonates, hydroxides and organic complexes from the decomposition of biomass and microorganism biomass immobilization (Adriano, 2001). However, organic materials such as sewage sludge and other manure forms are very quickly mineralized in most of arid and semi-arid areas due to high temperatures. The effect of organic matter also may be associated with the change in soil pH caused by different organic matter sources (Narwal and Singh, 1998).

7.5.5 Forms and Sources of Trace Elements

The sources of trace element inputs into arid soils strongly affect their bioavailability in soils. Chelated forms provide the highest bioavailability to plants, followed by inorganic salts, and organic wastes forms. It has been noted that corn took up more Zn from organic Zn sources (ZnEDTA and Zn-ligninsulfonate) than inorganic forms (ZnO and $ZnSO_4$) on calcareous soils (Gallagher et al., 1978), especially at low temperatures such as in the early spring season. Higher growth temperatures diminished the differences between organic and inorganic Zn sources. Various Zn sources are used for correcting Zn deficiency (Table 7.4). Adriano (2001) pointed out that Zn chelate was more readily utilized by the plants than Zn applied as an inorganic salt. Chelated sources of Zn were about 5 times as effective as inorganic sources in overcoming Zn deficiency (Adriano, 2001). Boawn et al. (1957) reported that bean plants took up 3.5 times more Zn from ZnEDTA than from $ZnSO_4$ in an alkaline soil with Zn applied. Chelated Zn in soils may have a great diffusion of Zn^{2+} into the root surface (Adriano, 2001). However, Rehm et al. (1980) reported that there was no advantage of the chelated forms (ZnEDTA) of Zn over inorganic forms ($ZnSO_4$ and ZnO) when they were applied in a suspension fertilizer to the side of and below the corn seeds at planting. Myttenaere and Mousny (1974) reported that CrEDTA was less absorbed and metabolized by plant roots than Cr^{3+} and CrO_4^{2-}, but it had a higher translocation rate in plants. Chelated forms of trace elements are less immobilized and adsorbed by the soil matrix. On reducing contact of Zn with the soil matrix and thus reducing immobilization, the band application of Zn fertilizers was found to be more effective in correcting Zn deficiency than the broadcast application. Band application requires lower rates of $ZnSO_4$ than broadcast application. In addition to band and broadcast application of Zn fertilizer, Zn is also applied to foliage of crops during a growing season.

Table 7.4. Zinc sources and application rates for correcting zinc deficiency on crops[a]

Crop	Source	Zn rate, kg/ha	Application Methods
Rice	$ZnSO_4$, ZnO	9	Broadcast
Corn	$ZnSO_4$, ZnO, Zn chelate ZnEDTA	0.6-17	Broadcast or band
Vegetables	$ZnSO_4$, ZnEDTA	0.6-9	Broadcast or band
Flax	$ZnSO_4$	1.0-11	Broadcast
Bean	$ZnSO_4$	1.0-17	Broadcast or band
Sweet corn	$ZnSO_4$, ZnEDTA	0.9-17	Broadcast or band
Sorghum	$ZnSO_4$	1.0-17	Broadcast or band
Potato	$ZnSO_4$	1.0-11	Broadcast or band

[a]Data from Adriano, 2001.

Organic wastes (animal manure, organic compost, and sewage sludge) are also important sources of trace elements in soils. It was found that the organic waste bound trace elements underwent decomposition/immobilization processes when they entered arid soils (Banin et al., 1990). A large portion (50–80%) of organic matter of stabilized sewage sludge was resistant to decomposition in soil and was similar to soil humic substances (Adriano, 2001). After the termination of sludge application, organic C content of the sludge-treated Californian soils decreased by approximately 40% over ten years (Hyun et al., 1998). Latterell et al. (1978) reported that the annual average rate of decomposition of organic matter in sludge-amended soils was in the range of 10–18%. Banin et al. (1990) reported that mineralization rates increased at higher application rates of sludge in Israeli arid soils. The decomposition rates of the organic matter in sludge-amended soils decreased with time as the easily decomposable organic matter was broken down. The supply of micronutrients Zn and Cu from these organic sources was a slow process, so that it was difficult to meet the requirements of plants for these micronutrients in the early growing seasons. However, they supplied the plants with enough micronutrients in the late growing seasons. Nevertheless, it also was observed that the bioavailaiblity of Cd in Californian sludge-mended soils did not change, even after the termination of the sludge application. In California sludge-treated soils, the soluble Cd concentration did not increase, as the organic C in these soils declined over ten years following termination of the sewage sludge applications (Hyun et al., 1998). The soluble Cd in the soils increased with the organic carbon content of the soils.

The plant tissue Cd concentrations increased with the soluble soil Cd. One possible explanation is that there were still adequate amounts of organic matter in these soils to complex Cd ten years after the termination of the sludge applications (Hyun et al., 1998).

Organic acid – heavy metal complexes had different stabilities and resistances to degradation. Most of the low molecular organic acids were generated from rhizodeposition and as microbial metabolites during soil organic matter decomposition. Citrate-metal complexes generally were more biodegradable than oxalate-metal complexes in calcareous soils with 10–15% $CaCO_3$ (Renella et al., 2004). The biodegrability of low molecular weight organic acid – metal complexes was also dependent upon the nature of the metal, following the order: $Mg > Zn > Cu \approx Pb > Cd$ (Renella et al., 2004).

In roadside soils, lead was present in the more soluble forms such as PbClBr and $PbSO_4$ from automobile emissions compared to soils near smelters or in mining sites (Adriano, 2001), which contained oxides, sulfides and carbonates (galena, anglesite and cerussite) with low solubility. However, after oxidation of sulfide into sulfate, the soils became very acidic, resulting in the increase in both solubility and bioavailability of the trace metals.

7.5.6 Salt Concentration

Water resources become scarce in arid and semi-arid zones. Due to the limited availability of fresh water, irrigation with saline water is practiced. In general, salts increase the ionic strength of the soil solution and the bioavailability of trace elements in arid soils. Salts increase both metal solution concentrations and metal concentrations in plants (i.e., increase the bioavailability of Zn, Pb, Cu, Fe and Mn) in arid and semi-arid soils. Helal et al. (1996) reported that NaCl accelerated the root mortality of maize and increased both the concentrations of Zn, Cu and Cd in the soil saturated extracts and their uptake by maize from a polluted Egyptian desert soil (with pH 8.5 and 2.3% of $CaCO_3$), while $CaCl_2$ increased the solubility of these trace elements, but the root mortality and plant uptake decreased. Kadria and Michel (2004) reported that salinity increased both the bioaccumulation of Zn and Pb in rye grass (*Lolium perenne* L.) in a soil with pH 7.6 and salinity up to 900 $mS.m^{-1}$. Above that level, Zn and Pb contents in the plants decreased with a salinity of 1200 $mS.m^{-1}$. Martinez et al. (2001) pointed out that concentrations of Cu, Fe, Mn and Zn in plant shoots on saline and/or sodic soils may increase, depending on the type of plants, salinity, trace element concentrations, and environmental conditions. The concentrations of Cu in plants were found to increase steadily with increasing electrical conductivity (Sillanpaa, 1982). In the surface soils of agricultural, industrial and urban regions of Isfahan, central Iran, DTPA-extractable Cd was strongly correlated to EC values (Amini et al., 2005). Weggler-Beaton et al. (2000) found that Cd concentrations in the soil solution and shoots of wheat and Swiss chard (*Beta vulgaris* cv. *Foodhook Giant*, Fig. 7.8) increased linearly with increasing Cl concentrations in the soil solution of biosolid-amended soil from southern Australia. Cd concentrations in shoots of Swiss chard with 120 mmol L^{-1} NaCl treatments were almost double those in the zero salt treatments. In contrast, increasing $NaNO_3$ concentrations either reduced shoot Cd concentration or had no effect. Speciation calculations predicted that the solution concentration of free metal ion Cd^{2+} was not significantly affected by NaCl or $NaNO_3$ treatments. McLaughlin et al. (1994) reported that soil salinity increased Cd concentrations in potato tubers.

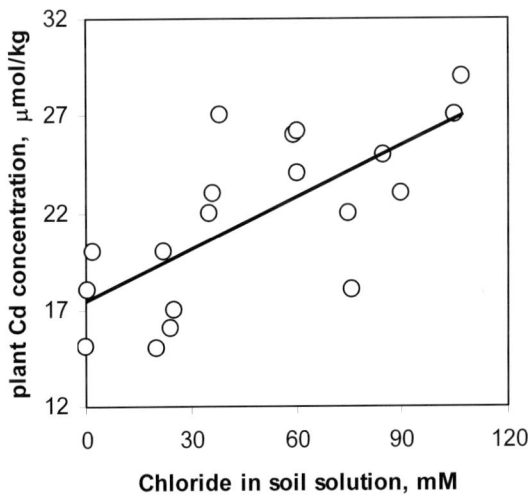

Figure 7.8. Swiss chard Cd concentrations as affected by chloride levels in soils (modified from Weggler-Beaton et al., 2000)

The enhancing effect of NaCl on Cd uptake was due to chloride complexation of Cd (Smolders et al., 1998). High salinity in soils increased the concentrations of chloride complexes of trace elements (such as $CdCl^+$ or $CdCl_2^0$) in soil solution, which increased and correlated best with Cd uptake of both plant species as discussed above. In addition, salinity also affected plant root function, and Na competition with Cd for sorption sites in soil may be a possible contributor.

7.5.7 Nitrogen and Phosphate Fertilizers and Trace Element Interactions

Most trace metals may be precipitated with phosphate into insoluble metal phosphates (Table 7.5). Most metal phosphates have low solubility. High localization of phosphates reduces the bioavailability of Zn in arid soils. The banded application of P near the seeds depresses Zn uptake by corn (Adriano and Murphy, 1970; Grant and Bailey, 1993), causing Zn deficiency. However, both N and P fertilizers increase Cd concentration in plants. Cadmium and Zn are antagonistic in root uptake and distribution within plants.

Table 7.5. Solubility product constants of common compounds/minerals of selected major and trace elements at 25 °C[a]

Element	Name	Formula	K_{sp}
Cd	Cadmium phosphate	$Cd_3(PO_4)_2$	2.53×10^{-33}
Co	Cobalt phosphate	$Co_3(PO_4)_2$	2.05×10^{-35}
Cu	Copper phosphate	$Cu_3(PO_4)_2$	1.4×10^{-37}
Pb	Lead phosphate	$PbHPO_4$	3.7×10^{-12}
	Lead hydroxypyromorphite	$Pb_5(PO_4)_3OH$	1.6×10^{-77}
	Lead chloropyromorphite	$Pb_5(PO_4)_3Cl$	4.0×10^{-85}
Ni	Nickel phosphate	$Ni_3(PO_4)_2$	4.74×10^{-32}
Zn	Zinc phosphate	$Zn_3(PO_4)_2$	9.0×10^{-33}
	Zinc pyromorphite	$Zn_5(PO_4)_3OH$	7.9×10^{-64}
	Hopeite	$Zn_3(PO_4)_2 \cdot 4H_2O$	4.0×10^{-36}

[a]From Lide D.R., 2001; McGowen et al., 2001.

Grant et al. (2002) and Grant and Bailey (1997) reported that fertilization with monoammonium phosphate consistently increased Cd concentration and decreased Zn concentration in durum wheat and flax seed. The effect was greater with banded as compared to broadcast placement of P fertilizers. Increased Cd concentration with the phosphate application was suggested to be related to high ionic strength, reduced pH, and enhanced root proliferation in the micro-region around the fertilizer granules. Enhanced root development in response to phosphate fertilization may increase the accumulation of Cd. Reduction of Zn accumulation associated with phosphate application may also contribute to the increase in Cd concentration in durum grain, possibly through the enhancement of Cd translocation to the grain. Mitchell et al. (2000) noted that increases in the Cd concentration in wheat with N fertilization were greatest immediately after fertilization. The concentration of Cd in both shoots and grains of wheat increased significantly with increasing N rate (up to 800 mg/kg as urea) for samples at maturity (Fig. 7.9). Both solution Cd concentration and DTPA-extractable soil Cd increased significantly with an increasing nitrogen rate. Cadmium was the element most affected by increasing the N rate. There were minor changes in the uptake of other elements, such as N, P, K, Ca, Mg, Mn, Zn and Cu, with N application rates and times. During the same time period, the concentrations of Cd in flax seed decreased as the seed concentration of Zn increased, across the soil types and fertilizer treatments. Cadmium and Zn antagonism had been observed in metal uptake by roots and distribution within plants. Jiao et al. (2004) observed that cadmium concentration was lower and Zn concentration higher in flax seed and wheat grain than in roots, shoots or straws of both species.

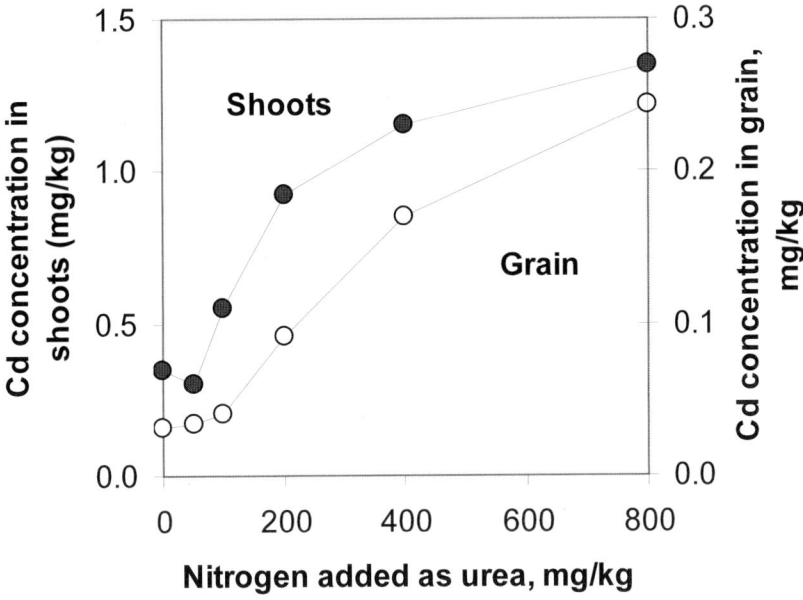

Figure 7.9. Effects of N rates on Cd concentrations in wheat shoots and grain (data from Mitchell et al., 2000)

7.5.8 Plant Species and Growth

Plants show different abilities of uptake of trace elements. Lettuce and rape usually have a higher capacity for the uptake of trace elements than do cereal plants. Grant and Bailey (1997) reported that flax tended to accumulate high concentrations of Cd relative to cereal crops, such as wheat grain and shoots.

Crop cultivar variation in trace element uptake had been observed with corn, rice, barley, wheat, and cocoa (Adriano, 1986). Wheat and soybeans showed variety-based variation of P-Zn interaction. Crop varieties with a low capacity for uptake of Zn in soils were more vulnerable to Zn deficiency by P application. Crops sensitive to Zn deficiency included beans, potatoes, peaches, flax, corn, rice, soybeans, and citrus (Adriano, 2001).

Plant species that release phytosiderophores take up high amounts of trace elements from calcareous soils amended with sewage sludge and inorganic metal salts, especially Zn. Awad and Romheld (2000a, 2000b) reported that Fe-deficient plants precultured for low Fe nutrition status (such as wheat and sorghum) took up high amounts of Zn, Ni and Cd. Phytosiderophores were released by the roots of Fe-deficient wheat plants, which increased the bioavailability of trace elements in arid soils. Phytosiderophores alone extracted higher amounts of Zn, Cu, Ni, Cd and Fe from arid soils directly. Awad and Romheld (2000b) found that concentrations of Zn, Ni, and Cd in shoots of wheat plants with low Fe nutritional status were 25–100% higher than in Fe-precultured plants.

In addition, plant growth rate also affects plant uptake and the bioavailability of trace elements. It was reported that the rate of As uptake by rice increased as the rate of plant growth increased (Onken and Hossner, 1995).

7.5.9 Other Soil Properties

In addition to soil pH and organic matter content, clay and carbonate contents, CEC also strongly affects the bioavailability to plants and bioaccessibility of trace elements in soils. Bioaccessibility is defined as the amount of a contaminant that is soluble due to simulated in-vitro gastric functions and has the potential to cross the intestinal wall (Hamel et al., 1998). Stewart et al. (2003) reported soils with higher quantities of clay and carbonates, higher pH, and a higher cation exchange capacity generally sequestered more Cr(III), posing less bioavailability and bioaccessibility. The bioaccessibility of Cr(III) in aridisol ranges from 17–39% and 10–33% after 1 and 100 day soil-Cr aging, respectively.

7.6 DISTRIBUTION OF BIOAVAILABLE TRACE ELEMENTS IN GLOBAL ARID AND SEMI-ARID SOILS

Bioavailable trace elements in world arid and semi-arid soils vary widely, depending upon the nature of the parent materials, soil pH, $CaCO_3$ and clay content, and soil texture. The contents of bioavailable trace elements in arid and semi-arid soils of selected countries are presented in Table 7.6. The data are recalculated from Sillanpaa (1982) and cited from Liu (1996), Han and Banin (1997, 1999) and many others.

7.6.1 Africa (Sillanpaa, 1982)

Hot water-extractable B and ammonium oxalate/oxalic acid-extractable Mo in 200 Egyptian arid soils ranged from 0.39–3.42 mg/kg and 0.030–0.395 mg/kg, respectively. DTPA-extractable Zn and Mn were from 0.377–11.93 and 7.28–51.87 mg/kg, respectively. EDTA-extractable Cu varied from 3.25–34.32 mg/kg. The average bioavailable B, Mo, Zn, Mn and Cu was 1.216, 15.02, 15.6, 0.185 and 1.53 mg/kg, respectively. These soils were sampled from both wheat and maize fields. The soils had 1–24.1% $CaCO_3$ and 7.57–8.95 pH. Bioavailable Zn varied considerably among the arid soils studied.

7.6.2 Asia

7.6.2.1 China (Liu, 1996)

Water-soluble B in loessial soils of the North Western region of China was in the range of 0.04–14.7 mg/kg with an average of 0.54 mg/kg (n = 968) (Table 7.7). About 62.7% of the soil samples contained less than 0.5 mg/kg water soluble B and 20.3% of the samples contained less than 0.25 mg/kg. Water-soluble B in alluvial soils of the North China Plain was 0.1–5.0 mg/kg with an average of 0.45 mg/kg. In soils of the Tibet region, water-soluble B varied from 0.1–2.66 mg/kg. Bioavailable B in saline soils of the arid regions varied from 0.19–6.70 mg/kg and the average values were from 0.79–2.08 mg/kg. Calcareous paddy soils contained 0.1–1.79 mg/kg (average 0.72 mg/kg). The arid and semi-arid regions in China, in general, had abundant B supplies to plants. However, meadow soils with high contents of organic matter and poor drainage in North China may have a B deficiency in wheat, maize, and soybean.

Table 7.6. Concentrations of bioavailable trace elements in arid and semi-arid soils of selected countries[a]

Country	Sample number		CaCO$_3$	pH (H$_2$O)	B	Cu	Mn	Mo	Zn
					\multicolumn{5}{c}{Extractant}				
					Hot water	EDTA	DTPA	Oxalate buffer	DTPA
			%		\multicolumn{5}{c}{mg/kg}				
Malta	25	Avg	58.6	7.91	1.48	9.36	9.75	0.15	13.4
		Stdev	5.7	0.1	0.69	3.25	2.60	0.07	16.3
		Min	52	7.78	0.82	6.50	6.37	0.03	2.63
		Max	68.4	8.08	2.86	16.8	14.7	0.28	51.2
Mexico	100	Avg	10.1	8.11	1.51	5.07	16.8	0.26	1.27
		Stdev	15.3	0.32	0.94	1.95	8.32	0.22	2.28
		Min	0.1	6.77	0.25	1.17	6.24	0.04	0.29
		Max	59.8	8.85	4.49	11.3	57.9	1.40	19.3
Pakistan	156	Avg	6.9	9.38	0.90	6.24	11.1	0.26	1.07
		Stdev	4	0.27	0.88	2.99	2.99	0.21	1.91
		Min	0.7	7.89	0.10	1.69	4.68	0.03	0.13
		Max	18.1	9.25	5.75	29.5	18.6	1.36	15.7
Egypt	100	Avg	4.4	8.01	1.13	15.7	17.4	0.19	1.70
		Stdev	1.8	0.19	0.48	5.07	5.85	0.06	1.20
		Min	1.3	7.57	0.39	4.29	9.49	0.05	0.60
		Max	16.3	8.43	3.56	34.3	51.9	0.40	9.75
Iraq	119	Avg	25.2	8.04	1.77	6.24	11.8	0.16	0.36
		Stdev	6.4	0.15	1.78	1.56	3.77	0.15	0.42
		Min	2.1	7.68	0.21	2.73	4.68	0.02	0.09
		Max	43.5	8.44	11.6	12.7	24.8	0.85	4.63
Syria	20	Avg	27.5	7.94	1.35	6.63	19.4	0.15	1.57
		Stdev	15.7	0.21	0.87	2.86	17.0	0.09	2.25
		Min	6.8	7.61	0.40	2.47	7.54	0.03	0.13
		Max	59.2	8.51	4.13	16.6	87.1	0.38	9.69
Lebanon[b]	16	Avg	17.1	7.8	0.70	7.80	23.3	0.10	1.11
		Stdev	20.7	0.37	0.21	2.60	15.0	0.09	0.55
		Min	0.4	7.1	0.44	5.33	8.84	0.02	0.27
		Max	67.1	8.1	1.26	15.0	49.3	0.33	2.17

[a]Recalculated from Sillanpaa M., 1982.

[b]All data from wheat field experiments except for Lebanon which were from pot study.

Loessial soils contained 0.01–0.32 mg/kg bioavailable Mo (extracted with an oxalate buffer at pH 3.3) with an average of 0.06 mg/kg (n = 419). Seventy-four percent of the soil samples contained less than 0.10 mg/kg. Some crops in this region showed Mo deficiency. In soils of the

North China Plain, the bioavailable Mo varied from a trace amount to 0.33 mg/kg, of which 84% of the soil samples contained less than 0.15 mg/kg of bioavailable Mo. The bioavailabe Mo in the North West region of China was in the range of 0.01–0.55 mg/kg, with the majority of soils below 0.15 mg/kg. Soils in the Tibet region contained 0.03–1.24 bioavailable Mo. Irrigated oasis soils contained the highest bioavailable Mo (0.88–1.24 mg/kg). Calcareous paddy soils had a trace to 0.44 mg/kg with an average of 0.12 mg/kg of bioavailable Mo. In soils of the North China Plain and loessial soils of the Loess Plateau, Mo deficiency was observed and Mo fertilizers was applied to soybeans, peanuts, and legumes.

Table 7.7. Concentrations of bioavailabile Zn (DTPA-extractable Zn), Cu (DTPA-extractable Cu), and B (water-soluble B) in arid and semi-arid soils of North China[a] (from Liu, 1996)

Soil	Range, mg/kg		
	Zn	Cu	B
Blown sand soils	0.25-0.82	0.01-0.8	0.16-1.43
Brown calc soils	0.14-2.48	0.18-3.6	
Chao soils	0.09-1.5		0.10-0.50
Chernozems	0.34-10	0.38-2.7	
Chestnut soils	0.12-3.88	0.18-1.93	0.15-2.46
Cinnamon soils	0.21-1.16	0.35-1.58	0.18-1.01
Heilu soils	0.2-0.99	0.28-1.7	0.14-1.09
Irrigation-warping soils	0.3-1.43	0.22-4.2	
Loessal soils	0.13-0.77	0.01-0.7	0.04-0.64
Shajiang black soils	0.04-1.22	0.07-9.16	
Sierozems	0.12-0.99	0.11-1.72	0.28-2.29
Solonchaks	0.2-2.66	0.29-1.92	0.40-14.7
Solonetz	0.18-1.2		
Tier soils	0.21-1.42	0.4-2.69	0.10-1.98

[a]Data from Liu, 1996. The data cover the North China Plain, Loess Plateau, and Xinjiang Province.

Loess soils in the Loess Plateau contained 0.9–34.5 mg/kg DTPA-extractable Mn with an average of 0.93 mg/kg. In the soils of the North China Plain, DTPA-extractable Mn was in the range of 0.5–120 mg/kg. The average Mn in the North China Plain was 8.71 mg/kg. In arid soils of the North West region, DTPA-extractable Mn varied from 0.38–41.6 mg/kg (average Mn, 7.93 mg/kg). In the calcareous paddy soils of China, the exchangeable Mn (by ammonium acetate) and easily reducible Mn (by ammonium acetate-hydroquinone) were in the range of a trace to 59.5 and 3–109 mg/kg, with an average of 4.18 and 177 mg/kg, respectively. There were large areas in the loess region showing a Mn deficiency in some crops. Most of the Mn-deficient soils were in the arid zones of the North China Plain and North West regions. When these calcareous soils contained less than 2–3 mg/kg DTPA-extractable Mn, Mn fertilizer was applied to wheat, barley, maize, oats, peanuts, soybeans, peas, legumes and fruit trees in the region.

Soils in the North China Plain and Loess Plateau regions contained 0.04–3.01 mg/kg DTPA-extractable Zn with an average of 0.44 mg/kg. The concentrations of DTPA-extractable Zn in northern China are presented in Table 7.7. In the loessial soils of the Loess Plateau, 64% of the soil samples had less than 0.5 mg/kg of bioavailable Zn. The bioavailable Zn in the arid soils of North China varied from 0.08–11.84 mg/kg with an average of 1 mg/kg, with 41% of the soil samples having < 0.5 mg/kg of bioavailable Zn. The average amount of bioavailable Zn in calcareous soils was 0.35 mg/kg (trace - 1.12 mg/kg). The North China Plain and Loess Plateau are major Zn-deficient regions in China. Calcareous paddy soils frequently displayed Zn deficiency in rice. Zinc fertilizers have been applied to rice, maize, sorghum, wheat, cotton and fruit trees where bioavailable Zn was less than 0.5 mg/kg.

Loessial soils in the Loess Plateau contained 0.01–4.20 mg/kg DTPA-extractable Cu with an average of 0.93 mg/kg (Table 7.7). Bioavailable Cu in the North China Plain varied from 0.07–9.95 mg/kg. In the North West region, soils contained 0.06–19.20 mg/kg DTPA-extractable Cu. The average bioavailable Cu was 1.83 mg/kg in the calcareous paddy soils with a range of trace to 6.85 mg/kg. Copper deficiency was not often observed in the arid and semi-arid soils of China.

Water soluble Se in the North West region varied from 0.0002–0.0429 mg/kg. Water soluble Se accounted for 2.13–6.34% of the total Se in the soils of North China. Selenium is an essential element to animals and humans. When water soluble Se in soils is less than 0.003 mg/kg, Se deficiency in animal and human beings may occur. EDTA-extractable Se in the alkali desert soils of North China was in the range of 0.011–0.090 mg/kg; this was about 5–11% of the total Se in the soils. Selenium deficiency was mostly found in the Loess Plateau and Tibet region. NH_4OAc-extractable Ni in soils from Beijing was 0.29 mg/kg.

7.6.2.2 India

In saline-alkali soils of the Punjab region of India, soluble B in saturation extracts was in the range of 3–11.8 mg/kg (Kanwar and Singh, 1961). In the vertisol of the Gujarat region, water-soluble B was in the range of 0.5–0.7 mg/kg (Raychaudhuri and Datta, 1964). In the arid soils of the Rajsthan region, water soluble B varied from 0.89–10.2 mg/kg, with an average of 3.22 mg/kg (Moghe and Mathur, 1966).

7.6.2.3 Pakistan (Sillanpaa, 1982)

In 156 arid soils sampled from the wheat fields of Pakistan, hot water-extractable B and ammonium oxalate/oxalic acid-extractable Mo were in the range of 0.104–5.746 and 0.029–1.364 mg/kg, respectively. DTPA-extractable Zn and Mn varied from 0.13–15.73 and from 4.68–18.59 mg/kg, respectively. EDTA-extractable Cu was from 1.69–29.51 mg/kg. The average and standard deviation of bioavailable B, Mo, Zn, Mn and Cu were 0.897 ± 0.884, 0.257 ± 0.213, 1.066 ± 1.911, 11.05 ± 2.99, and 6.24 ± 2.99 mg/kg, respectively. These arid soils contained 0.7–18.1% $CaCO_3$ with soil pH from 7.89–9.25. In addition, in another 86 arid and semi-arid soils sampled from maize fields, higher bioavailable Mn (15.86 ± 9.49 mg/kg) was found, but bioavailable Zn, Mo and Cu were also lower than those in the soils from arid wheat fields. Bioavailable Zn, Mo and Cu were 0.78 ± 0.74, 0.190 ± 0.103, and 5.72 ± 1.95 mg/kg, respectively. However, the soils sampled from the maize fields had similar bioavailable B concentrations compared to those samples taken from the arid wheat fields.

7.6.3 Europe, Australia, and Former Soviet Union

In Malta, 25 arid soils contained 7.78–8.08 mg/kg of water-extractable B with an average of 1.482 mg/kg (Sillanpaa, 1982). The $CaCO_3$ contents in these soils varied from 56%–68.4% with an average of 58.6%. The soil pH varied from 7.78–8.08. These soils had 6.5–16.77 mg/kg EDTA-extractable Cu, with an average of 9.36 mg/kg. DTPA-extractable Zn and Mn were in the range of 2.63–51.2 and 6.37–14.69 mg/kg, respectively. The average DTPA-extractable Zn and Mn were 13.38 mg/kg ± 16.3 mg/kg and 9.75 ± 2.6 mg/kg, respectively. The bioavailable Zn in these arid soils had a greater variation than did the bioavailable Cu, Mn and B. Ammonium oxalate/oxalic acid-extractable Mo varied from 0.027–0.282 mg/kg with an average of 0.146 ± 0.074 mg/kg.

Hot water-extractable B in the chernozems of Bulgaria was in the range of 0.7–1 mg/kg (Pavaleyev, 1958). In the steppe soils of the Crimea region of the former Soviet Union, hot water-extractable B was 2–2.57 mg/kg, while solonetses and solonchacks in the Krasnodar region had 35–37 mg/kg of hot water-extractable B (Aubert and Pinta, 1977). However, in the chernozem of the Armenia region, bioavailable B was 0.53–1.05 mg/kg. In the solonchacks of Turkmenistan, bioavailable B was 32.4 mg/kg. In soils near a lead smelter at Pot Pirie, South Australia, EDTA-extractable Zn, Cd and Cu ranged from <5 to >40, <0.15 to >1.0 and < 2 to 10 mg/kg, respectively (Cartwright et al., 1976).

7.6.4 North America

7.6.4.1 Mexico (Sillanpaa, 1982)

In 100 arid and semi-arid soils of Mexico, hot water-extractable B varied from 0.247–4.485 mg/kg. The average hot water-extractable B was 1.508 ± 0.936 mg/kg. These soils contained 0.1–59.8% $CaCO_3$ with an average of 10.1 ± 15.3%. The soil pH in these soils was in the range of 6.77–8.85. The EDTA-extractable Cu was 5.07 ± 1.95 mg/kg with a range of 1.17–11.31 mg/kg. These soils contained 1.27 ± 2.75 of DTPA-extractable Zn and 16.77 ± 8.32 mg/kg of DTPA-extractable Mn. The bioavailable Mo as extracted by ammonium oxalate/oxalic acid was in the range of 0.036–1.404 mg/kg, with an average of 0.26 mg/kg.

7.6.4.2 United States

Soils in the semi-arid grass land from New Mexico contained the following amounts of DTPA-extractable B, Cu, Cd, Pb and Zn: 0.02, 1.3, 0.01, 0.8 and 0.2 mg/kg, respectively (Fresquez et al., 1990). In two uncontaminated California soils (coarse-loamy, mixed, thermic Typic Haploxeralfs and fine-loamy, thermic Xerollic Calciorthids), DTPA-Zn was in the range of 4.26–5.47 mg/kg (LeClaire et al., 1984). The amounts of DTPA-extractable Cu, Ni, Pb, Se and Zn in an uncontaminated semi-arid soil from south-western Wyoming were 2.3, 0.4, 1.5, 0.02 and 1.3 mg/kg, respectively (Johnson and Vance, 1998). In a background Typic Torrifluvent from Arizona with pH 7.7, the amounts of DTPA-extractable Zn, Cu, Pb, Ni and Cr were 0.799, 2.98, 2.39, 0.103 and 0.406 mg/kg, respectively (Brendecke et al., 1993). Barbarick and Ippolito (2003) reported that soils (fine, smectitic, mesic Aridic Paleustolls and fine, smectitic, mesic Aridic Argiustolls) from Colorado contained 1.9–4.0 mg/kg DTPA-extractable Zn and 3.2–5.1 mg/kg DTPA-extractable Cu. Another Aridic Argiustolls soil from Colorado had 0.5 mg/kg DTPA-extractable Zn and 2.1 mg/kg DTPA-extractable Cu (Barbarick and Workman, 1987).

7.6.5 Middle East

7.6.5.1 Iraq (Sillanpaa, 1982)

Arid and semi-arid soils (n = 119) in Iraq had soil pH from 4.68–8.44. The $CaCO_3$ content in these soils varied from 2.1–43.5%, with an average of 25.2%. These soils contained 0.208–11.57 mg/kg hot water-extractable B, 0.091–4.628 and 4.68–24.83 mg/kg DTPA-extractable Zn and Mn, respectively. The average bioavailable B, Zn and Mn in these soils were 1.768 ± 1.781, 0.364 ± 0.416 and 11.83 ± 3.77 mg/kg, respectively. The average bioavailable Mo (extracted by ammonium oxalate/oxalic acid) was 0.160 ± 0.148 mg/kg, with a range of 0.025–0.853 mg/kg. The bioavailable Zn was very low compared to other arid and semi-arid soils. Bioavailable Zn, Mo and B had a greater variation than did bioavailable Cu and Mn in these soils.

7.6.5.2 Iran

In silt clay soils (0–30 cm) of Isfahan, Central Iran, the amount of EDTA-extractable Zn, Cu, Pb, Ni, Cd, Co and Cr were 3.2, 1.8, 2.6, 0.6, 0.16, 0.6 and 0.8 mg/kg, respectively (Khoshgoftarmanesh and Kalbasi, 2002). Concentrations of these trace elements increased in subsoils (30–60 cm) and increased with applications of municipal waste leachate. In the surface soils of agricultural, industrial and urban regions of Isfahan, central Iran, the average DTPA-Cd was 0.09 mg/kg, and about 80% of the soil samples had less than 0.1 mg/kg DTPA-extractable Cd (Amini et al., 2005). DTPA-Cd was strongly correlated with EC in the soils.

7.6.5.3 Israel

In alluvial and loessial soils, hot water-extractable B was in the range of 0.6–1.2 mg/kg. Desert alluvial soils contained 1.1–1.6 mg/kg of bioavailable B. The bioavailable B in brown soils of the semi-arid regions varied from 0.4–0.6 mg/kg. The bioavailable B in brown red sandy soil was in the range of 0.3–0.5 mg/kg. The Rendzinas Valley had 0.7–1.1 mg/kg bioavailable B. Muck soils had a very high soluble B content (5–24 mg/kg) (Ravikovitch et al., 1961). In 45 soils with 0.5–68% $CaCO_3$ and pH 7.0–8.25, the average concentrations of NH_4NO_3-extrctable Zn, Ni, Cu, Cr and Mn were 0.48 ± 2.16, 0.47 ± 0.95, 0.67 ± 1.16, 0.03 ± 0.09 and 4.93 ± 4.94 mg/kg, respectively (Banin et al., 1997a; Han and Banin, 1997).

7.6.5.4 Lebanon (Sillanpaa, 1982)

In 16 soils sampled from maize fields, hot water-extractable B and ammonium oxalate/oxalic acid-extractable Mo were 0.702 ± 0.208 and 0.100 ± 0.088 mg/kg, respectively. Bioavailable B and Mn varied from

0.442–1.261 and 0.018–0.325 mg/kg, respectively. The DTPA-extractable Zn and Mn and EDTA-extractable Cu were in the range of 0.273–2.17, 8.84–49.27 and 5.33–14.95 mg/kg, respectively, with an average of 1.105 ± 0.546, 23.27 ± 14.95 and 7.8 ± 2.6 mg/kg, respectively. These soils contained 0.4–67.1% $CaCO_3$ (average: 17.15%) with soil pH 7.1–8.1. The bioavailable Zn varied less than in the other arid soils of the Middle East.

7.6.5.5 Syria (Sillanpaa, 1982)

In 20 arid and semi-arid soils of Syria, the $CaCO_3$ content was from 6.8–59.2% with an average of 27.5%. The soil pH in wheat fields was in the range of 7.66–8.44 mg/kg. The DTPA-extractable Zn and Mn in these soils were 1.573 ± 2.249 and 19.37 ± 17.03 mg/kg, respectively. The EDTA-extractable Cu varied from 2.47–16.64 mg/kg with an average of 6.63 ± 2.86 mg/kg. Ammonium oxalate/oxalate acid-extractable Mo varied from 0.031–0.377 mg/kg. The average bioavailable Mn was 0.148 ± 0.095 mg/kg. The hot water-extractable B was in the range of 0.403–4.134 mg/kg with an average B of 1.352 ± 0.871 mg/kg. The bioavailable Zn and Mn were extremely unevenly distributed with a large variation.

7.7 MICRONUTRIENT DEFICIENCY IN ARID AND SEMI-ARID SOILS

7.7.1 Zn Deficiency

A high soil pH in arid soils frequently causes some micronutrient deficiencies. Since Zn and Cu are micronutrients to plants, severe Zn and Cu deficiencies have been observed in arid and semi-arid soils due to a high pH and the high binding of carbonate. In calcareous soils, Zn deficiency is prevalent. Zinc deficiency is reported to be widespread in Asian neutral, alkaline and calcareous soils. In the calcareous soils of northern China, the average DTPA-extractable Zn is 0.37 mg/kg, less than the threshhold value to avoid deficiency (0.5 mg/kg). Zinc deficiency in these areas has been reported in most crops and trees, such as rice, maize, sorghum, wheat, cotton, rape, apples, pearl, peach trees, soybeans, and other crops in the legume families (Table 7.8) (Liu, 1996). The North China Plain is the major area with a Zn deficiency in China. Liu (1996) pointed out that a Zn deficiency may occur in calcareous soils with a soil pH > 6.5, a soil organic matter < 1% and with a DTPA-extractable Zn < 0.5 mg/kg, but mostly in the Zn-sensitive crops listed above. In addition, overliming acid soils decreases the bioavailability of most trace metals, thus leading to Zn deficiency. It has been reported that extractable Zn, Zn activity in the soil solution, and plant uptakes of Zn decreased with increases in soil pH after liming (Friesen et al., 1980; Pepper et al.., 1983). Zinc deficiency can also occur in soils with a high content of soil phosphate due to possible Zn phosphate precipitation and in peat soils due to a lower total Zn content and strong organic complexation. Zinc deficiency has been also observed in newly leveled soils due to the low availability of Zn in these subsoils.

Table 7.8. Plants/animals and countries susceptible for micronutrient deficiency

Trace Element	Plants/Animals	Country
Zn	crops and trees such as rice, maize, sorghum, wheat,	North China Plain, North West China
	cotton, rape, apple, pearl, peach trees, soybean,	Turkey, India, Irag, Mexico, and Pakistan
	some crops in legume family	South/Western Australia
Cu	wheat, oats, barley, alfalfa, sunflower, water melon,	South /Western Australia, China,
	onion, spinach, lettuce, carrots, lucerne, Sudam grass,	Americas, Africa, Europe
	citrus, conifers, sheep	
Mn	oats, rye, wheat, rice, maize, peas, soy beans	South Australia, Indian, Syria, Pakistan,
	potatoes, cotton, tobacco, sugarbeet, tea, sugar-cane,	North China Plain, North West China
	pineapples, pecan, peaches, spinach, citrus,	
	a number of forest trees	
Co	sheep	Australia, China
Se	human beings	North China, Tibet Plateau of China

A common critical limit for Zn deficiency in soils has been 0.5 mg/kg DTPA-extractable Zn for different crops (maize, wheat, and rice) (Sillanpaa, 1982). DTPA-extractable Zn concentration of 0.5–1.0 mg/kg has been marginal for sensitive crops (Sillanpaa, 1982). Brown isohumic calcareous soils of New South Wales in Australia with 35–60 mg/kg of total Zn showed Zn deficiency due to the low bioavailability of Zn. Zinc deficiency has been reported to occur in wheat on solidized solonetz and solodic soils and other calcareous soils of South Australia. Zinc application at the rate of 0.6–28 kg/ha to cereals, pastures, and maize of Australia has been reported (Reuter, 1975). Zn deficiency occurs in Turkey, India, Iraq, Mexico, and Pakistan (Table 7.8). Zinc deficiency is frequently observed in rice on calcareous paddy soils.

Zinc fertilizer is used to supply plants to correct for Zn deficiency. In 1987, Zn fertilizers were used in an area of more than 2,670,000 ha in China (Zhu, 1996). Zinc inorganic salts, chelate forms, and biosolids such as sewage sludge have been applied to arid and semi-arid soils to correct for Zn deficiency.

7.7.2 Copper Deficiency

Copper deficiency in grain crops is known as reclamation disease, wither-tip, yellow-tip or blind ear, and in woody species such as in citrus, known as dieback or exanthema (Adriano, 2001). The EDTA-extractable Cu concentrations in global soils for Cu deficiency have been suggested to be in the range of 1.0–1.3 mg/kg (Sillanpaa, 1982). The critical DTPA-extractable Cu concentration in most soils is 0.2 mg/kg. Copper is an essential micronutrient to plants. Below this critical concentration, Cu deficiency may occur. The critical Cu concentration in plant shoots has been reported to be between 2.5–5.2 mg/kg in wheat and other crops and vegetables (Sillanpaa, 1982; Adriano, 2001). Copper deficiency has been reported to occur in wheat on solidized solonetz and solodic soils and other calcareous soils of South Australia as well as in the America, Africa, Asia, and Europe (Adriano, 2001). The critical concentrations of EDTA-, DTPA- and $Ca(NO_3)_2$-extractable Cu in these soils were reported in the range of 0.4–0.8, 0.1–0.2 and 0.005–0.015 mg/kg, respectively (King and Alston, 1975). Plant and soil analyses were able to predict Cu deficiency in wheat in the region. Furthermore, copper deficiency in cereals was more obvious during drier years, especially drier springs (Reuter, 1975). This was because the root system of Cu-deficient plants was reduced during drier seasons. In addition, copper deficiency was also widely observed in sheep during seasons with lush pasture growth in the spring (Reuter, 1975). The duration of Cu application at the rate of 0.7–23 kg/ha to cereals of Australia has been reported from 2 to 10 years (Reuter, 1975).

7.7.3 Manganese Deficiency

Manganese deficiency occurs principally in soils with a high pH or calcareous soils since Mn in these soils is mostly present in insoluble oxides. Manganese deficiency has been found for more than 20 crops including oats, rye, wheat, rice, maize, peas, soy beans, potatoes, cotton, tobacco, sugar beets, tea, sugar-cane, pineapples, pecans, peaches, spinach, citrus, and a number of forest trees (Table 7.8) (Sillanpaa, 1982). The critical DTPA-extractable Mn for Mn deficiency has been suggested to be 1.6–3.9 mg/kg, and soils with up to 5.2–6.5 mg/kg DTPA-extractable Mn has been considered to indicate susceptibility (Sillanpaa, 1982).

Manganese deficiency has been reported to occur on the coast calcareous soils of South Australia. Manganese deficiency may also occur in the soils of India, Syria, and Pakistan (Table 7.8). In China, most Mn-deficient soils are in the arid zones of the North China Plain and North West regions.

7.7.4 Cobalt Deficiency

An adequate supply of Co in the pasture is important to the health of grazing animals. Ruminants are reported to suffer from Co deficiency, recognized locally under different names such as Bush Sickness, Denmark Wasting Disease, Coasty Diseas, and Pining (McKenzie, 1975). In plants, Co is required by legumes for optimal functioning of symbiotic nitrogen fixation. Rumen microbes require Co for the synthesis of vitamin B12. Soil is the main source of Co for grazing grass and crops.

Cobalt deficiency has been reported in ruminant animals from feed in Australia with inadequate Co. When Co concentration in plants was less than 0.04–0.07 mg/kg, Co deficiency occurred (Kubota, 1980). The bioavailability of Co in soils primarily controls Co in plants. Cobalt deficiency in sheep is widely reported in South Australia. However, Co deficiency varies greatly from year to year. This may be related to plant uptake and the bioavailability of Co in soils as affected by rainfall. Han et al. (2002b) reported that short saturation of arid soils increased the bioavailability of Co in arid soils. The solid-phase binding and forms of Co in arid soils are sensitive to changes of redox potential. Submergence affects redox potential, pH, and the breakdown of organic matter with the production of CO_2. These processes influence the the multi-equilibrium of Fe and Mn in soils. Therefore, other metals including Co may be transformed from one component to another in the soil. Extractable Co and plant uptake of Co were found to increase significantly in waterlogged soils (Beckwith et al., 1975; Berrow et al., 1984). The duration of a cobalt application at the rate of 0.06–0.24 kg/ha to a pasture in Australia has been reported to last from 2–5 years (Reuter, 1975).

In the United States, the lower Atlantic Costal Plain is a region with low soil Co, producing less than 0.07 mg/kg in legumes. Cobalt deficiency in animals has not been reported in the rest of the U.S.

7.7.5 Selenium Deficiency

Selenium deficiency in humans occurs in the loessial region of the North China and Tibet Plateau of China (Liu, 1996). The arid and semi-arid soils in these areas of China contain low total Se in their parent materials.

7.8 TOXICITY

The toxicity of trace elements in plants depends upon the nature of the trace elements, soil properties, and plant tolerance. In general, Cd is the most toxic heavy metal, followed by Ni, Cu and Zn (Mitchell et al., 1978). Nickel, Cu and Zn have equal toxicity to wheat in calcareous soils. Nickel is more phytotoxic to lettuce and wheat grown in acid soils than in calcareous soils, whereas Cd, Cu and Zn toxicity depend on the plant species and metal concentration ranges (Table 7.9) (Mitchell et al., 1978). Sensitive plant species have lower critical soil concentrations that cause adverse effects on both plant growth and yields. In a California Domino silt loam soil amended with Cd-enriched sewage sludge with a pH of 7.5 and 1.0% of $CaCO_3$, critical soil Cd concentrations were much lower for Cd-sensitive plants, such as spinach, soybean, curly cress, and lettuce than for tolerant plants, such as rice (Bingham et al., 1975, 1976) (Table 7.9). Critical soil Cd concentrations producing a 25% yield decrease were between 4–13 mg/kg for Cd-sensitive plants (spinach, soybeans, curly cress, and lettuce), while soil Cd concentrations ranged from 15–146 mg/kg for sudan grass (*Sorghum halepense* Pers var. *Sudanese Hitche*.), alfalfa (*Medicago sativa* L.), clover (*Trifolium repens* L.), fescue (*Festuca elatior* L.) and Bermuda grass (*Cynodon dactylon* Pers), and were approximately 170 mg/kg for tomato and cabbage plants. Paddy rice was tolerant even at 640 mg/kg soil Cd concentration. The corresponding tissue Cd concentrations producing a 25% yield reduction were 9–43 mg/kg for sudan grass, alfalfa, clover, fescue and Bermuda grass. The critical Cd leaf concentrations for rice, soybean and field beans were 3, 7 and 15 mg/kg respectively, while the critical Cd concentrations in their edible fruits/seeds were 2, 1.7 and 7 mg/kg, respectively. However, the critical Cd leaf concentrations increased to 121, 125, and 160 mg/kg for turnips, tomatoes, and cabbage, respectively. Most agricultural crops and vegetable crops (lettuce, corn, carrots, radishes, wheat, and squash) have 32–75 mg/kg critical Cd leaf concentrations for a 25% yield reduction. In a study with four different arid soils from southern California with pH 7.4 – 7.8 and 1.2 – 28.3% of $CaCO_3$, Mahler et al. (1978) reported that the critical soil Cd concentrations producing 50% yield reduction were 139 and 250 mg/kg for lettuce (*Lactuca sativa* var. *longifolia*) and Swiss chard (*Beta vulgaris* car. *cicla*) respectively, with corresponding tissue concentrations of 160 and 203 mg/kg, respectively (Table 7.9)

In a hydroponic experiment, Page et al. (1972) reported that the growth of beets, beans, and turnips was reduced by 50% at Cd solution concentrations of 0.2 mg/L, while solution concentrations increased to 1 mg/kg for corn and lettuce. Tomato and barley produced 50% growth reductions at a 5 mg/L solution concentration, whereas with cabbage it was at 9 mg/L. This indicates that the tolerance of plants to trace elements (Cd) varies.

Table 7.9. Diagnostic soil and plant tissue Cd criteria for a 25% and 50% yield decrements[a]

Plant	Soil pH	DTPA Extract from Soil mg/kg	Saturation extract from Soil mg/L	Shoots/leaf mg/kg	Yeild component
			25% Yield Decrement		
Sudangrass	7.5	11	4	9	Shoots
Alfalfa	7.5	22	6	24	Shoots
White clover	7.5	29	7	17	Shoots
Tall fesue	7.5	71	9	37	Shoots
Bermudagrass	7.5	107	9	43	Shoots
Spinach	7.5	2.4		75	Shoots
Soybean	7.5	3		7	Dry bean
Curlycress	7.5	4.8		70	Shoots
Lettuce	7.5	7.8		48	Head
Corn	7.5	10.8		35	Kernal
Carrot	7.5	11		32	Tuber
Turnip	7.5	16.8		121	Tuber
Field bean	7.5	24		15	Dry bean
Wheat	7.5	30		33	Grain
Radish	7.5	57.6		75	Tuber
Tomato	7.5	96		125	Tipe fruit
Zucchini squas	7.5	96		68	Fruit
Carrot	7.5	102		160	Head
Rice	7.5	>384		3	Grain
			50% Yield Decrement		
Lettuce	7.4	135	6	200	Shoots
	7.5	55	10	190	Shoots
	7.7	15	4	121	Shoots
	7.8	25	5	120	Shoots
Chard	7.4	215	9	190	Shoots
	7.5	70	13	214	Shoots
	7.7	150	8	180	Shoots
	7.8	175	7	178	Shoots

[a]From Bingham et al, 1976; Mahler et al., 1978.

Besides yield reduction, the other symptoms of Cd toxicity in plants (sweet corn, lettuce, red beets, field beans, turnips, peppers, and tomatoes) include chlorosis, reddish-orange coloration of the leaf margin, necrosis, and wilting (Page et al., 1972). Even at the levels of trace elements almost to the USEPA limits, potential toxicity of these trace elements might occur to some crops and plants. Yield reductions at Cu, Zn and Ni concentrations far below the USEPA-503 limits in nonacid soils have been observed in sludge field experiments, particularly with dicotyledonous crops (McBride, 1995). Based on long-term field studies with a variety of crops in soils amended with sewage sludge, McBride (1995) concluded that the ultimate impact of toxic metals from sewage sludge at levels approaching the proposed USEPA limits on various soil-crop systems was potentially harmful. The soil adsorption capacity for trace elements increased in sludge-amended soils, but this protection may not be permanent or effective for all toxic metals. The USEPA limits have strongly relied on field observations for metal uptake by corn, which has led to an underestimation of phytotoxicity thresholds applicable to a wide range of crops, partly because corn is able to root deeply and is metal-tolerant (McBride, 1995). He further pointed out that the decision to use 50% yield reduction and plant shoot concentrations of heavy metals as phytotoxicity indicators may have obscured incipient toxicity. Many long-term field experiments in arid soils have shown high bioavailability of heavy metals in sludge-amended soils. This will be discussed in detail in the later chapters.

Chapter 8

TRACE ELEMENT POLLUTION IN ARID ZONE SOILS

The input of toxic trace elements into soils is increasing as a result of the reuse of reclaimed sewage water for irrigation, the disposal of wastewater sludge and municipal refuse, industrial activities, and atmospheric fallout. Compared to air and water pollution, the clean up of soil pollution will take a much longer time. When pollution sources are removed, the atmosphere is cleansed immediately, rivers generally within a year or so and only small, enclosed water bodies may present a long-term problem. However, soil pollution with toxic trace elements including heavy metals may be virtually "permanent" (Davies, 1980). Soil is a perfect sink for toxic trace elements which may rapidly accumulate in soils, but they are only slowly depleted by leaching, plant uptake, and erosion. Widespread increases of trace metals including Cd, Hg and Pb have been estimated to be on the order of 10–15% since the turn of the century in agricultural soils in Europe (McBride, 1995). The global perspective of post-industrial age pollution will be discussed in detail in the next chapter. This chapter addresses the sources and pathways of pollution in arid and semi-arid regions. Monitoring and remediation of polluted soils will be discussed as well.

8.1 SOURCES OF POLLUTION IN ARID AND SEMI-ARID SOILS

8.1.1 Irrigation with Sewage or Reclaimed Sewage Water

Heavy metals may enter soils through irrigation with reclaimed sewage water. This is especially important in arid and semi-arid areas where water supplies are limited and fresh water is very valuable. The agricultural sector is the largest user of water in the arid and semi-arid areas. Irrigation with reclaimed sewage water is the most readily available and economically feasible way to supplement the fresh water in these areas. In Israel, sewage irrigation on a large scale was initiated in 1972, and about 66% of the effluents were reused for irrigation or recharging the aquifer, as compared to about 2.4% of the reclaimed municipal waste water in the U.S., and about 11% of the total annual flow in Australia (Avnimelech, 1993). In fact, Israel has gained valuable experience from the practice of irrigating with recycled sewage water for the past 30–40 years (Avnimelech, 1993; Banin et al., 1987). In the U.S., there have been more than 3000 land application sites (Chang et al., 2002), and in California alone, 496 x 10^6 m^3 of reclaimed water is beneficially used annually. Approximately 70% of the total reclaimed water is used in agricultural and landscape irrigation (Chang et al., 2002). In Mexico City, Mexico, a metropolis of 15 million inhabitants, more than 70,000 hectares of crop land are irrigated with reclaimed waste water. In Athens, Greece, 7 x 10^5 m^3/day of primary effluents are reused with a majority of the applications used for crop land and landscape irrigation (Chang et al., 2002). Sewage effluents have been used in irrigation for over 60 years in Cairo, Egypt (Chang et al., 2002). In addition, waste water reclamation and reuse in irrigation has been reported in Saudi Arabia, South Africa, The Netherlands, Germany, and Italy. However, in some developing countries, untreated waste water and even raw sewage has been used for agricultural irrigation. In China, at least 1.33 x 10^6 hectares of agricultural land are irrigated with untreated or partially treated waste water from cities (Chang et al., 2002).

Sewage effluents contain plant essential nutrients. Typical Israeli sewage effluents were reported to have 20–100 mg/L total N, 6–20 mg/L total P and 10–50 mg/L K in raw sewage, and have 30–60 mg/L total N, 8–18 mg/L total P and 14–45 mg/L K in oxidation ponds (Avnimelech, 1993). The supply of nutrients in sewage water irrigation is similar in range to that for the conventional fertilization of corn, cotton, citrus, and Rhodes grass. For the irrigation of 2000–4600 m^3/ha of corn, 90–300 kg N/ha and 16–100 kg P/ha are supplied through sewage water applications, while irrigation with sewage water provides 4000–8000 m^3/ha of cotton with 120–360 kg N/ha and 32–100 kg P/ha. Nutrient supplies are exceptionally variable with sources of sewage and seasonal changes even within the same source and the same reservoir. Plant nutrient deficiency may occur at the beginning of the season if nutrients are supplied only through irrigation water. Additionally, irrigation with N rich water may encourage undesirable vegetative growth, e.g., cotton, later in the season.

The general amounts of trace elements in sewage treatment plant effluents are dependant upon a number of conditions including the source, location, regional industrial activities, and sewage water treatment processes such as the separation of biomass as sludge from the effluents. Conventional waste water treatment includes a primary and secondary treatment process for removing the organic matter and suspended and colloidal solids in the waste water. In general, primary and secondary treatments may remove an overall 50%, 65%, 76%, and 79% of the Ni, Zn, Cr and Cu, respectively (Chang et al., 2002). With the advancement of treatment technologies, trace element concentrations in treated effluents have decreased. As a result, the concentrations of trace elements in the treated effluents are generally low and usually meet the USEPA requirements for agricultural irrigation (Table 8.1). However, these elements are elevated in sludge through the subsequent stabilization and volume reduction processes.

Table 8.1. Agricultural irrigation water standards

Element	US Irrigation Standards[a, b]	
	Long-term	Short-term
	mg/L	
B	0.75	2.0
Cd	0.01	0.05
Co	0.05	
Cu	0.2	5.0
Cr	0.1	20
Ni	0.2	2.0
Mo	0.01	0.05
Pb	5.0	20
Zn	2.0	10

[a] National Research Council, 1996
[b] Avnimelech Y. 1993.

Typical concentrations of trace elements in treated sewage effluents in the U.S., southern California, and Israel are listed in Table 8.2. Data from bimonthly analyses of sewage treatment plant effluents from seven local treatment plants located in inland southern California showed high B concentrations, while all other trace elements were generally low (Bradford et al., 1975). The average values of Cd, Co, Cu, Cr, Ni, Pb and Zn were 0.01, 0.005, 0.021, 0.007, 0.04, 0.023 and 0.05 mg/L, respectively. Typical sewage effluents from Dan and Haifa, two sewage plants in Israel, contained less As, Cd, Pb, B, Mo and Co, but had high concentrations of Cr, Cu and Zn (Avnimelech, 1993). However, concentrations of trace elements varied with treatment plants. Cadmium in sewage effluents from Natanya, Givat Brenner, and Tirat Hacarmel, three sewage treatment plants in Israel, was between 0.01–0.02 mg/L (Banin et al., 1981). Nickel, Cu, Pb, Zn and B in Givat Brenner sewage effluents were 0.029, 0.044, 0.051, 0.002 and 0.062 mg/L, respectively (Hardiman et al., 1984a, b). Most of the trace elements and heavy metals in sewage water were below the long-term irrigation standards suggested by USEPA (Table 8.1). Only Cd in the sewage water from California and Cd and Cr from Israel were close to or at the limit. All of these elements were also lower than the amounts permissible for sludge application. In Israel, limited industrial heavy metal inputs and their effective removal from sewage have resulted in low heavy metal content of sewage effluents. Presently in the U.S., most reclaimed municipal waste water meets current irrigation water quality criteria.

Table 8.2. Trace element composition of sewage treatment plant effluents from the U.S., southern California, and Israel

Element	US[a]		Southern California[b]		Israel[c,d]
	1996		1975		1981
	Range	Mean	Range	Mean	Range
			mg/L		
B	0.0003-0.0025	0.0005	0.3-2.5	1.00	0.0005
Cd	<0.005-0.22	<0.005	<0.005-0.22	0.01	0.01-0.02
Co	<0.005-0.05	0.02	<0.001-0.01	0.005	<0.004
Cu	0.005-0.05	0.02	0.006-0.053	0.021	0.031-0.056
Cr	<0.001-0.1	0.001	<0.001-0.1	0.007	0.063-0.073
Ni	0.005-0.5	0.01	0.003-0.6	0.04	0.022-0.038
Mo	0.001-0.02	0.005	0.001-0.018	0.008	<0.004
Pb	0.001-0.2	0.005	0.003-0.35	0.023	0.01-0.022
V			<0.001-0.009	0.004	
Zn	0.01-0.4	0.04	0.004-0.35	0.05	0.107-0.541
Hg	<0.002-0.01	0.002			
As	<0.005-0.02	<0.005			<0.005

[a]National Research Councial, 1996
[b]Bradford et al., 1975
[c]Banin et al., 1981.
[d]Avnimelech Y. 1993.

The long-term application of reclaimed sewage for irrigation purposes results in the accumulation of trace elements in arid soils. Banin et al. (1981) found that prolonged irrigation (28 years) with treated sewage effluents from rural sources led to the accumulation of heavy metals (Cd, Cu, Ni and Pb) in the top 10–15 cm layer of the coastal plain soils of Israel (Table 8.3). Statistically significant differences were found for the total content and saturated paste extractable fractions of Cd in all three soils and for Cu, Ni and Pb in clay soils when comparing these fields with adjacent ones irrigated with normal water. However, chromium concentrations were not statistically different for any of the soils. This results in the increased uptake of certain metals by crops, and their possible introduction to animals and humans through the food chain. Long-term irrigation with sewage effluents increased the total amount of Cd per year in the plow layer of some soils in Israel by 5–10% (Avnimelech, 1993). Hardiman et al. (1984a, b) reported that long-term irrigation with reclaimed sewage effluents increased the trace element concentrations in soil solutions as well. Cadmium, Pb and Cu in soil solutions (saturated paste extracts) ranged from 1.5–2.4, 2.0–3.1, and 11.4–23.4 µmol/kg (ppb), respectively. Both Cd and Cu concentrations, in soil solutions of soils with long-term irrigation, were 10 times higher than those in control soils while Pb was 2–3 times higher than those in control soils.

Table 8.3. Concentrations of trace elements in Israeli arid soils (0-30 cm) irrigated with secondary sewage effluents (from Banin et al., 1981).

Extraction	Element	Years of irrigation with secondary sewage effluents					
		12		18		28	
		Sewage Effluents	Normal water	Sewage Effluents	Normal water	Sewage Effluents	Normal water
Total mg/kg	Cd	2.8	2.0	1.3	1.0	2.2	1.3
	Cr	119	114	74	68	86	75
	Cu	43	35	24	21	35	31
	Ni	63	51	21	20	39	31
	Pb	19	14	7.1	6.0	5.0	4.9
	Zn	82	70	44	40	76	52
Saturated paste extract mg/L	Cd	2.4	1.7	1.9	1.2	3.1	0.9
	Cr	6.6	5.7	7.1	5.9	3.2	3.0
	Cu	62	49	22	14	43	27
	Ni	47	32	20	11	35	20
	Pb	10.3	7.4	5.7	4.1	4.9	4.3

Bradford et al. (1975) reported that the concentrations of trace elements (B, Cd, Ni, Pb, Cu and Zn) in saturated extracts were greater in Californian soils irrigated with sludge treatment plant effluents than non-irrigated soils. On the other hand, long-term irrigation with sewage effluents in sandy or cracked limestone regions (Karst areas) resulted in the build–up of these trace elements in ground water through either colloidal particle enhanced leaching or through cracks (Avnimelech, 1993). Yet, in clay-rich calcareous soils of arid and semi-arid regions, the mobility of trace elements and heavy metals was limited.

In Central Iran, irrigation with sewage increased EDTA-extractable metals (Ni, Cr, Co, Pb, Cu, Zn and Mn) in soils. At high irrigation rates (600 tons/ha), sewage irrigation increased the concentrations of some heavy metals in rice. However, these concentrations remained below the critical levels of these metals (Khoshgoftarmanesh and Kalbasi, 2002).

In China, an estimated 8.4% of the 2.1×10^6 hectares of water-irrigated farm lands are seriously polluted (Chang et al., 2002). Almost 50% of the total areas showed accumulation of pollutants in soils (Chang et al., 2002). Some arid and semi-arid areas around big municipals in North China have received long-term irrigation with waste water heavily polluted by industrial waste discharge. As a result, cadmium and other heavy metals have accumulated in these soils making rice grown on these soils inedible due to the very high concentrations of Cd.

Treated sewages are used for irrigation in agricultural lands in Mexico, Jordan, and Tunisia. But in India, China, and Pakistan, a high proportion of the untreated raw sewages used for irrigation contain disease-causing pathogens and toxic waste from industry. Sewage is probably the biggest source of water for urban farming, especially in the arid and semi-arid regions. Sewage irrigation provides an estimated one fifth of the world's food.

Selenium is also a problem common with irrigation. Selenium has been reported in the arid and semi-arid soils of the western U.S. (17% of western U.S. land) comprising some 1.5 million acres of farm land on Cretaceous sediments (Thompson et al., 2003). In the Central Valley of California, high levels of Se are found in irrigated agricultural lands. Some drainage water Se concentrations can reach 1.4 mg/L. In California, the origin of Se in the sedimentary rocks of the Coast Range Hills comes from marine algae (Horne, 2000). Elevated levels of Se can lead to mortality, developmental defects, and reproductive failure in migratory aquatic birds and fish. Similar Se toxicity problems occur in other arid and semi-arid regions of the world with alkaline, seleniferous soils derived from marine sediments. Selenium problems have been reported in Colorado, the Dakotas, and Northwestern China. Naturally occurring selenium in these sediments and soils is assumed to be incorporated in pyrites in marine sedimentary rocks such as shales (Brown et al., 1999). When these soils are irrigated for agricultural purposes, indigenous Se is solubilized and moves by agricultural drainage waters where it accumulates in ponds and reservoirs. Selenium can be concentrated in waterborne plants and animals (3000 mg/kg). Serious Se problems have been reported in the Kesterson National Wildlife Refuge in Merced Country and the San Joaquin Valley of California (Brown et al., 1999). Selenium in these contaminated soils from the Kesterson Reservoir area is present as both selenate and selenite in the top few cm of soil but is reduced to elemental selenium at lower soil levels (Pickering et al., 1995; Tokunaga et al., 1997). Reduction of selenate is both biotic and abiotic (Oremland et al, 1989; Losi and Frankenberger, 1998; Myneni et al, 1997) and the reduction process is rapid (Tokunaga et al., 1996, 1997). Reduced elemental Se can be also reoxidized to the selenate form during irrigation, resulting in higher mobility since selenate is transported as an aqueous complex in drainage water.

8.1.2 Land Application of Sewage Sludge

Sewage sludge is another potential source of heavy metals in soils. Sludge production in the United States reached 4.5×10^6 tons in the early 1980's, and the European Economic Community exceeded 6×10^6 tons of dry solids annually during that same period (Lake et al., 1984). An estimated 5.3×10^6 tons per year (dry weight–DW) of sludge are currently produced in the United States from publicly owned treatment plants (NRC, 1996). Israel produced about 60,000–70,000 tons of sludge (DW) annually in sewage water treatment plants (Banin et al., 1990). This has no doubt increased substantially along with a growth in population.

Arid and semi-arid soils are low in organic matter as well as many other essential macro and micronutrients needed by plants. The application of N and P fertilizers to arid and semi-arid lands significantly increases yield and improves crop and forage quality. However, the use of these fertilizers is costly and their effectiveness is dependent upon the precipitation, whereas organic amendments are relatively inexpensive. Sewage sludge that is either incorporated or surface-applied improves soil physiochemical conditions, such as soil texture and water capacity by building up organic matter. Infrared analysis found polysaccharidic and proteinaceous materials in the soluble and insoluble light organic sludge components (Steinhilber and Boswell, 1983). This indicates that aromatic compounds were in the insoluble light fraction. Pascual et al. (1997) reported that soil biological quality (soil biomass carbon, basal respiration and others) was improved by the organic amendment of arid soils with sewage sludge and other municipal solid waste. Another study showed that four years of applications of sewage sludge to an arid soil of Arizona had no significant adverse effect on soil microbial populations and activity (Brendecke et al., 1993). Banin et al. (1990) reported that the application of sewage sludge in arid Israeli soils increased soil (0–30 cm) organic matter content which increased with the application rates of the sludge. The recovery of organic matter in Israeli arid soils after the application of sludge decreased with higher amendment rates. Also, the large amounts of organic matter had a significant effect on the bulk density of the top ten cm of the soils.

Sewage sludge also provides essential nutritional elements (N, P, K, and some micronutrients – Fe, Zn and Cu) for agricultural crops, especially in many calcareous, arid and semi-arid soils deficient in major- and micronutrients. In the Upper Rio Puerto watershed of west-central New Mexico, one of the most severely eroded and degraded watersheds in the U.S., soils with a pH of 7.8 are low in soil organic matter and plant-available N. These soils are low in fertility with a low productivity of plant biomass. Fresquez et al. (1990) reported that the application of dried anaerobically digested sewage sludge linearly increased soil N, P and K as well as increased the levels of DTPA-extractable micronutrients (Cu, Mn and Zn). Total forage yields increased two- to three-fold in sewage-sludge amended soils compared to the unamended treatment. The levels of tissue N, P and K, and the crude protein in blue grama (*Bouteloua gracilis* (H.B.K.) Lag. ex Steud), galleta (*Hilaria jamessi* (Torr.) Benth.), and bottle brush squirrel tail (*Sitanion hystrix* (Nutt.) J.G. Sm.) increased linearly with sludge applications. All micronutrients increased to the levels found in efficient plant tissues.

The application of sewage sludge to agricultural land is considered a suitable method for its disposal. Agricultural use of sewage sludge consumes approximately 44%, 20–25%, and more than 31% of sludge production in the U. K., Germany, and the United States, respectively (Feigin et al., 1991; Lake et al., 1984; Bilitewski et al., 1994). In California, over 60% of the publicly owned treatment works in the state employs land application in sewage sludge management (Chang et al., 2002). In 14 western European countries, an average of 38% of sewage sludge has been used for agricultural land application, while 43% has been used for landfills and 10% for incineration (Chang et al., 2002).

However, sewage sludge has elevated levels of potentially toxic trace elements (Zn, Cd, Cu, Pb, Cr and Ni), which may exceed natural soil concentrations by two orders of magnitude or more (Jung and Logan, 1992; Lake et al., 1984; Essington and Mattigod, 1991). In general, sewage sludge contains high concentrations of toxic trace elements and heavy metals (Table 8.4). The bulk of the total element mass is associated with light-density materials (Essington and Mattigod, 1990). The heavy-density fractions (> 2.96 $Mg\ m^{-3}$) of the sludge and amended soil contain barite, celestite, lead silicate, lead phosphate, chromium oxide, or oxyhydroxide, sphalerite and chalcopyrite (Essington and Mattigod, 1991). Moreover, trace element concentrations in sewage sludge are highly variable depending on the sources and process treatments. When industrial waste contributes significantly to the waste water flow, the sewage sludge contains much higher contents of trace elements. In non-industrial communities, the trace elements concentrations are lower. The median values for Cr, Cu, Pb, Ni and Zn in digested sewage biosolids in the U.S. were reported at 500, 800, 50, 80 and 1,700 mg/kg, respectively (Mortvedt, 1996). In general, sewage sludge contained less As, Cd, Hg, Mo and Se as compared to Zn, Cu, Ni, Cr and Pb. The median values for As, Cd, Hg, Mo and Se in sewage biosolids were 10, 10, 6, 4 and 5 mg/kg, respectively. However, some samples had 3000–5000 mg/kg of Cd and Ni. Mercury in sewage sludge of the U.S. was less than 56 mg/kg, Se less than 17 mg/kg, and Mo and As less than 210–230 mg/kg. Typical sewage sludge (both composted and liquid) from Los Angeles County Sanitation Districts contained 30–106, 275–599, 712–1,702, 655–1,775 and 1,785–4,643 mg/kg Cd, Ni, Cu, Pb and Zn, respectively (Sposito et al., 1982). Metals in composted sewage sludge were in the following order: Zn > Pb > Cu > Ni > Cd (Sposito et al., 1982). In Israel, sewage sludge from Dan, Natania, Herzeliyah, and Haifa had average values of 2,527; 249; 55.6; 338; 12.76; 270; 7.45; 5.67; 1.65 and 2.49 mg/kg Zn, Cu, Ni, Cr, Cd, Pb, Co, Mo, Se and As, respectively (Banin et al., 1990).

Table 8.4. Trace element composition of sewage sludge from the U.S. and Israel

Element	US[a,b]	Southern California[c]		Israel[d]
	Total	Total	Saturation extract	Total
	mg/kg	mg/kg	mg/L	mg/kg
B	4-757	74-680	2.1-17	
Cd	1-3410	1-140	0.05-1.2	2-38
Co	1-18	3-230	0.02-0.2	2-16
Cu	84-17000	200-1050	0.14-24	66-630
Cr	10-99000	60-400	<1.00.18-18	110-993
Ni	2-5300	10-2140	0.1-0.37	15-243
Mo	1-214	2-25	0.13-2	2-10
Pb	13-26000	15-1700	0.03-0.1	36-680
V		265-1200	0.5-1.5	
Zn	101-49000	265-1150		940-7200
Se	2-17			0.3-5.3
As	1-230			0.4-5.9

[a]Chaney R.L., 1983.
[b]National Research Council, 1996.
[c]Bradford et al., 1975
[d]Banin et al., 1990

The solubility of trace elements in sewage sludge is strongly dependent upon the nature of the elements and the sludge properties. Water soluble (water and 0.06 M $CaCl_2$-extractable) Pb, Cu, Cd and Zn in air-dried and digested sludge are < 1%, 2%, 11% and 36%, respectively (Lagerwerff et al., 1976). Anaerobic processes result in low water solubility of these elements. Anaerobic incubation of sludge for 1 week had < 0.1%, < 0.03%, < 0.1–6.2%, 1.5–9.2% and < 0.01–0.8% of Cd, Pb, Cu, Ni and Zn, respectively (Stover et al., 1976). Long-term anaerobic incubation for three months decreased the water solubility of Cu and Ni, but increased Cd, Cr and Pb (Bloomfield and Pruden, 1975). Subsequent aeration increased the water solubility of Cr, Cu and Zn. Besides water extraction, other single extractants such as EDTA, acetic acid, and diluted HCl also have been used to indicate the bioavailability of these trace elements in sludge (Lake et al., 1984).

In sewage sludge, heavy metals are predominantly in the carbonate bound, the residual, and the organically bound fractions. However, the major forms of metals in sludge are dependent upon the sources of the waste water and treatment processes.

Cadmium in sewage sludge is primarily present in the carbonate fraction. Banin et al. (1990) reported that Cd in sludge from Israel was mainly bound to the carbonate fraction (46%), followed by the easily reducible oxide (17%), the organically bound (15%), and the exchangeable fractions (11%). Sposito et al. (1983) found that most of the Cd in the sludge was in the carbonate fraction, followed by the organically bound and the residual fractions. Emmerich et al. (1982) found that the Cd in sludge from Los Angeles was primarily present in the carbonate bound fraction (56%), and to a lesser extent, in the residual (36%) and the organically bound fractions (22%). McGrath and Cegarra (1992) reported that the Cd in sludge was mostly present in the carbonate, the residual, and the exchangeable fractions. Candelaria and Chang (1997) found that the Cd in sewage sludge predominately resided in the EDTA-extractable fraction (mainly the carbonate bound).

Copper in sludge was predominately found in the organically bound fraction. The Cu in sludge from Israel prevailed in the organically bound portion (80%) (Banin et al., 1990). Sposito et al. (1983) and Emmerich et al. (1982) also reported that Cu mainly occurred in the organic fraction, followed by the carbonate and residual fractions.

Iron in the sludge was mainly (> 90%) present in the residual fraction according to the findings of Knudtsen and O'Connor (1987). In the sewage sludge from Israel, Fe was found predominately in the reducible oxide bound (38.8%) and the residual (35%) fractions (Banin et al., 1990).

Generally speaking, Ni was found to be present in approximately equal amounts in the carbonate (32%), the residual (26%), and the organically bound fractions (24%) of sludge (Emmerich et al., 1982). Likewise, McGrath and Cegarra (1992) reported that Ni found in sludge was mostly in the residual, the organically bound, and the carbonate portions.

Banin et al. (1990) reported that most of the Pb in Israeli sewage sludge was bound in the reducible oxide fraction (46%), followed by the readily reducible oxide (20%) and the carbonate (18.7%) fractions. McGrath and Cegarra (1992) found that Pb mainly existed in the residual and the carbonate fractions.

Zinc in sludge from Israel had been reported to occur in the carbonate (34.5%), the easily reducible oxide (26.5%), and the organically bound (31%) fractions (Banin et al., 1990). Emmerich et al. (1982) reported that Zn in sludge predominately existed in the carbonate bound (57%) and the organically bound fractions (28%). Knudtsen and O'Connor (1987) found that 53% of the Zn in sludge occurred in the carbonate fraction. McGrath and Cegarra (1992) reported that Zn in sewage sludge mainly resided in the carbonate, the organically bound, and the residual fractions. In addition, Cr in sludge mostly occurred in the residual and the carbonate fractions (MaGrath and Cegarra, 1992).

Long-term applications of sewage sludge and irrigations with recycled sewage water result in a substantial accumulation of heavy metals in the soils and the crops grown in them (Fig. 8.1) (Chang et al., 1984; Bingham et al., 1976; Jung and Logan, 1992; Sposito et al., 1982; McGrath and Cegarra, 1992; Lake et al., 1984; Essington and Mattigod, 1991; Beckett, 1989; Villarroel et al., 1993). Soil amendment with municipal sewage sludge is permitted by the USEPA Clean Water Act 503 Regulations (1993). This regulation allows the accumulation of trace elements such as Cr, Cd, Cu, Pb, Hg, Ni, Se and Zn from levels of 10– > 100 times the levels present in the background concentrations of these metals in most soils (McBride, 1995). The continued application of sewage sludge to arid soils in California has significantly increased soluble Cd concentrations (extracted by 0.015M $CaSO_4$ + 0.03M NaCl) and organic matter accumulation in the soils (Hyun et al. 1998). Furthermore, the sustained applications of sludge in the second decade increased the soluble Cd concentrations in the soils by two–three times as compared to that after the first nineteen years of application.

Organic matter added to arid soils in the forms of sewage sludge and other solid waste is decomposed following the model $C = C_0 (1-e^{-kt}) + C_1$ (Pascual et al., 1998). The decomposition is initially a rapid process of mineralization, followed by a second slower phase. With decomposition, trace elements originally bound in organic materials are released into the soils and soil solution, and they become available to plants.

However, ground water, as opposed to plants, could be in danger of potential contamination in the long-term, especially in light textured soils. Nickel, Cu, Zn, Cd and Pb have increased in Californian sludge-amended soils proportionally to the sludge applications (Sposito et al., 1982; Chang et al., 1984). The addition of the sludge to Israeli soils increased the concentrations of Cu, Zn and Cd in the top soils by up to 183%, 349%, and 305%, respectively (Banin et al., 1990). After five years of land applications with sludge in a semi-arid site of south-western Wyoming, the total Cu, Ni and Pb levels significantly increased in the upper soil layers (Johnson and Vance, 1998). Also, Cu, Se and Zn concentrations in 15 grass species significantly increased (Johnson and Vance, 1998). In the arid southwestern desert soils of Arizona, four years of annual sludge applications increased DTPA-extractable Zn, Cu, Pb and Ni even though the applications did not significantly increase the total metal concentrations (Brendecke et al., 1993). Fortunately, there are very few observations of pollution in groundwater due to long term land applications of sewage sludge.

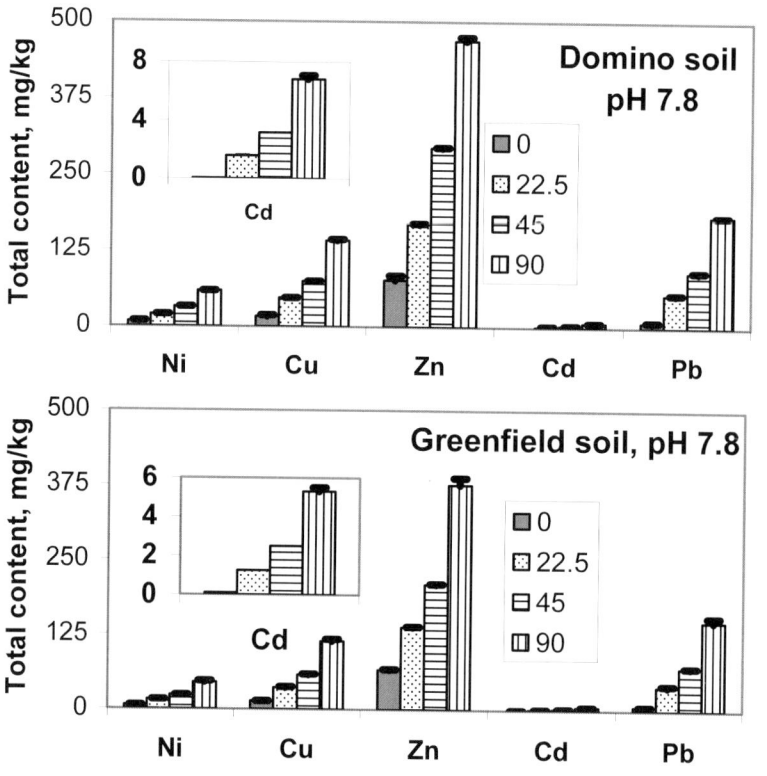

Figure 8.1. The accumulation of trace elements in two Californian soils. The soils received four-year sludge applications at the rate of 0, 22.5, 45, and 90 tons ha^{-1} year^{-1}. Domino loam soil, fine-loamy, mixed, thermic Xerollic Calciorthid; Greenfield sandy loam soil, coarse-loamy, mixed, thermic Typic Haploxeralf (Data extracted from Sposito et al., 1982)

Therefore, land application of sewage sludge at reasonable rates does not significantly increase the solubility of soil heavy metals and other toxic elements. Fresquez et al. (1990) reported that sewage sludge applied at the rate of 45 Mg ha^{-1} in the arid soil of west-central New Mexico gave the most favorable soil and vegetation results. Zinc, Fe, Mn and Cu in sludge-amended soils increased the supply to forage growth, but Cd and Pb did not significantly increase. Levels of Cd and Pb in all forage tissues did not increase significantly as a result of the sludge amendment. It was concluded that the application of sludge after the second growing season did not significantly increase the solubility of Cd and Pb. Brendecke et al. (1993) reported that the application of sludge to arid southwestern desert soils had no significant effects on the microbial populations or activity in the soils. Both the annual and total cumulative trace element loading rate limits and maximum permissible concentrations for land application of sewage sludge in several countries are listed in Tables 8.5, 8.6, and 8.7. These limits in many European communities are considerably lower than those regulated by the USEPA. McBride (1995) pointed out that even at the levels for trace elements that approach the USEPA limits, potential toxicity of heavy metals to some crops and plants might occur. Yield reductions in nonacidic soils at Cu, Zn and Ni concentrations far below the USEPA-503 limits have been observed in sludge field experiments, particularly with dicotyledonous crops (McBride, 1995). The USEPA limits strongly rely on data from field studies for the metal uptake by corn, but corn is a deeply rooted plant and is generally tolerant to many heavy metals. Therefore, the current USEPA limits lead to an underestimation of the phytotoxicity of these toxic elements in regard to a wide range of crops (McBride, 1995). He further pointed out that the decision to use a 50% yield reduction and plant shoot concentration of heavy metals as phytotoxicity indicators may have obscured the incipient toxicity. Actually, the availability of metals is best measured from the contents of these metals in the plant roots rather than the tops (McBride, 1995). Many long-term field experiments in arid soils have shown high bioavailability of heavy metals in sludge-amended soils.

Table 8.5. Cumulative loading rates of trace elements for land applications of sewage sludge in selected countries[a]

Element	US	European Union[b]	Spain	China	Canada
		kg ha^{-1}			
As	41				
Cd	39	1.4-5.4	1.4	1.2	2.6
Cr	3000	60-260	60		248
Cu	1500	60-240	60	200	140
Pb	300	60-560	60	700	86
Hg	17	1.8-2.8	1.8		0.6
Mo	18				
Ni	420	20-110	20	120	72
Se	100				
Zn	2800	200-5200	200	600	326
B					

[a]Data derived from Chang A.C. et al., 2002.
[b]Recommended level and mandatory level, respectively.

Table 8.6. Maximum annual loading rates of trace elements for land applications of sewage sludge in selected countries[a]

Element	US	European Union	Spain
		kg ha^{-1} yr^{-1}	
As	2		
Cd	1.9	0.15	0.15
Cr	150	3	
Cu	75	12	12
Pb	15	15	15
Hg	0.85	0.1	0.1
Mo	0.9		
Ni	21	3	3
Se	5		
Zn	140		30

[a]Data derived from Chang A.C. et al., 2002

Table 8.7. Maximum permissible trace element levels of sewage sludge for land applications in the U.S. and China[a]

Element	Maximum permissible mg kg^{-1} (dry weight)		
	US	China	
		Acid Soil[b]	Neutrla/Alkaline soil
As	75	75	75
Cd	85	5	20
Cr	3000	600	1000
Cu	4300	250	500
Pb	840	300	1000
Hg	57	5	15
Mo	75		
Ni	420	100	200
Se	100		
Zn	7500	500	1000
B		150	150

[a]Data derived from Chang A.C. et al., 2002
[b]Soil pH < 6.5.

Currently, there are two hypotheses proposed to describe the phytoavailability of trace elements in sewage sludge amended soils: the plateau theory and the sludge time bomb hypothesis (Chang et al., 1997). The plateau theory states that the metal adsorption capacity added with sewage sludge will persist as long as the metals of concern persist in the soil and the metals will remain in chemical forms not readily available for plant uptake (Chang et al., 1997). Thus, metal concentrations in plants will reach a plateau after the termination of sludge application. The sludge time bomb hypothesis proposes that the metal adsorption capacity of the soil is augmented by organic matter added as sewage sludge. This capacity will revert back to its original background level with time following the termination of the sewage sludge applications as the mineralization of the organic matter releases metals into more soluble forms, thus becoming a time bomb (Chang et al., 1997). Chang et al. (1997) found that the conditions necessary for the plateau and time bomb theories to take place may be found from current long-term experiments. However, an actual plateau or time bomb is not evident from the ten years of experimental data in the soils of California, U.S. in which the sewage sludge applications reached 2,880 Mg/ha. Hyun et al. (1998) found that when sewage sludge applications were terminated, the soluble Cd concentrations of the sludge-treated soil were higher than those in the control soils and did not decline significantly over the next ten years. Yet, the soluble Cd concentrations or the phytoavailability of Cd in the sludge-treated soils did not increase as the organic C in these soils declined over the ten years following the termination of sewage sludge application. However, with continuous sludge applications, the soluble Cd concentration of sludge-treated soils increased with each incremental addition of sludge for the same ten years.

8.1.3 Atmospheric Inputs and Traffic Contributions

Atmospheric inputs have been reported as one of the major contributors of heavy metals and other trace elements in soils. Since there are less forests and land plant covers in arid and semi-arid regions than in humid and temperate regions, wind erosion increases the role of transport of trace elements and heavy metals in these regions. Hinkley et al. (2002) reported that the finer-grained components of atmospheric dust in the southwestern U.S. were significantly enriched in many trace elements (Pb, Cd, Cu, Se, Zn, As, Sb and Bi). Commonly these trace elements were found to be at levels of 10–100 times above the average levels found in the earth's crust. The degree of enrichment in the dusts was about the same as that in pre-industrial dusts preserved in Antarctic ice and in European peat bogs, which have natural sources of enrichment. The similar levels of trace element enrichment of the dusts appear to indicate that naturally-occurring high amounts of trace elements, as seen in the pre-industrial samples, are possibly not a significant component of the current dust in the southwestern U.S. High enrichment in current dust may be related to increasing anthropogenic pollution.

Analysis of the Greenland ice core covering the period from 3,000–5,000 years ago (Greek, Roman, Medieval and Renaissance times) revealed that Pb was present at concentrations four times as great as the natural values from about 2,500–1,700 years ago (500 B.C to 300 A.D.) (Hong et al., 1994). These studies show that Greek and Roman lead and silver mining and smelting activities polluted the middle troposphere of the Northern Hemisphere on a hemispheric scale two millennia ago, long before the industrial revolution. Cumulative Pb fallout during these eight centuries was as high as 15% of that caused by the massive use of Pb alkyl additives in gasoline since the 1930s (Hong et al., 1994). Pronounced Pb pollution also was observed during Medieval and Renaissance times. Furthermore, the accumulative deposition from anthropogenic sources in pre-industrial times (600 BC to AD 1800) was at least as large as the cumulative deposition during the Industrial period (AD 1800 to the present) (Renberg, 1994). This will be further discussed in the next chapter.

The elements Cd, Zn and Pb are considered to be atmophile elements which, on the global scale at present, have greater mass transport through the atmosphere than through streams (Banin et al., 1987; Stumm and Morgan, 1982). Their input into forest floors and forest soils may, therefore, be strongly influenced by atmospheric fallout. Many studies found that Cd, Zn and Pb were enriched in the forest floor and in the top layers of the mineral soils in forests, and the accumulation rate had increased over the last few decades (Heinrichs and Mayer, 1980). The point sources emitting trace elements included mining and smelting, refineries and factories, chemical plants, power stations and incinerators.

Another important source of nonpoint source pollution emitters is traffic sources. Banin et al. (1987) reported that a significant accumulation of the atmophile elements, Cd, Zn and Pb was observed in the uppermost thin layers of an Israeli soil. The soil was beneath the canopy of pine trees and surrounding the tree trunks, located in a roadside forest site. These elements were non-homogeneously present in the top few millimeters of the soil with concentrations as high as 1.5, 6, and 15 times the natural soil background concentrations of Zn, Cd and Pb, respectively. A zone highly contaminated with these elements was also present in the mineral soil immediately adjacent and underneath the tree trunks. This was possibly due to stemflow conduction of the air-deposited metals into the soil. High concentrations of these elements in the local environment of trees may create a zone of long-term, high metal availability. The adsorption capacity of the soil may be saturated, and the excesses of these elements will be mobilized in the soil solution and be bioavailable to trees. In forests and other natural ecosystems, where no mechanical soil-mixing occurs, the steady-state distribution with depth of these elements will show higher concentrations of these elements at locations where atmospheric fallout is concentrated on the soil surface. The increased Pb accumulation in the roadside soil of Israel was due to the use of leaded gasoline for two decades. The Cd accumulation was also most likely related to vehicle traffic, resulting from wear and tear of car-tire treads containing Cd and the combustion of gas and motor oil (Banin et al., 1987). Moreover, the accumulation of heavy metals from traffic source increased with decreases in distance from the roads. The deposition of air-borne metals from vehicles is highly directional. For roadside sites in the U.S. near the Santa Ana Freeway in southern California, Page et al. (1971) had reported fall-out of 592, 351, 184, 108 and 71 mg Pb m^{-2} $year^{-1}$ at the leeward side of the road at distances of 15, 30, 46, 107 and 168 m, respectively. Banin et al. (1987) estimated that 14075 mg Pb m^{-2} had accumulated in the 0–20 cm layer of the arid soil in Israel over a 20-year period (from 1967 to 1987) with an average range of 700 mg Pb m^{-2} $year^{-1}$ near and beneath the tree trunks, where the distance from the trees to the road was 26 m and was virtually parallel to the road direction. However, in the soils outside the margin of the canopy, only 1774 mg Pb m^{-2} had accumulated in the 0–29 cm soil layer with an average of about 90 mg Pb m^{-2} $year^{-1}$. This indicated that deposits from traffic sources are highly variable and highly directional in relation to the deposition of air-borne trace elements. However, with the increasing use of clean gasoline, such as the elimination of tetraethyl lead as a gasoline additive beginning in the 1970's in the U.S. and some other countries, the contribution of air-borne trace elements from traffic sources has dropped significantly as compared to the 1960s-1970s.

Banin et al. (1987) proposed a thin-horizon sampling approach to study the effects of traffic sources and atmospheric fallouts on soils in the arid zone of Israel. They pointed out that the large and systematic variability in the concentrations of the atmophile elements in the soil would be masked if a more conventional and less-detailed sampling scheme was used. The measured Pb concentration varied between 209 mg/kg in the top layer of an arid soil near the road in Israel and 66 mg/kg at a depth of 20 cm (Banin et al., 1987). If the profile had been sampled as one 0–20 cm horizon, the weighed average concentration observed would have been 76.3 mg/kg Pb. If it had been sampled in two 10–cm thick horizons, the concentrations would have been estimated to be 86.6 and 66 mg/kg in the top and bottom horizons, respectively (Banin et al., 1987). This distribution would strongly affect the bioavailability in arid soils.

When the toxic element is at low concentrations in the soil, it is more likely that it will be fixed to the solid matrix of the soil. As the total concentration increases, the fixation capacity becomes saturated. Thus a larger proportion of the total metal will be found in the soil solution and will be more available to plants. The gradual decrease in the distribution of the elements in the top soils and usage of the traditional one-top-layer sampling may underestimate the potential toxicity and bioavailability of air-deposited elements in the ecosystems. The difference between the actual concentration profiles in the soil and those obtained by analysis of samples from the thick layers in non-disturbed soils in natural ecosystems will be maximal for those metals that are air-transported and deposited from the atmosphere. This is particularly true if their mobility in the arid soils is low due to intense chemical interaction, including adsorption, precipitation or chelating.

8.1.4 Agricultural Management

Agricultural activities such as the application of fertilizers, organic compost and animal wastes, as well as pesticide or herbicides increase the concentrations of many trace elements. Cadmium, As and Cr are the three most relevant elements in agricultural management. Commercial phosphate fertilizers have small amounts of trace elements, including Cd, since phosphate rock contains various trace elements (Table 8.8). On a global basis, phosphate rocks contain the ranges of 1–225 mg/kg Cr, 3–300 mg/kg V, 2–37 mg/kg Ni, 0.1–60 mg/kg Cd, 1–15 mg/kg As, 3–35 mg/kg Pb and 0.01–0.06 mg/kg Hg. The global averages of 91% of the phosphate rock reserves have 188, 88, 25, 29, 11, 10 and 0.05 mg/kg of Cr, V, Cd, Ni, As, Pb and Hg, respectively. However, some phosphate rocks from the western U.S. contain as much as 130 mg/kg Cd (Mortvedt et al., 1981). Phosphate rocks from North Africa and Morocco have the highest concentrations of Cd (30–60 mg/kg), while those from Russia and South Africa have the lowest Cd (0.1–0.2 mg/kg). Cadmium concentrations in some Australian phosphate rocks ranges from 4–109 mg/kg (Mortvedt et al., 1981). Phosphate rocks from Morocco also have the highest Cr (225 mg/kg) while those from South Africa and Russia the lowest Cr (1–13 mg/kg). The highest Pb (35 mg/kg) is found in phosphate rocks from South Africa and the highest V is found from North Africa (300 mg/kg). Phosphate rocks from Russia contain the lowest Ni, As (1–2 mg/kg) and Hg (0.01 mg/kg).

The transfer of trace elements in phosphate rocks to P fertilizers is dependent upon the manufacturing processes. Triple superphosphate fertilizer contains 60–70% of the Cd present in phosphate rocks (Wakefield, 1980). The transfer coefficients may be similar for most other elements and heavy metals even though there are little data on the transfer of other elements from phosphate rocks to P fertilizers. In general, based on some long-term (> 50 years) soil fertility experiments in the U.S., annual Cd rates from the application of phosphate fertilizers are estimated to range from 0.3 to 1.2 g per ha. The addition of Cd to soils as a contaminant from P fertilizers at these rates does not appear to result in increased Cd levels in plants as a result of long-term P fertilization (Mortvedt, 1987). Moreover, the bioavailability of trace elements in P fertilizers is lower in calcareous arid soils than in acidic soils.

Table 8.8. Average concentrations of trace elements in phosphate rocks (PR)[a]

Country	As	Cd	Cr	Pb	Hg	Ni	V
				mg/kg			
Russia	1	0.1	13	3	0.01	2	100
USA	12	11	109	12	0.05	37	82
South Africa	6	0.2	1	35	0.06	35	3
Morocco	11	30	225	7	0.04	26	87
North Africa	15	60	105	6	0.05	33	300
Middle East	6	9	129	4	0.05	29	122
Average of 91% of PR reserves	11	25	188	10	0.05	29	88

[a]Kongshaug et al., 1992.

On the other hand, the long-term application of pesticides and other agricultural chemicals containing trace elements, such as As, Pb and Cu, increase their concentrations in soils (Han et al., 2000, 2001b, 2001c, 2004a). Compounds containing arsenic have been widely used as pesticides, insecticides, herbicides, soil sterilants, silvicides and dessicants over the past century. Inorganic arsenicals including As_2O_3, $NaAsO_2$, $Pb_3(AsO_4)_2$, CaH_2AsO_4 and Paris green have been used extensively in agriculture and forestry. Agricultural use of these arsenicals peaked in the 1940's, when it accounted for more than 90% of the domestic arsenical use. Currently, agricultural use of As accounts for 15% of arsenical use (Loebenstein, 1994). In addition, methylarsonic acid, dimethylcalciumpropylarsonate, calciummethylarsonate and dimethylarsenic acid have been applied to crops, orchards, turf and have been used in silviculture. From about 1965 to 1992, arsenic acid had been used to desiccate cotton plants in Texas and Oklahoma to remove leaves for mechanical cotton picking (Loebenstein, 1994). Surely these agricultural practices increased As concentrations in soils. Moreover, local disposal of these inorganic arsenic pesticides along with cotton waste containing arsenical pesticides contaminated the groundwater of Texas. Arsenic concentrations in groundwater occurring near Knott, Texas were found in the range of < 0.05 to < 2,500 mg/L (Welch et al., 2000). In addition, many formulations of Cu-containing fungicides, such as Bordeaux mixture and copper oxychloride, are used on many fruits and vegetables. Some soils in South Australia (peach orchards) contain 210 mg/kg Cu due to fungicide sprays (Tiller and Merry, 1981). The average Cu concentration in 95 Australian orchard soils was 99 mg/kg with a range of

18–320 mg/kg. Most of the Cu accumulated in the surface horizons (Tiller and Merry, 1981).

The long-term application of animal wastes and organic composts increases both the total contents and the bioavailable fractions of trace elements in soils. Trace elements are found in animal manures since these trace elements are added to their diets as additives. Swine manure usually contains high concentrations of Cu, Mn and Zn. Copper is regularly used to treat lameness in dairy cattle and is used as a growth promoter in swine. In swine and poultry manures, Cu concentrations have been reported primarily in the range of 899–1,550 and 19–1,196 mg/kg, respectively (Payne et al., 1988; Bolan et al., 2003). The total Cu concentration ranged from 0.1–1.55 mg/L and from 0.5–10.5 mg/L in swine and diary effluent, respectively (Bolan et al., 2003). Most of the Cu in both the effluent and solid materials was organically complexed. In swine manure, the major portions of Cu, Mn and Zn were in the organic, the oxide and the carbonate bound fractions (Hsu and Lo, 2000). Synthetic acid rainwater and neutral-pH extractable fractions of these trace elements from swine manure were low (< 10%) (Hsu and Lo, 2000). In soils with pH 7.8–8.9 from the North Eastern Spain regions, the application of composted cattle manure increased the total Zn and EDTA- and DTPA-extractable Zn in soils (Ramos and Lopez-Acevedo, 2004). All of these fractions of Zn increased with the contents of organic matter in soils. Long-term applications of poultry litter have been reported to increase accumulation of trace elements, such as Cu and Zn, even in the humid soils (Han et al., 2000). Chicken litters have high As and other trace elements. Arsenic is added in various organic forms, such as *p*-arsanilic acid and roxarsone, to poultry diets for the control of coccidial intestinal parasites and to improve feed efficiency. Han et al. (2004a) found that As in chicken litter is bioavailable, and water soluble As ranged from 5.3–25.1 mg/kg, representing 36–75% of the total As in chicken litter.

Agricultural activities and management do not significantly affect the solubility of other trace elements. In a tropical semi-arid soil, long-term cultivation and management history did not affect the concentrations of DTPA-extractable Zn, water-soluble Zn, or exchangeable and organically complexed Zn (Agbenin, 2003).

8.1.5 Mining and Industrial Activities

Mining and industrial activities produce large quantities of volatiles and dust particles and increase concentrations of trace elements and heavy metals in soils, waters, and vegetation. When mineral deposits containing concentrated trace elements and heavy metals are exposed at the earth's surface, trace elements are released. Trace elements enriched in coal and carbonaceous materials include Hg, F, Cr, Mo, Pb, As, Se, Cd, and B. Grinding, smelting, processing and refining of ores and metal products pollute the surroundings of smelters through metal enriched effluents and dust or particulates. Crushing ore and waste materials increases the surface area of minerals exposed to weathering processes. The oxidation of sulfide minerals containing many trace elements results in acidic mine drainage, enhancing the mobility and bioavailability of these elements. In southeastern Spain under subtropic arid and semi-arid conditions with scarce precipitation, the city of Cartagena has experienced mining and metallurgic activities since ancient times. Cartagena was once a primary center for the extraction and exploitation of lead and silver for the citizens of Carthago and Rome. It is one of the most important lead reserves in Europe. In the atmosphere of Cartagena, high concentrations of Pb, Cd and Zn are found in both dust-fall and suspended particulate matter. The soils in the vacant industrial areas (abandoned lead smelter locations) contain $150,000 \pm 173,740$ mg/kg Pb, $26,809 \pm 25,580$ mg/kg Zn, $9,300 \pm 15,910$ mg/kg Cd, and $2,939 \pm 7,100$ mg/kg Cu (Martinez Garcia et al., 2001). Soils in this old town have $5,179 \pm 6,590$, $2,133 \pm 7,880$, $36,907 \pm 67,860$ and $3,048 \pm 9,960$ mg/kg Pb, Cd, Zn and Cu, respectively. But soils in the adjacent agricultural land outside the city contain $1,256 \pm 3,100$; 14 ± 20; $2,122 \pm 5310$ and 37 ± 30 mg/kg Pb, Cd, Zn and Cu, respectively. Soil pH in abandoned smelter soils is lower than soils in the town and in agricultural soils. These data indicate that mining and lead smelter activities resulted in the extreme accumulation of Pb, Zn, Cd and Cu in the Cartagena soils. Even agricultural soils around the region were strongly contaminated by these heavy metals during the long-term civilization. Soils in these regions not only acted as sinks for various trace elements and heavy metals, but also acted as strong polluted sources to adjacent areas.

The Tharsis mine is another mining area located in the semi-arid Huelva province of southwestern Spain. This region is part of the Iberian Pyrite Belt, one of the largest belts of massive sulfide deposits in the world. The sulfide mineralization occurred in the lower unit of the volcanic-sedimentary complex, composed of mafic and felsic volcanic rocks intercalated with marine schists (Chopin et al., 2003). The major type of minerals is the massive sulfide ore bodies (95% pyrite, FeS_2; chalcopyrite, $CuFeS_2$; sphalerite, ZnS; and galena, PbS) and a Cu-rich stockwork zone which holds most of the remaining mineralization. The mining at Tharsis dates back to 2500 BC with significant mining operations during the Roman era (201 BC–475 AD for Ag, Au, Pb and Cu) and from 1850 to the present day (for Fe, S and Cu) (Chopin et al., 2003). Pre-Roman smelting produced two million tons of As-rich slag, some containing up to 23% As. Roman smelting produced 18 million tons of slag enriched in Fe, Cu, Zn, Co, Au, Ag and Ni. During the period from 1850 to present day, about 49 x 10^6 tons of ores have been extracted (Chopin et al., 2003). Large scale mining with a continuous mining industry has resulted in the production of large quantities of mining waste. The mining waste has an enormous potential for contaminating the surrounding areas with metals and trace elements. In the soils at Tharsis, As, Cu, Pb and Zn are in the ranges of 6–6,300; 5–693; 14–24,800 and 16–420 mg/kg, respectively (Table 8.9). Lichen and shrubs grown in these soils contain elevated levels of trace elements (Table 8.9). Lichen was found to contain high Pb concentrations (118 mg/kg) with an average of 58 mg/kg Pb, which was much higher than those in uncontaminated soils (Chopin et al., 2003). Soils in the mining areas had the maximum concentrations of these metals, such as 24,800 mg/kg Pb and 6,300 mg/kg As. Concentrations of these metals decreased with increasing distances from the mining area. In soils 1.5–13 km away from the mine, As, Cu, Pb and Zn decreased to 20–22, 20–28, 32–103 and 79–94 mg/kg, respectively. Soils in the locations underneath the wind direction had higher concentrations of these elements since the wind brings all of the dust and aerosols containing high concentrations of these elements from the mining areas' deposits. The soil pH decreased to the range of 3.7–8.27 in the polluted areas due to the oxidation of metal sulfides. Lower soil pH values increased the mobility and bioavailability of these trace elements.

Table 8.9. Concentrations of trace elements in vegetation and soils at Tharsis mined areas of semi-arid South West Spain[a]

Samples		As	Cu	Pb	Zn
			mg/kg		
Lichen	range		12-35	9-118	43-68
	mean		20	58	53
Cistus ladanifer	range	0.18-9	7-37	2-195	36-173
	mean	5	13	35	88
Soil	range	6-6300	5-693	14-24800	16-420
	mean	302	123	1001	102

[a]Chopin et al., 2003.

Soils around the Pb-Zn Port Pirie smelter in the semi-arid region of South Australia were found to contain 2,220 mg/kg Pb (about 3 km away from the mine) (Cartwright et al., 1976). Lead concentrations decreased to 140 mg/kg at a distance of 16 km and were further reduced to 32 mg/kg at 33.5 km. In tailing piles of the Leadville mining area of Colorado, U.S., Pb concentrations ranged from 6,000–10,000 mg/kg (Brown et al., 1999). XAFS spectroscopy showed that 50% of the total Pb occurred as adsorbed complexes on iron (hydro)oxides in selected carbonate-buffered tailings with a near-neutral pH, whereas Pb in sulfide-rich low pH samples were dominated by Pb-bearing jarosite. There was no Pb adsorption in sulfide-rich low pH tailings. Total metals as digested with HNO_3 in Aspen (Colorado) silver mine dump materials ranged from 135–21,700 mg/kg Pb, 13.3–233 mg/kg Cd, and 144–20,000 mg/kg Zn (Boon and Soltanpour, 1992). Sulfide minerals in a large tailings dump in northern Idaho, U.S. were reported to be oxidized by bacteria, and the dissolution of ore minerals and metals into their sulfates occurred, discharging into the nearby rivers (Davies, 1980). In soils from Balya Maden, Turkey, As, Pb and Zn were 1,510; 31,650 and 45,440 mg/kg, respectively (Reeves and Baker, 2000).

Selenium toxicity occurring in the western U.S. was found to be associated with known seleniferous geological formations or areas of historic uranium exploration and mining (Boon, 1989). The mining of coal, bentonite, and uranium in the western U.S. has increased the potential for Se contamination in soils, surface water, and groundwater. Extremely high Se, As, and Mo values have been reported in areas of uranium enrichment in Utah, North Dakota, South Dakota, Colorado, Wyoming, New Mexico, and Arizona (Boon, 1989). Many of these areas were watersheds for reservoirs and irrigated agricultural land. Brown et al. (1999) used X-ray absorption fine structure (XAFS) spectroscopy to determine the oxidation state, local coordination environment and the relative proportion of different As species in three Californian mine wastes in the Mother Lode District. They included a fully oxidized tailing, a partially oxidized tailing and a roasted sulfide ore. Arsenic in fully oxidized tailings (Ruth Mine) was bound to the ferric oxyhydroxides and aluminosilicates (clay), while in the partially oxidized tailings (Argonaut Mine) 20% of the reduced As was bound in arsenopyrite (FeAsS) and arsenical pyrite ($FeS_{2-x}As_x$) and 80% As(V) was in a ferric arsenate precipitate such as scorodite. Roasted sulfide ore of the Spenceville Mine contained As(V) substituted for sulfate in the crystal structure of jarosite ($KFe_3(SO_4)_2(OH)_6$) and sorbed to hematite surfaces (Brown et al., 1999).

8.1.6 Serpentine and Seleniferous Soils

In addition to anthropogenic pollution, some serpentine soils derived from Fe and Mg-rich ultramafic rocks are enriched in Ni, Cr and Co. In North America, ultramafic rocks form two discontinuous bands along the east and west side of the continent. The largest area of ultramafic terrain is in the Klamath Mountains province of northern California and southern Oregon (Lee et al., 2001). Serpentinite is a metamorphic rock formed from low temperature (300–600 °C) hydrothermal alteration of igneous ultramafic rocks. The major minerals are olivine, orthopyroxene, clinopyroxene and chromite. Most ultramafic rocks have been partially hydrated or completely serpentinized. Soils formed from serpentinite contain an abundance of Fe, Mn, Cr, Ni and Mg and low concentrations of the plant-essential nutrients Ca and K (Lee et al., 2001). Some serpentine soils in California contain 1,691 mg/kg Cr and 4,483 mg/kg Ni (Reeves and Baker, 2000).

Seleniferous formations occur in the Great Plains region from Canada to Mexico, accounting for $> 700,000$ km^2 of the western U.S. Seleniferous soils are frequently associated with Se-containing geological formations (Boon, 1989). As discussed above, seleniferous formations occur in North Dakota, South Dakota, Montana, Wyoming, Colorado, Kansas, and New Mexico. Some soils derived from Se-rich parent materials, such as Cretaceous shales of the middle-western United States, have > 10 mg/kg Se and sometimes exceed 50 mg/kg (Reeves and Baker, 2000).

8.2 MONITORING POLLUTION IN ARID AND SEMI-ARID SOILS

8.2.1 Plants

Perennial plants, especially trees, may act as environmental monitors and historical records. The *Sequoia gigantea* of California (U.S.) can reach ages up to 2,000 years old, while the bristle cone pine (*Pinus longaevia*) forests have some specimens up to 4,900 years old (Davies, 1980). The annual growth cycles of trees produce annual rings. After years, the tree ring is metabolically inactive and stable in its composition. The annual rings in tree trunks contain the records of trace elements, which reflect the local environment in which it was formed. Sheppard and Funk (1975) reported that the historical changes of trace elements recorded in the rings of trunks of Ponderosa pine growing on the banks of the Spokane river, Idaho (U.S.) were in agreement with the trace element data as revealed by core data for an adjacent lake sediment. Both records of trace elements (in the tree rings and in the core samples from the lake) reflected the historical activities of the active Coeurd' Alene mining production.

8.2.2 Soil Microorganisms and Soil Enzymes

High concentrations of trace elements and heavy metals inhibit soil microorganisms and certain soil enzymes. The effects of trace elements on microorganisms include changes in the microbial population (microbial biomass), microbial community structure, microbial activities such as the mineralization of carbon and nitrogen, and enzyme activities. Pollution may lead to a decrease in microbial diversity. Cadmium can have deleterious effects on membrane structure and function by binding to the ligands, such as phosphate, and the cysteinyl and histidyl groups of proteins (Collins and Stotzky, 1989). Soil bacteria are more sensitive than fungi to Cd (Vig et al., 2003). Moreno et al. (1999) reported that high Cd concentrations in sewage sludge decreased soil microbial biomass carbon and stimulated the metabolic activity of the microbial biomass in sewage sludge amended arid soils from Spain with a pH of 8.1. The metabolic quotient (qCO_2) is a relationship between soil respiration and microbial biomass and is expressed as ng CO_2-C h^{-1} mg biomass C. The qCO_2 is used as a marker of the environmental stress of the microbial population. The metabolic quotient is a very sensitive index of the stress caused by the incorporation of Cd-contaminated sewage compost in the arid soil. Heavy metals at low levels do not significantly affect C/N mineralization, but at high levels they reduce the mineralization rates (Dar and Mishra, 1994). The effects of cadmium on soil microbial processes and soil enzyme activities have been extensively studied in soils with pH > 7.0 (Table 8.10). The microbial biomass (including bacteria and actinomycete biomass), nitrification and N-mineralization are inhibited by high levels of Cd in arid and semi-arid soils (Table 8.10). The addition of 10 µg/g Cd to Indian soils with a pH of 7.7–8.1 reportedly had no remarkable effect on C and N mineralization and microbial biomass, whereas significant decreases in these parameters were observed at 25 and 50 µg/g Cd in soil, irrespective of the sludge addition (Dar and Mishra, 1994). DTPA-extractable Cd exhibited a significant negative correlation with the microbial biomass ($r = -0.58^*$ to -0.86^*; $p < 0.05$).

Moreover, heavy metals were found to reduce enzyme activities by interacting with the enzyme-substrate complex by denaturing the enzyme protein and interacting with its active sites (Nannipieri, 1994). Cadmium bound to the active sites of the enzymes, such as alkaline phosphatase, which was inactivated by the disruption of its metabolism (McGrath, 1999). Several soil enzymes were proposed to be important to organic decomposition and agricultural ecosystem health. These included dehydrogenase, urease, phosphatase, protease, ß-glucosidase and others (Moreno et al., 1999). Dehydrogenase activity was proposed as an indicator of the total metabolic activity of soil microorganisms. Phosphatase was reported to be an important enzyme which catalyzes the hydrolysis of organic P to produce inorganic P, making it available to plants. Protease hydrolyzed N-α-benzoil-L-argininamide while ß-glucosidase was involved in the C cycle and hydrolyzed ß-glucosidic bonds of the carbohydrate chains. Researchers have found that Cd had a negative effect on the synthesis of the endocellular enzyme, dehydrogenase, but had a positive effect on urease, phosphatase (Moreno et al., 1999), as well as fungal biomass (Megharaj et al., 2003) (Table 8.10). In addition, protease and ß-glucosidase activities seemed to be unaffected by Cd concentrations in sewage sludge. Dar (1996) reported a decrease in both dehydrogenase and alkaline phosphatase activity at levels of 50 mg/kg Cd in different soils with pH 7.7–8.1 from India.

Table 8.10. Summary of the effects of Cd on microbial processes and enzyme activities in arid and semi-arid soils

Processes/enzymes	Soil pH	Reference
Inhibitory Effects		
Biomass	7.7, 7.8, 7.9, 8.1	Griffiths et al., 1997; Moreno et al., 1999; Dar, 1996; Dar and Mishra, 1994
Bacteria, actinomycetes	5.7-8.5, 7.7, 7.9, 8.1	Hattori, 1989; Dar, 1996
Respiration	5.7-8.5, 7.7, 7.9, 8.1	Dar and Mishra, 1994; Hattori, 1989
N-mineralization	5.7-8.5, 7.7, 7.9, 8.1	Dar and Mishra, 1994; Hattori, 1989
Nitrification	5.8-7.8	Liang and Tabatabai, 1978
Dehydrogenase,	7.7, 7.9, 8.1	Moreno et al., 1999, 2001; Dar, 1996
Urease	8.1	Moreno et al., 2001
Protease	5.7-8.5	Hattori, 1989
Arylsulfatase	7.6	Bewley and Stotzky, 1983
Alkaline phosphatase	7.9, 8.1, 7.7	Dar, 1996
Stimulation Effects		
Fungal	5.7-8.5	Hattori, 1989
N-mineralization	8.52	Hassen et.al, 1998
b-glucosidase	8.1, 5.7-8.5	Hattori, 1989; Moreno et al., 1999
Hydrolyzing N-a - benzoil-L-argininamide	8.1	Moreno et al., 1999
Protease	8.1	Moreno et al., 1999

A number of basic criteria have been proposed for a microbiological property to be an indicator in monitoring soil pollution by heavy metals (Brookes, 1995): (1) The property needs to be accurately and precisely measurable across a wide range of soil types and soil conditions; (2) Because a large number of samples usually have to be analyzed, it is preferable that the property can be easily and economically measured; (3) The property needs to be of a nature that control or background measurements also can be made so that the effects of the pollutant can be precisely determined; (4) The property needs to be sensitive enough to indicate pollution but also sufficiently robust not to give false alarms; (5) The property needs general scientific validity based upon reliable and contemporary scientific knowledge; and (6) Reliance upon a single property may be unsafe. Based on these criteria, three groups of microbiological properties might be considered as indicators of soil pollution by heavy metals. The first group of indicators are those that measure the activity of the whole microbial population (e.g., microbial respiration and soil N mineralization as discussed above). The second group measures the size of the microbial population at the single organism level, the functional group level or the whole population level. The third group is a combination of both activity and biomass data giving specific activities of the microbial population. Combining microbial activity and population measurements (e.g., biomass specific respiration) appears to provide a more sensitive indication of soil pollution by heavy metals than either activity or population measurements alone (Brookes, 1995).

8.2.3 Other Media

Other potential monitors of historical trace elements are core samples of ice, peat, and lacustrine deposits. Each individual stratus reflects the environment in which it formed. Analyses of these core samples could reconstruct the past local and regional environmental history.

8.3 REMEDIATION OF POLLUTED ARID AND SEMI-ARID SOILS

8.3.1 Physical and Chemical Remediation

Remediation approaches may generally be classified as physical, chemical, and biological approaches. The first two are referred to as engineering strategies; the latter as bioremediation. Physical remediation refers to soil excavation and encapsulation, "dig-and-dump", with subsequent disposal in landfills, solidification, and isolation. Solid waste on land can be encapsulated by use of impermeable base liners and surface covers. Limestone or clay liners for landfills may be an economical chemical barrier used to precipitate or adsorb soluble metals in waste. However, this does not address the issue of soil decontamination. In some instances, physical remediation is combined with *in situ* immobilization with chemicals. Other methods, such as soil washing with water, diluted acid or complexing agents (EDTA), have an adverse effect on biological activity, soil structure and fertility. In addition, these practices can lead to significant engineering costs, as well as possible leaching into groundwater.

Chemical remediation refers to the application of various minerals or chemicals to adsorb, bind, precipitate or co-precipitate trace elements and heavy metals in soils and waters thereby reducing their bioavailability, toxicity, and mobility. *In situ* immobilization refers to the treatment of contaminants in place without having to excavate the soils or waste, often resulting in substantial cost savings. However, *in situ* immobilization or extraction by these physicochemical techniques can be expensive and are often only appropriate for small areas where rapid and complete decontamination is required.

Sorption on mineral surfaces is one of the major processes that can bind and sequester trace elements. Sorption significantly reduces the mobility of trace elements in groundwater. The transformation of elements between oxidation states (through reduction or oxidation reactions) and redistribution among solid-phase components decreases the mobility and bioavailability of these trace elements (Han and Banin, 1997). The effectiveness of an adsorption reaction in binding an ion is determined by the following variables: pH, the charge on the mineral surface as a function of pH and the type of sorption complex formed. In addition, different ions can compete for the same types of reactive surface sites, and organic and/or inorganic ligands may inhibit or enhance sorption of a trace element. Similarly, surface coatings such as biofilms may block reactive sites and/or create new sorption sites (Brown et al., 1999).

The addition of minerals or materials of high sorption capacity (such as synthesized or natural zeolites, apatites, dolomitic lime, phosphate, Fe and Mn oxides, or organic matter residues) can increase the adsorption of these minerals and reduce the bioavailability of trace elements (Table 8.11). Lime, zeolite, and apatite significantly reduced the Zn concentration in tissues of maize (*Zea mays*) and barley (*Hordeum vulgare*) from spiked soils (Chlopecka and Adriano, 1996). The addition of synthesized zeolite pellets (at the application rate of 1% by soil weight) significantly reduced the concentrations of Cd in the roots and shoots of a range of crop plants (lettuce, *Lactuca sativa*; oats, *Avena sativa*; and rye grass, *Lolium perenne*) in soils with a pH of 7.8 (Gworek, 1992a). Lettuce leaves, rye grass, oat shoots, beet (*Beta vulgaris*) leaves, and beet roots grown in Pb contaminated soils amended with zeolites were 49–73%, 47–77%, 58–68%, 62%, and 26–83% lower in Pb content, respectively, than those grown in soils without the addition of zeolites (Gworek, 1992b).

Apatite is used to remediate Pb contaminated soils because apatite dissolution releases phosphate, which combines with Pb to form highly insoluble Pb-phosphate minerals. Apatites follow linear (zero-order) dissolution kinetics (Manecki et al., 2000) with rates of Pb uptake by the apatites decreasing in the same order as the apparent dissolution rate constants ($K_{Ap}°$): hydroxyapatite > chlorapatite > fluorapatite (Manecki et al., 2000). This suggests that Pb uptake is controlled by the total amount of dissolved phosphate in the system, and the rate-controlling step is the apatite dissolution (Manecki et al., 2000). However, it has been reported that chloropyromorphite can be formed from galena (PbS) in the presence of hydroxyapatite (Zhang and Ryan, 1999) where dissolution or oxidation of galena was the rate-limiting step (Zhang and Ryan, 1999). Galena was not stable when exposed to oxidizing conditions. The formation was favorable with increasing pH. The effects of Ca, K, Na, calcite and EDTA on Pb immobilization by hydroxyapatite have been investigated (Ma, 1996). Aqueous Ca and calcite slightly inhibited Pb immobilization, whereas EDTA significantly reduced the effectiveness of hydroxyapatite to immobilize Pb. Hydroxyapatite effectively immobilized > 71% of the aqueous Pb in $PbHAsO_4$ contaminated soils from Washington State (Ma, 1996).

Table 8.11. Chemicals and minerals for the remediation of trace element contaminated arid and semi-arid soils

Chemicals / Minerals	Elements	References
Apatite	Pb, Cd, Zn	Ma 1996 ; Chen et al., 1997
Zeolites	Zn, Cd, Pb	Gworek, 1992a,b; Chlopecha and Adrinao, 1996
Mn oxides	Cd, Pb	Mench et al., 1994
Phosphate rocks and salts	Pb, Cd, Zn	McGowen et al., 2001; Ma et al., 1995
Organic matter	Cr(VI)	Tokunaga et al., 2003; Oliver et al., 2003; Cifuentes et al., 1996
Nitrate	Cr(VI)	Tokunaga et al., 2003; Oliver et al., 2003; Cifuentes et al., 1996

In addition to Pb, mineral apatite has been used to remediate Cd and Zn contaminated semi-arid soils in Idaho, U.S. (Chen et al., 1997). Idaho sites were contaminated from mine tailings and smelter emissions. It is one of the largest and most complexly contaminated, abandoned hazardous waste sites in the United States. The heavy metal-bearing components in the soils are primarily sulfides and oxides including PbS, $PbSO_4$ and PbO. Due to the long-term oxidation of these sulfides, the soil pH has significantly decreased. Chen et al. (1997) found that the mineral apatite from North Carolina, U.S. was extremely effective in reducing the Pb leached from contaminated soils in Idaho by 99.5–99.9% (when using a pH 5–10 solution) and reduced Cd and Zn by 70–98%. They found that hydroxyl fluoropyromorphite was formed in the reactions of the apatite with aqueous Pb solution and contaminated soil leachates. Minor amounts of otavite were also formed in the reaction of the apatite with aqueous Cd solution. Surface complexation and coprecipitation were the most important mechanisms, with possibly ion exchange and solid diffusion also contributing to the overall sorption processes of Zn^{2+} and Cd^{2+} on hydroxyapatite surfaces (Xu and Schwartz, 1994). This indicates that mineral apatite could be a cost-effective option for the remediation of trace element-contaminated soils, wastes, and water.

In addition to apatite, diammonium phosphate and phosphate rocks have been directly used to remediate metal contaminated soils. Diammonium phosphate was found to effectively reduce heavy metal solubility and transport in smelter-contaminated soils (McGowen et al. 2001). Phosphate rocks had been found to be effective in immobilizing Pb from aqueous solutions (with a minimum Pb removal of 39–100%) and reducing water-soluble Pb from a $PbHAsO_4$-contaminated soil (by 57–100%) from Washington State (Ma et al., 1995). The primary mechanism of Pb immobilization was via dissolution of the phosphate rocks and precipitation of a fluoropyromorphite-like mineral (Ma et al., 1995). The addition of the phosphate rock reduced $Ca(NO_3)_2$ and HOAc-extractable Pb from the Pb-contaminated soils but had little effect on EDTA-extractable Pb (Ma et al., 1997a). The phosphate rocks converted Pb from the water soluble, the exchangeable, the carbonate, the Fe-Mn oxides and the organic fractions to the residual fractions (Ma et al., 1997b).

Manganese oxides have been used to remediate contaminated soils as well. Hydrous manganese oxides decreased Cd and Pb in rye grass (*Lolium* spp.) and tobacco (*Nicotiana tabacum* L.) from soils with pH 7.4–7.8 (Mench et al., 1994). Cadmium concentrations decreased in water and in 0.1 M $Ca(NO_3)_2$ and acetic acid extracts of treated soils (Mench et al., 1994).

Another important process is to precipitate or co-precipitate trace elements by raising soil pH. Liming is commonly used to raise the soil pH and precipitate trace elements to reduce their bioavailability and mobility. This method is especially suitable for the removal of heavy metals in soluble and ionic forms. Mitigation and remediation of acidic mining drainage include biologically enhancing the acidic mining drainage to remove sulfide, collecting the leachate and removing the metals, covering the mine wastes with lime or limestone, diverting water inflow to the waste or from the waste, capping the wastes with clay, using bactericides to prevent bacteria causing the oxidation of sulfide, and applying phosphates to the surface. However, these procedures are very costly and have varying degrees of success. In addition, electroreclamation can be applied for the removal of ionic species in contaminated soils.

8.3.2 Biological Remediation and Phytoremediation

Biological remediation is defined as "the use of living organisms to reduce or eliminate environmental hazards resulting from accumulations of toxic chemicals and other hazardous wastes." Chromium(VI) is more soluble, mobile, and toxic in soils, while reduced Cr(III) is nontoxic and stable in vadose zones of arid and semi-arid regions. Bioremediation has been successfully applied to remediate Cr(VI) contaminated vadose soils in arid and semi-arid regions. Chromium (VI) contaminated vadose soils are remediated by an *in situ* reduction process of Cr(VI) to Cr(III). This process is accelerated by the addition of organic carbon and nitrate (Tokunaga et al., 2003; Oliver et al., 2003; Cifuentes et al., 1996). Cifuentes et al. (1996) found that yeast extract amended arid soils from New Mexico removed more Cr(VI) than grass and manure amendments. Tokunaga et al. (2003) successfully remediated Cr(VI) contaminated Californian calcareous soils (with 1–4% calcite at a pH of 8.3) by adding organic carbon. Microbially dependent processes are largely responsible for Cr(VI) reduction in the thick vadose zones of arid and semi-arid regions (the efficiency could approach 100%) (Oliver et al., 2003). The rate constant of Cr(VI) reduction is linearly correlated with added organic carbon in soils and is slower in alkaline conditions. The effective first-order reduction rate constants range from 1.4×10^{-8} to 1.5×10^{-7} S^{-1} (Tokunaga et al., 2003). The rate constant increases when the initial Cr(VI) levels are lower and the organic carbon content is higher.

On the other hand, organic materials directly adsorb Cr(VI) and reduce Cr(VI) in arid soils. Many forms of organic carbon directly reduce Cr(VI), such as phenols, organic acids, humic substances, grass biomass, and organic manure. But in general, abiotic reduction in bulk sediment is minimal (Oliver et al., 2003). Tokunaga et al. (2003) reported that ferrous iron, native humic substances, and added organic carbon supported abiotic Cr(VI) reduction, which occurs at a high initial Cr(VI) concentration (10,000 mg/L). However, under slightly alkaline-oxidizing conditions, microbially dependent pathways dominated over abiotic reduction in soils contaminated initially with 1,000 mg/L Cr(VI).Other remediation processes for Cr(VI) contaminated soils include H_2S injection, aqueous Fe(II) injection, and the use of reduced Fe solids. Aqueous-phase Cr(VI)-Fe(II) redox reactions may be significant if Fe^{2+} concentrations are in equilibrium with relatively soluble, ferric hydroxide-like phases (Tokunaga et al., 2003). The overall interactions involving microbial activity, organic carbon degradation, Fe^{2+}, and mineral surfaces control the net rates of Cr(VI) reactions in soils.

Reduced Cr(III) may be re-oxidized into Cr(VI) by Mn oxides in soils. Chung et al. (2001) reported that native Cr(III) in subsurface materials of arid and semi-arid regions can be oxidized to soluble chromate by native manganese oxides. If subsurface materials contain a low content of Mn oxides, the re-oxidization of reduced Cr(III) is not significant.

The new trend in bioremediation is the use of phytoremediation as a cleanup tool. Phytoremediation technologies (i.e., processes using plant-based remediations) is an emerging system of strategies that uses various plants to degrade, extract, contain or immobilize contaminants from soil and water (Raskin and Ensley, 2000). Plant species for remediation of trace element contaminated soils are summarized in Table 8.12. Phytoextraction is the predominant process employed where metals and trace elements are concentrated in the stems and leaves of metal accumulating-plants (Raskin and Ensley, 2000). It can be employed *in situ* for the clean up of contaminated sites with widespread and low to medium level contaminants in near-surface groundwater or in shallow soils over long periods of time. Furthermore, the technique can be utilized with relatively low costs. The development of phytoremediation is being driven primarily by the high cost of other soil remediation methods as well as the desire to use an environmentally benign process.

Table 8.12. Plant species for the phytoremediation of trace element contaminated soils

Plants	Trace elements
Brassica juncea	Pb, Cr(VI), Cd, Cu, Ni, Zn, ^{90}Sr, B, and Se
Thlaspi caerulescens	Ni, Zn
Thlaspi rotundifolium ssp. *cepaeifolium*	Pb
Alyssum wulfenianum	Ni
Buxacease and Euphoribiacease families	Ni
Brassica napus	Se, B
Festuca arundinacea Schreb cv. Atla	Se
Hibiscus cannabinus L.cv. Indian	Se
Populus	As, Cd
Helianthus annuus	Cs, Sr, U

Potential obstacles for the large-scale application of phytoremediation technologies, however, include the time required for remediation, the pollutant levels tolerated by the plants used, the disposal of the contaminated plants, and the fact that only the bioavailable fraction of the contaminants will be treated. This means that phytoremediation does not achieve 100% removal or reduction of the contaminants. From the ecological, toxicological, and medical (health) points of view, the bioavailable fraction should be the most important point of consideration (Van der Lelie et al., 2001). Only the bioavailable trace elements in the soils can be taken up by plants. Therefore, successful phytoremediation requires the proper selection of hyperaccumulator plants with respect to the climatic conditions, soil type, and tolerance for the concentration of trace elements, as well as the pre-evaluation of the bioavailability of trace elements in soils. In addition, most of the hyperaccumulator plants grow very slowly with lower biomasses. This means that phytoremediation will take a longer period of time (up to several years or decades) to clean up contaminated sites. Moreover, hyperaccumulator plants have not been found for many trace elements and heavy metals. Finally, many highly contaminated sites contain extremely elevated amounts of trace elements and heavy metals, which deliver high levels of toxicity to these plants (Raskin and Ensley, 2000).

It has been reported that a Chinese fern, *Pteris vittata*, was an effective As hyperaccumulator plant, while *Thlaspi caerulescens* hyperaccumulated Zn, Ni, Cd, and Cu (Ma et al., 2001; Zhao et al., 2002). Indian mustard (*Brassica juncea*), a high-biomass plant, accumulated Pb, Cr(VI), Cd, Cu, Ni, Zn, Cr, B and Se (Salt et al., 1995; Han et al., 2004b, c). *Astragalus* and *Stanleya* species were capable of accumulating and tolerating thousands of micrograms of Se per gram of leaf tissue even in the presence of high soil sulfate (Goodson et al., 2003). Oat (*Avena sativa*) and barley (*Hordeum vulgare*) tolerated high Cu, Cd and Zn concentrations and accumulated elevated concentrations of these metals in the plant shoots (Ebbs and Kochian, 1998). Hybrid poplar trees took up As and Cd while *Alyssum bertolonii*, (an Italian serpentine plant), *A. mural,* and *A. serpyllifolium* subsp. *lusitanicum* accumulated up to 1% Ni (Reeves and Baker, 2000). *Alyssum*, of the *Brassicaceae* family, had been found to hyperaccumulate Ni (1,280–29,400 mg/kg Ni) in many arid and semi-arid regions of Mediterranean Europe and the Middle East including Turkey, Iraq, and Syria. *Bornmuellera, Cochlearia aucheri, C. sempervivum* and *Thlaspi* (the *Brassicaceae* family) were found in Turkey to accumulate high levels of Ni (2,000–31,200 mg/kg). *Streptanthus polygaloides* of the *Brassicaceae* family and *Arenaria of Caryophyllaceae* were found in California, U.S. to accumulate 14,800 and 2,330–2,370 mg/kg of Ni, respectively (Reeves and Baker, 2000). Many species of the *Asteraceae* family (*Haplopappus condensata, H. fremontii, Machaeranthera glabriuscula, M. parryi, M. ramosa, M. venusta), Stanleya pinnata* and *S. bipinnata* in the *Brassicaceae* family, *Atriplex confertifolia* in the *Chenopodiaceae* family, *Castilleja chromosa* in the *Scrophulariaceae* family, and many species in the *Leguminosae* family (*Astragalus bisulcatus, A. osterhoutii, A. pattersonii, A. pectinatus*, and *A. racemosus*) in the Midwest U.S. accumulated high concentrations of Se (1,190–14,920 mg/kg) (Reeves and Baker, 2000).

Sunflowers and Indian mustard accumulated more Cr than other agricultural plant species (Shahandeh and Hossner, 2000). To date, there are approximately 400 known metal hyperaccumulator plants in the world (Baker and Walker, 1989). However, some non-hyperaccumulator plants such as corn, sorghum and alfalfa may be more effective than the above mentioned hyperaccumulator plants at removing large quantities of trace elements from soils due to their fast growth rate and large biomass. Soil conditioning, such as pH adjustment, acidification, and the addition of a chelating agent (EDTA) increases the bioavailability of trace elements and heavy metals in soils, especially in arid and semi-arid soils, resulting in increasing the efficiency of remediation.

Phytoremediation was applied to remediate Se contaminated soils in the western United States. *Astragalus* and *Stanleya* species accumulated 2–37% of the total soil Se from five high Se calcareous soils from California with pH ranges of 7.1–8.4 and 1.3–42% $CaCO_3$ (Goodson et al., 2003). The biological labile pools of soil Se as plant uptake was consistently less than the chemical labile pool of soil Se (4–73% of total soil Se) as extracted in 0.1 M KCl from soils. Wetlands were very successful in trapping and volatizing 60–90% of the selenite and removing 40–66% of the selenate from Se enriched water with 15% lost through volatilization (Allen, 1991; Hansen et al., 1998; Thompson et al., 2003). Vegetated wetland (*Typha angustifolia* L. and *polypogon monspeliensis* L. Desf.) removed about 40% of the Se (selenate) mass versus about 23% removal by the unplanted controls (Thompson et al., 2003). The sediment and the detritus layers were major contributors with the removal of Se. Higher retention times resulted in greater Se (selenate) mass removal (66%) and lower effluent Se concentrations. The success for phytoremediation of Se in arid and semi-arid regions, such as in central California, was highly dependent upon the following factors: (1) soil salinity and high concentrations of toxic elements, (2) the presence of competitive ions affecting Se uptake, (3) adverse climatic conditions, (4) water management strategies that produce less effluent, (5) unwanted consumption of plants high in Se by wildlife and insects, and (6) acceptance of phytoremediation as a remediation technology by the public and growers in the regions (Banuelos, 2000).

Indian mustard (*Brassica juncea Czern L.*), tall fescue (*Festuca arundinacae*), birds foot trefoil (*Lotus corniculatus*), and kenaf (*Hibiscus cannibinus*) have been used to reduce soil Se levels near Los Banos, California. Total Se concentrations in soil depths from 0 to 60 cm were lower in all cropped plots than in the bare plots after four years (Banuelos, 2000). The efficiency of Se lost from the soil after four years for each of the four crop rotations was in the range of 17–60% (Banuelos, 2000). The cropped plots with only tall fescue had 25% lower soil Se concentrations after four years in depths from 0–45 cm (Banuelos, 2000). However, the amount of soil Se removed by the above-ground portion of a crop did not exceed 10% of the lost soil Se in the fields. The majority of Se lost from the fields may have been due to biological volatilization and leaching and other processes.

Another application or example of phytoremediation is phytostabilization by reforestation, such as the reclamation of metalliferous mine wastes. Phytostabilization is the stabilization of contaminants in surface soils (especially in root zones) by preventing them from leaching down profiles or entering surface runoff. The additional benefits of reforestation in reclamation include supplying local timber needs, the provision of employment in rural areas, the enhancement of the surrounding scenery, the establishment of perimeter wind breaks and shelter belts, the provision of food and shelter for wildlife, and the reduction of surface water and resulting erosion (Williamson and Johnson, 1981).

8.3.3 Volatilization Processes for Remediation of Se and As Contaminated Soils

Microorganisms transform inorganic and organic arsenic compounds into volatile methylarsines, ionic mercury into elemental mercury, and selenium into H_2Se and methylated Se compounds under anaerobic conditions. Selenium is biologically reduced and volatilized from selenate (SeO_4^{2-}) and selenite (SeO_3^{2-}) to elemental selenium (Se^0) and dimethylselenide. Dungan and Frankenberger (2000) found that optimum Se volatilization occurred at pH 6.5 and temperature at 35 °C and EC at 11dS m^{-1}. Volatilization of Se increases with increasing SeO_3^{2-} concentration. The methylation of Se is a biological process. Fungi and bacteria are the predominant Se-methylating organisms in soils and sediments. Some algal species are also Se-methylating. Dimethylselenide is the major biological metabolite of Se methylation (Dungan and Frankenberger, 2000). Microbial transformations of Se to less toxic volatile forms may be an approach to remediate seleniferous soils and sediments.

Arsenic is transformed and volatilized from inorganic and organic arsenic compounds into methylarsines by microorganisms such as fungi. The high volatility of methylated arsines indicates significant fluxes into the atmosphere. The global cycling of arsenic shows that the natural input to the atmosphere is approximately 45 thousand tons of As $year^{-1}$, whereas anthropogenic sources are 28 thousand tons of As $year^{-1}$ (Chivers and Peterson, 1987). Approximately 35% of the atmospheric arsenic flux is from the low-temperature volatilization of arsenic from soil (Frankenberger, 1997). Cobalt stimulates arsenic volatilization. Arsenite stimulates trimethylarsince production while arsenate inhibits the process. On the other hand, methylated arsines are readily oxidized by air to their oxide and acidic forms. In addition, transgenic technology has been used to phytoextract mercury from contaminated soils and sediments, and mercury in plant leaves is then transformed into elemental mercury, releasing it into the atmosphere (Meagher, 2000; Bizily et al., 2000).

Chapter 9

GLOBAL PERSPECTIVES OF ANTHROPOGENIC INTERFERENCES IN THE NATURAL TRACE ELEMENT DISTRIBUTION

Industrial Age Inputs of Trace Elements into the Pedosphere

9.1 INDUSTRIAL AGE ANNUAL WORLD PRODUCTION OF TRACE ELEMENTS

Eight trace elements of greatest environmental concern are chosen, which are arsenic (As), mercury (Hg), lead (Pb), cadmium (Cd), chromium (Cr), nickel (Ni), copper (Cu) and zinc (Zn). These eight trace elements will be discussed in this chapter in the order of their production and level of environmental concern, as presented above. Of these, copper and lead are known to be the earliest metals utilized by humans. Lead was used by humankind at least 5000 years ago (Settle and Patterson, 1980; Adriano, 1986). The production of these eight elements has increased considerably since the dawn of the industrial age in the 1850s (Table 9.1).

Table 9.1. Global industrial age annual production of selected trace elements and heavy metals in 1880, 1900, 1950, 1990 and 2000 (million tons) (Data extracted from Han et al., 2002a, 2003b)

Year	Cd	Cr	Cu	Hg	Ni	Pb	Zn	As
1880	0.001	0.003	0.41	0.002	0.007	0.60	0.37	0.002
1900	0.002	0.01	0.73	0.003	0.02	0.86	0.64	0.010
1950	0.009	0.22	2.25	0.007	0.14	1.50	2.30	0.044
1990	0.017	3.51	9.04	0.004	0.94	3.33	7.20	0.066
2000	0.020	4.32	13.2	0.002	1.25	3.10	8.73	0.052
1900/1880	2.0	3.3	1.8	1.5	2.9	1.4	1.7	5.0
1950/1900	4.5	22	3.1	2.3	7.0	1.7	3.6	4.4
2000/1900	10	432	18	0.7	63	3.6	14	5.2
2000/1950	2.2	20	5.9	0.3	8.9	2.1	3.8	1.2

We will first reconstruct the *annual* production rates and develop functions for the production change with time from the start of the industrial age to the present (1850–2000). Records of world annual mine production for arsenic and mercury date from 1900, for zinc, copper and lead from 1915–1920 and for chromium and nickel from 1930. In a following section, the *cumulative* production for the entire period will be presented and discussed. Reconstruction of estimated cumulative production of metals during the industrial age was done on the basis of production functions fitted to the annual production data. The global trace element production data were collected from Adriano (1986), Kelly et al. (2002), Woytinsky and Woytinsky (1953), U.S. Bureau of Mines (1978–1980), U.S. Department of the Interior, and U.S. Geological Survey (1994–2000). Chromium production was calculated from chromite production assuming an average concentration of 27.03% Cr. Cadmium data before 1963 were obtained from world annual zinc production (Cadmium mainly as by-product of zinc ore and assuming 0.37% of zinc (Adriano 1986)). Since emission from coal-fired electric power plants represents one of the largest sources of mercury emissions to the atmosphere (Lindqvist et al., 1991; Nriagu and Pacyna, 1988), mercury release from global coal consumption is also estimated. An average concentration of 0.17 mg mercury kg^{-1} in coal was used to calculate the mercury contribution from world coal production records (Bragg et al., 1998; U.S. Dept. of Interior and U.S. Geol. Survey, 2002).

Arsenic (As): Arsenic has become an increasingly important environmental concern due to its potential carcinogenic properties (Goyer et al., 1995). Both natural and anthropogenic As sources pose threats to human health and global agricultural and social sustainability (Adriano, 1986). Many developing countries in South and Southeast Asia and in Central and South America (Bangladesh, India, Cambodia, Thailand, Vietnam, China, Pakistan, Argentina, Brazil, Chile, and Mexico) face increasing As crises with high As concentrations found in drinking water and severe arsenic toxicity in human and livestock. Globally, millions of people are at risk of As-induced cancer and other arsenic-related diseases. Recently, the U.S. Environmental Protection Agency (USEPA) announced a decrease in the allowable As standard for drinking water from 50 to 10 $\mu g\ L^{-1}$ (USEPA, 2002a). Meeting this standard level requires highly efficient purification technologies and also limiting As contamination sources. However, As is highly dispersed during mining, smelting, processing, recycling, disposing of wastes by the construction and pest-control industries and applying As in agriculture and forestry (Han et al., 2003b, Loebenstein, 1994). Nriagu and Pacyna (1988) have estimated contamination of air, water, and soils by trace metals, based on known pathways. More recently, Matschullat (2000) reviewed As cycles and fluxes in the geosphere and adjacent environmental compartments. The release of As from subsoil by groundwater extraction and from other natural sources of As indeed pose an acute threat to the health of people around the world, especially in Bangladesh, India, and Europe. Natural As inputs into the world soils are important arsenic sources, which were reported to be 1.5 times anthropogenic As inputs (Matschullat, 2000). However, a discussion of the natural As inputs is beyond the scope of this chapter, which focuses on anthropogenic As sources.

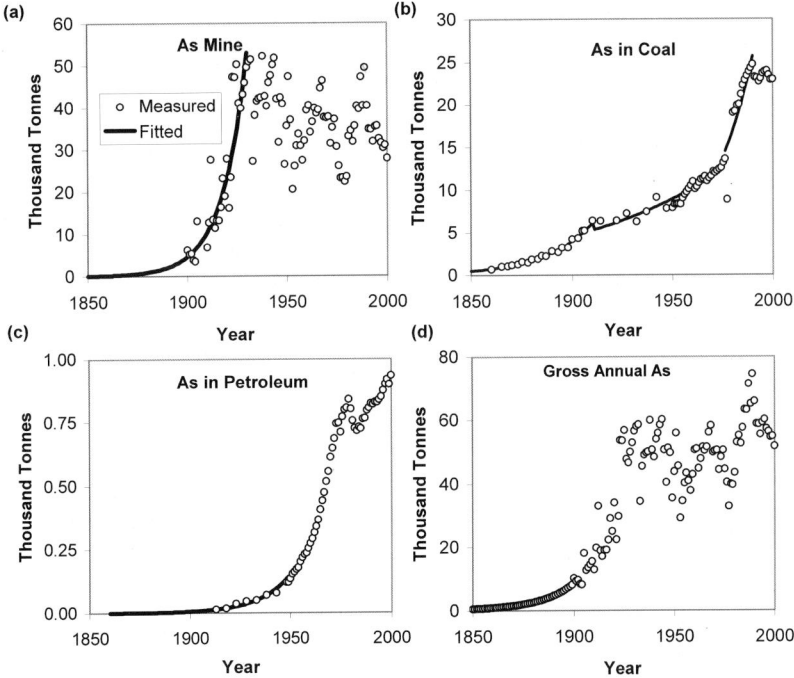

Figure 9.1. Measured and fitted global annual industrial age As production. (a) by mining; (b) released by coal burning; (c) released by petroleum burning; and (d) gross annual release into the environment (See explanations in the text) (after Han et al., 2003b. Reprinted from Naturwissenschaften, 90, Han F.X., Su Y., Monts D.L., Plodinec M.J., Banin A., Triplett G.B., Assessment of global industrial-age anthropogenic arsenic contamination, pp 396–397, Copyright (2003), with kind permission of Springer Science and Business Media)

The earliest records of world annual As production from mining date from 1900 (Kelly et al. 2002) (Fig. 9.1a). World As production from mining increased steeply from 1900 to 1930. The change in production is described by the exponential best-fit equation: $Y = 63.66\ e^{0.083\ X}$ where Y is world annual As mine production in tons and X is number of years since 1849 ($R^2 = 0.846$, significance at 5% probability level). From 1850 to 1900 and between 1906 and 1909, measured records were not available and they are estimated by extrapolation and interpolation, respectively, using the exponential equation derived for the data between 1900 and 1930 (Fig. 9.1a). After the 1930s, world As production fluctuated considerably and, as a general trend, has decreased (Fig. 9.1a). This may be related to the decrease in agricultural use of As as a pesticide, which peaked in the 1940s and accounted for more than 90% of arsenic use (Loebenstein, 1994). Still, global As annual mining production in 2000 was 4.5 times that for 1900 (Fig. 9.1a).

Most As was produced and recovered from flue dusts, speiss, and sludge associated with the smelting of copper, lead, gold, and silver ores that contain As (Loebenstein, 1994). Since As is a byproduct of the mining and processing of nonferrous metals, estimates of world production quantities have a high degree of uncertainty (Loebenstein, 1994) compared to heavy metals. Based on the assumption that there is approximately 6.5 kg of As per ton of copper (Edelstein, 1985) and estimated total accumulative post-industrial age global Cu production (Table 9.2) (Han et al., 2002a), the alternative accumulative post-industrial age As production until 2000 is estimated to be around 3.0 million tons. This Cu-based As estimate (3.0 million tons) is in excellent agreement with the total cumulative As production (3.3 million tons) based mainly on estimates from the U.S. Bureau of Mines and the model discussed above.

The other major anthropogenic sources of As in the environment are from the burning of coal (Nriagu and Pacyna, 1988) and petroleum. Arsenic is volatilized from burned coal, but can be condensed downstream in fine particulate materials (Clarke and Sloss, 1992). Since early world coal production records are not continuous (Woytinsky and Woytinsky, 1953; United Nations, 1976, 1978; Energy Information Administration, 2001, 2002), the three best fitting exponential equations ($Y = 93.35\ e^{0.0421X}$, $R^2 = 0.988$; $Y = 473.76\ e^{0.0131X}$, $R^2 = 0.897$; $Y = 16.35\ e^{0.0407X}$, $R^2 = 0.844$, where Y is production in tons and X is number of years since 1849), were used to estimate the world annual coal production when it was not available: for 1850–1910, 1911–1975, and 1976–1990, respectively. Arsenic concentration in coal varies within a wide range from 0.3 to 93 mg kg^{-1} (Bowen, 1979) or from 0.34 to 130 mg kg^{-1} (Piver, 1983). Some coal from Europe, New South Wales, New Zealand, and the U.S. contains As from trace amounts to 200 mg kg^{-1} (Baur and Onishi, 1969). Some coals from China even have a few thousand mg As kg^{-1}. Annual As contributions from the world-wide coal production were estimated (Fig. 9.1(b)), assuming that the mean As concentration in world coal is 5 mg kg^{-1} (Bowen, 1979).

Petroleum was known in Asia and the Caucasus even before the Christian era. However, commercial production did not begin until 1857 in Romania and in the United States until 1859 (Woytinsky and Woytinsky,

1953). World-wide petroleum production has increased exponentially since the 1950s. As concentration in petroleum is relatively low, ranging from 0.0024 to 1.63 mg kg^{-1} (Pacyna, 1987). An average As concentration of 0.26 mg kg^{-1} (Bowen, 1979) was used to estimate the As contribution to emissions into the environment resulting from petroleum burning (Fig. 9.1(c)).

World annual As release from coal and petroleum accounted for 80–86% of the relatively low world gross As production in the 1850s. When As mining increased steeply between 1900 and 1930, the fraction of inputs from fossil fuels decreased and reached 10–15% of the total in 1925–1945, and then increased to 46% in 2000 (Fig. 9.2a). Note that arsenic generated from petroleum was less than 6% of total As contributions from both coal and petroleum. Combined As production from mining and the world-wide coal and petroleum industries reached in 2000 was 51.8 thousand tons, 5.2 times that in 1900 (Fig. 9.1 (d), Table 9.1). Perhaps incidentally, this estimate is very similar to the lower boundary of annual worldwide emission of As into soils (52 x 10^6 kg per year in 1988), estimated by Nriagu and Pacyna (1988) by summing As inputs from various known pathways. Upon the complete dissipation of the total As anthropogenically produced before 2000, the average annual As dissipation rate into the pedosphere over the industrial age is 26.5 x 10^6 kg per year, which falls close to the previous estimate of anthropogenic emission of As (23.6–28 x 10^6 kg per year) (Hutchinson and Meema, 1987; Chilvers and Peterson, 1987) and anthropogenic inputs into the pedosphere (28.4 x 10^6 kg As per year, Matschullat, 2000). The pedosphere, as the sphere interfaced between the lithosphere, atmosphere, and hydrosphere, is where human beings primarily live. However, the total As input to soils (94 x 10^6 kg per year) estimated by Nriagu and Pacyna (1988) is higher than the estimates provided here with the sources-based approach. This discrepancy may be due to other natural sources as discussed above, which is not addressed here.

Mercury (Hg): Mercury has been used for the past 2500 years due to its unique chemical and physical properties. Mercury is released into the environment in significant amounts by both natural processes and anthropogenic activities, and it has been proven to be a potent neurotoxin and a global toxic pollutant. Mercury has been used in many industrial processes and commercial products, including laboratory equipment for temperature and pressure measurements, electrodes, ultraviolet lamps, diffusion pumps, dental preparations, batteries, explosives, catalysts, and precious metal recovery by amalgamation. Mercury also has been used extensively in antiseptics, slimicides, fungicides and antifouling paints. It is widely distributed in the environment. The large input of Hg into the environment has resulted in widespread occurrence of Hg in the entire food chain (Adriano, 2001).

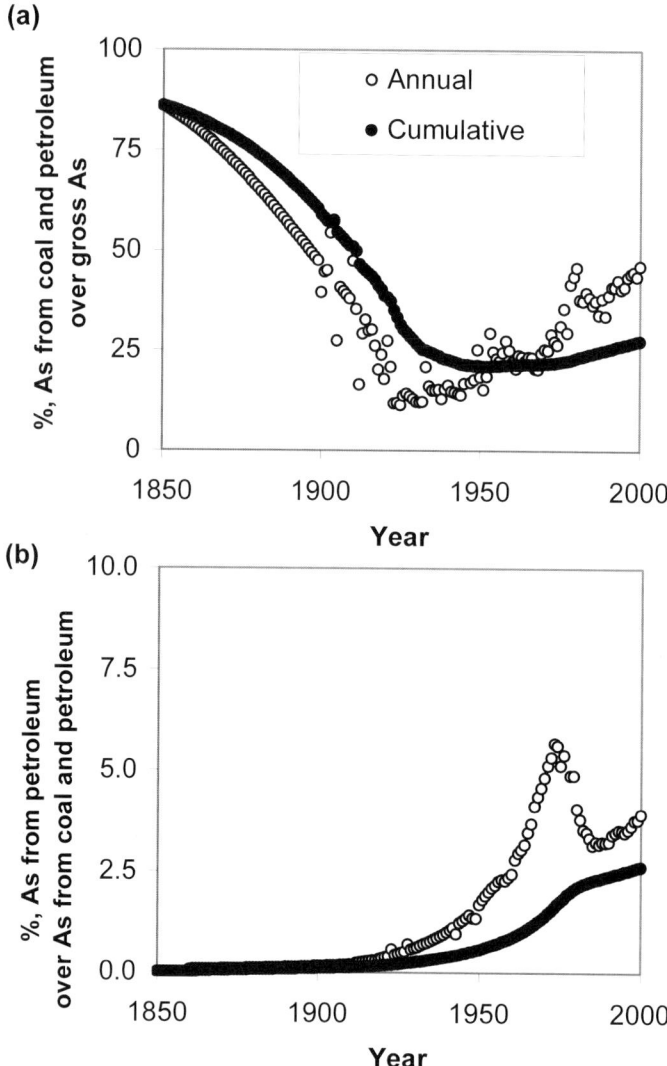

Figure 9.2. (a) The percentages of annual and cumulative As generated from the world-wide coal and petroleum industries over the global gross As production. (b) The percentages of annual and cumulative As generated from the world-wide petroleum industry over As generated from both coal and petroleum industries (after Han et al., 2003b. Reprinted from Naturwissenschaften, 90, Han F.X., Su Y., Monts D.L., Plodinec M.J., Banin A., Triplett G.B., Assessment of global industrial-age anthropogenic arsenic contamination, pp 396–398, Copyright (2003), with kind permission of Springer Science and Business Media)

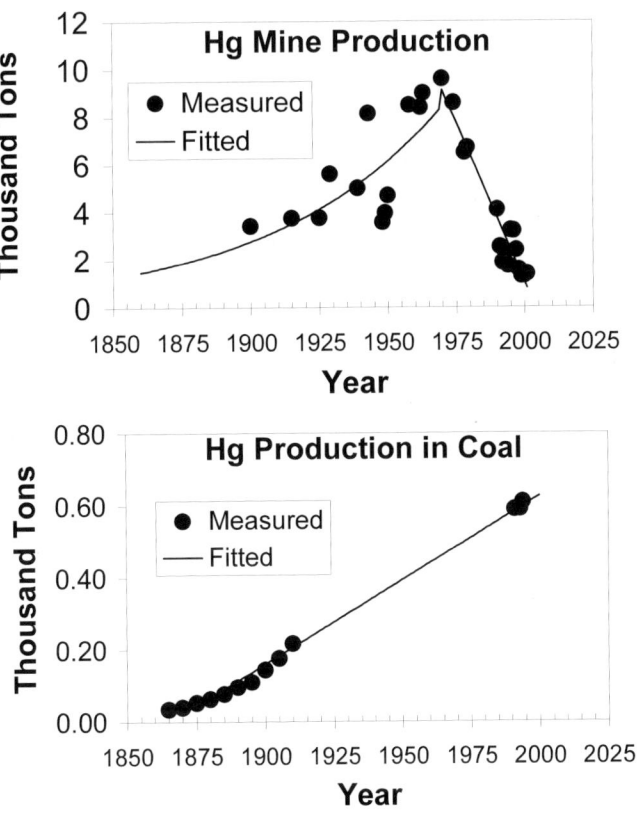

Figure 9.3. Measured and fitted global annual industrial age Hg mine production and Hg production from world coal industry (after Han et al., 2002a. Reprinted from Naturwissenschaften, 89, Han F.X., Banin A., Su Y., Monts D.L., Plodinec M.J., Kingery W.L., Triplett G.B., Industrial age anthropogenic inputs of heavy metals into the pedosphere, p 500, Copyright (2002), with kind permission of Springer Science and Business Media)

It has been estimated that the annual anthropogenic input of Hg into the environment is as high as 6×10^6 kg/yr (Nriagu and Pacyna, 1988). The anthropogenic Hg emission is in the range of 50–75% of the total annual Hg emission into the atmosphere (Ebinghaus et al., 1999; Fitzgerald, 1995). Combining both anthropogenic input and natural sources, a total of about 741×10^6 kg Hg has been released into the atmosphere, 118×10^6 kg released into water, and 806×10^6 kg released into the soil (Nriagu, 1979; 1991). Mining and refining operations account for 12% of the global anthropogenic emissions of Hg into the air (Van Horn, 1975). The atmospheric Hg loading has increased by a factor of three during the last 100 years (Fitzgerald, 1995).

Long-range atmospheric transport of Hg from fossil fuel combustion and solid waste incineration has increased Hg in freshwater and biota. In the United States, combustion of fossil fuels for power generation is estimated to generate about 30% of the total release of Hg into the atmosphere (Harriss and Hohenemser, 1978). One in every three lakes in the United States and nearly one-quarter of the nation's rivers contain various pollutants, including Hg (CNN, 2004). Forty States in the U.S. have issued advisories for methylmercury on selected water-bodies, and 13 states have statewide advisories for some or all sportfish from rivers or lakes (USGS, 2000). Fish consumption advisories for methylmercury account for more than three-quarters of all fish consumption advisories.

Exposure to Hg may cause serious harm to human health, at high doses it can even be fatal. Mercury toxicity usually involves the kidneys and/or nervous system disorders (central nervous system and neurobehavioral changes). Mercury contamination has been found in more than 70 federal sites with Hg-contaminated wastes.

World Hg production increased early in the 20th century (Fig. 9.3, Table 9.1). World Hg production after the 1970s started to decline (Fig. 9.3). However, the Hg contribution from the world coal industry has rapidly increased since the 1860s due to steady increases in world coal production (Fig. 9.3). The Hg production from the global coal industry in 1983 is estimated at 5.46×10^5 kg (Fig. 9.3). Mercury is highly volatile and exists almost exclusively in the vapor phase of combustion and gasification flue gases (Galbreath and Zygarlicke 1996). Therefore, global Hg emissions into the atmosphere contributed from coal burning accounted for 12% of the total anthropogenic Hg emissions into the atmosphere (45×10^5 kg /yr) estimated by Nriagu and Pacyna (1988).

Lead (Pb): Lead, a corrosion-resistant, dense, ductile, and malleable metal, is one of the trace elements used by humans for at least 5000 years. Lead was historically used for pigments for glazing ceramics, building materials, and pipes for transporting water. Prior to the early 1900's, Pb was primarily used for ammunition, brass, burial vault liners, ceramic glazes, leaded glass and crystal, paints or other protective coatings, pewter, and water lines and pipes (USGS, 2006). From the 1970s to the early 2000s, Pb has been widely used as in the production of bearing metals, cable coverings, caulking lead, solders, lead-acid storage batteries, radiation shielding in medical analysis and video display equipment, and, until the 1980s, as an additive in gasoline. By the mid-1980s, a significant reduction and/or elimination occurred in the used of Pb in nonbattery products, including gasoline, paints, solders, and water systems. By the early 2000's, the total demand for Pb in all types of lead-acid storage batteries represented 88% of apparent U.S. Pb consumption (USGS, 2006). Other significant uses included ammunition (3%), oxides in glass and ceramics (3%), casting metals (2%), and sheet lead (1%).The remainder was consumed in solders, bearing metals, brass and bronze billets, covering for cable, caulking lead, and extruded products.

Lead is a highly toxic metal. Lead can cause lesions in the central nervous system and apparently can damage the cells making up the blood-brain barrier that protects the brain from many harmful chemicals. It is absorbed by the red blood cells and circulated through the body where it becomes concentrated in soft tissues, especially the liver and kidneys. Lead may cause a range of health effects, from behavioral problems and learning disabilities, to seizures and death. Chronic Pb toxicity is a major public health problem in the United States affecting millions of children and adults, especially those who lived in old houses with Pb-containing paints and urban residential areas with high traffic. In 1978, there were nearly three to four million children with elevated blood Pb levels in the United States (USEPA, 2006). By 2002, that number had dropped to 310,000 children, and it continues to decline. Children 6 years old and under are most at risk, because their bodies are growing quickly. The primary sources of Pb exposure for most children are: deteriorating Pb-based paint, Pb contaminated dust, and Pb contaminated residential soil.

World Pb production continued to increase in the early 20th century (Fig. 9.4), but then slowed in the latter part of the 20th century due to discontinuing Pb additives to gasoline.

Cadmium (Cd): Cadmium is produced mainly as a byproduct from mining, smelting, and refining sulfide ores of zinc, and to a lesser degree, Pb and copper (USGS, 2006). Small amounts of Cd, about 10% of consumption, are produced from secondary sources, mainly from dust generated by recycling of iron and steel scrap (USGS, 2006). About three-fourths of Cd consumption is used in batteries; the remaining one-fourth is used for pigments, coatings and plating, and as stabilizers for plastics (USGS, 2006). Cadmium is also used in photography, lithography, process engraving, rubber curing and as fungicides. Cadmium, a potential toxic heavy metal with no known biological function, is present in a wide variety of consumer goods, and virtually all households and industries have products that contain some Cd.

World Cd production increased significantly after the 1950s. The annual world Cd production stabilized after the middle 1970s (Fig. 9.4, Table 9.1).

Chromium (Cr): Chromium is very widely used in modern industrial societies. Chromium has been used in iron, steel, and nonferrous alloys because it enhances hardenability and resistance to corrosion and oxidation. Production of stainless steel and nonferrous alloys with Cr are

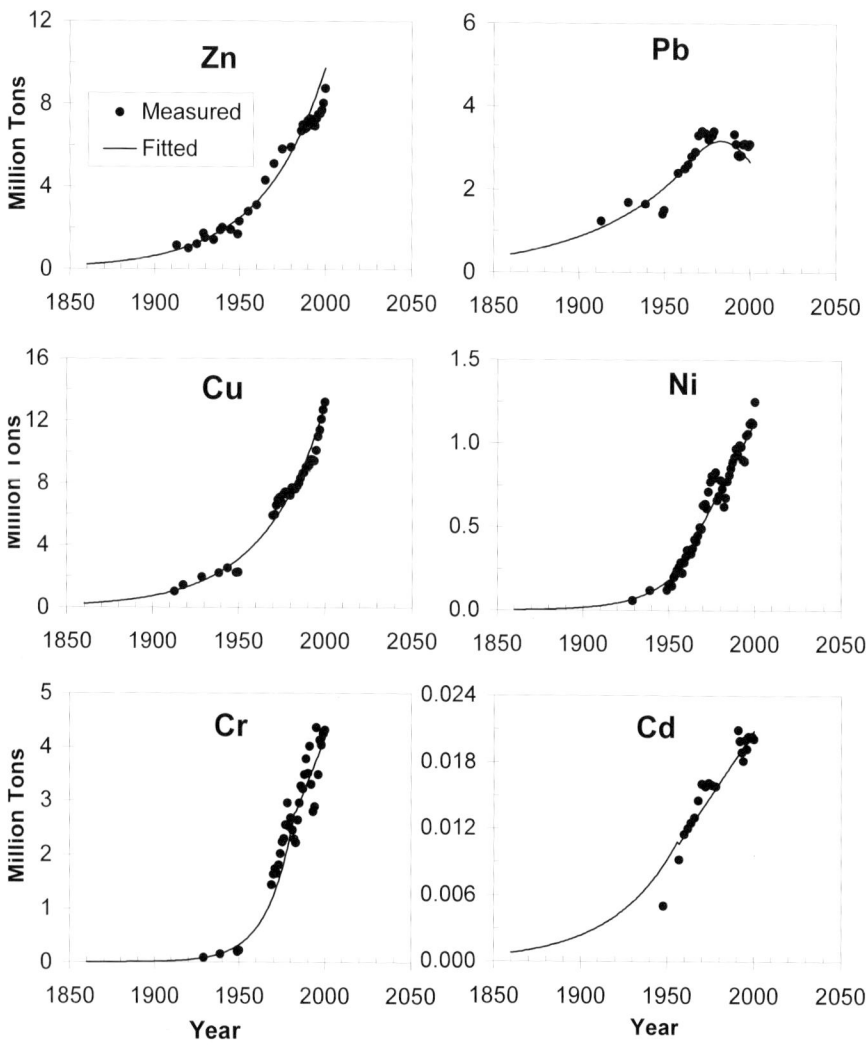

Figure 9.4. Measured and fitted world annual metal production. The measured world annual metal production data were collected from U.S. Geological Survey-Minerals Information, 1997; Adriano, 1986; Woytinsky and Woytinsky, 1953. Chromium production was calculated from chromite production assuming an average of 27.0% Cr. The fitted Cd data before 1963, were calculated from the world annual Zn production (Cd mainly as a by-product) (after Han et al., 2002a. Reprinted from Naturwissenschaften, 89, Han F.X., Banin A., Su Y., Monts D.L., Plodinec M.J., Kingery W.L., Triplett G.B., Industrial age anthropogenic inputs of heavy metals into the pedosphere, p 499, Copyright (2002), with kind permission of Springer Science and Business Media)

two of its most important applications (USGS, 2006). Chromium is also used in alloy steel, plating of metals, pigments, leather processing, catalysts, surface treatments, and refractories (USGS, 2006). Therefore, Cr has become a common constituent of diverse waste materials, and is found in varying concentrations at approximately one-third of Superfund sites in the U.S.

Anthropogenic Cr sources include ore refining, production of steel and alloys, pigment manufacturing, plating/metal, corrosion inhibition, leather tanning, wood preservation, and combustion of coal and oil. Other contamination sources are associated with nuclear reactor operation, irradiated fuel processing and fuel fabrication. Since hexavalent Cr is less adsorbed by soil and minerals and is very soluble in soils and aquifers, it easily enters source waters and waterways, posing high risks to humans and livestock. Hexavalent Cr is classified as a primary toxic and mutagenic contaminant.

World Cr production increased after early in the 20th century and significantly increased from 1975 to 1990 (Fig. 9.4) mainly due to large scale military use.

Nickel (Ni)/Copper (Cu)/Zinc (Zn): Nickel exhibits a mixture of ferrous and nonferrous metal properties, and Ni-based alloys are characterized by corrosion resistance. Therefore, Ni has been widely used in stainless steel (about 65% of the Ni consumed in the Western World) and superalloys/nonferrous alloys (12%). Turbine blades, discs and other critical parts of jet engines and land-based combustion turbines are fabricated from superalloys and Ni-based superalloys. The remaining 23% of consumption is applied in alloy steels, rechargeable batteries, catalysts and other chemicals, coinage, foundry products, and plating (USGS, 2006).

Copper is one of the oldest metals ever used and has been one of the important materials in the development of civilization (USGS, 2006). Because of its high ductility, malleability, and thermal and electrical conductivity, and its resistance to corrosion, Cu has become a major industrial metal after iron and aluminum in terms of quantities consumed. Copper has been widely used in power transmission and generation, building wiring and construction, telecommunication, and electrical and electronic products. In addition, Cu is extensively used in agriculture as a component in fertilizers, bactericides, fungicides, and algicides, and animal feeds (Han et al., 2000, Han et al., 2001b, c).

Zinc world production stands fourth among all metals after iron, aluminum, and copper. Zinc has been extensively used ranging from metal products (coating and alloying), rubber, chemical, paint, agricultural industries, to medicines. Zinc as a coating is used to protect iron and steel from corrosion (galvanized metal), as alloying metal to make bronze and brass, as a Zn-based die casting alloy, and as rolled Zn. Both Cu and Zn are also essential micronutrients for proper growth and development of humans, animals, and plants.

World Ni, Cu and Zn production increased early in the 20th century and Ni and Cu significantly increased from 1960 to 2000 (Fig. 9.4). After the 1980s, world Zn production then decreased.

In general, the world production for the eight trace elements studied increased exponentially. The annual world trace element production is in the following order: Cu > Zn > Pb > Cr >> Ni >> As >> Cd > Hg (Figs. 9.1–9.4). Of these, Cr had the most rapid increase in annual production since the early 20th century, while Cd had the slowest increase. Lead production stabilized and Hg production declined after the 1970s. Gross As production fluctuated from 1925 to 2000. The rates of increase in annual world trace element production since early 20th century are in the order: Cr >> Cu > Zn > Pb > Ni > As > Cd > Hg (Figs 9.1–9.4).

9.2 ESTIMATE OF INDUSTRIAL AGE CUMULATIVE WORLD PRODUCTION OF TRACE ELEMENTS

The measured trace element production data are sporadic and discontinuous (Figs. 9.1, 9.4–5). Due to the lack of early records of measured global production, the continuous annual production of the eight trace elements is estimated only from 1860. The cumulative world production of these elements is calculated by best-fitting (least square fitting) annual world metal production. The fitted continuous annual world Cu and Zn productions from 1860 to 2000, Cr before 1983, Cd before 1957, Ni before 1951, and Hg before 1970 are obtained from exponential modeling ($Y = Ae^{Bt}$, where Y is the output, t is the year, A and B are constants) based on actual metal productions (Han et al., 2002a). Due to rapid increases in Cr, Cd, and Ni production in the late 20th century, linear modeling is selected to estimate their continuous annual world production after 1983, 1957, and 1951, respectively (Han et al., 2002a). But after the 1970s, due to the decrease in world Pb and Hg production, a polynomial modeling (a 2nd order polynomial) is used to simulate the annual Pb production after 1968 and a linear modeling for annual Hg production after 1971.

Based on estimated continuous world annual metal production, the cumulative world production of these trace metals was calculated (Table 9.2, Figs. 9.5–9.7). However, the cumulative production of Pb and of Cu are greater than the estimates presented here since both metals were utilized by

Global Perspectives

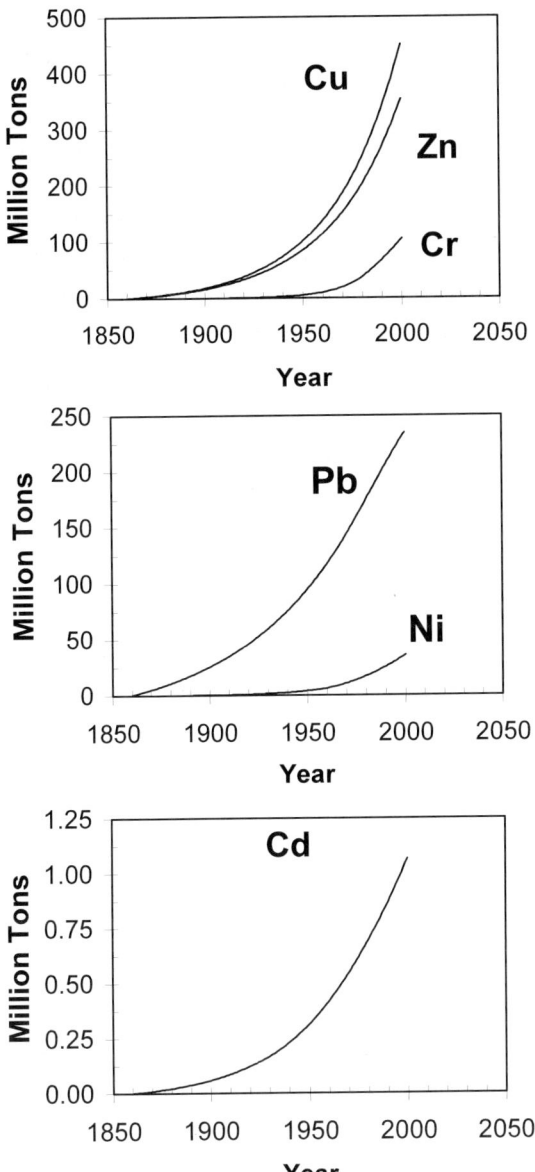

Figure 9.5. World cumulative metal productions since 1860 based on fitted world annual metal productions (after Han et al., 2002a. Reprinted from Naturwissenschaften, 89, Han F.X., Banin A., Su Y., Monts D.L., Plodinec M.J., Kingery W.L., Triplett G.B., Industrial age anthropogenic inputs of heavy metals into the pedosphere, p 499, Copyright (2002), with kind permission of Springer Science and Business Media)

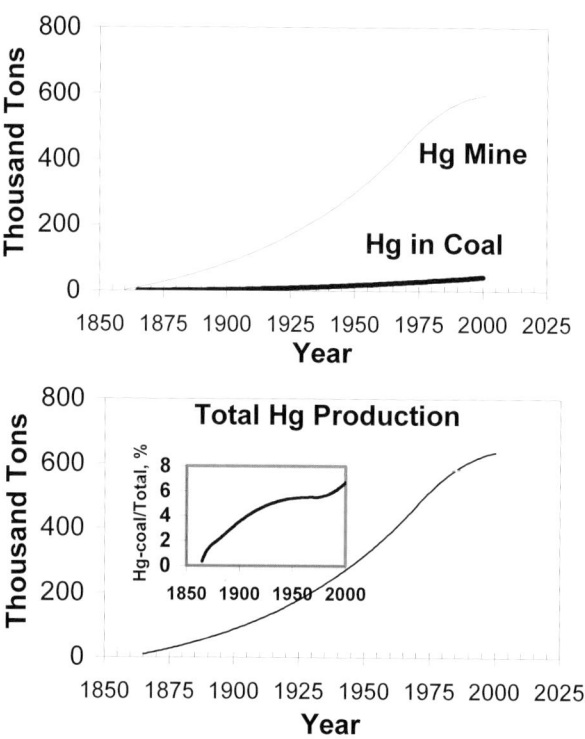

Figure 9.6. Global cumulative industrial age Hg mine production, Hg production from coal, and total global Hg production since 1860 based on estimated world annual Hg mine and coal production (inset: The percentage of Hg production from coal over the total global Hg production) (after Han et al., 2002a. Reprinted from Naturwissenschaften, 89, Han F.X., Banin A., Su Y., Monts D.L., Plodinec M.J., Kingery W.L., Triplett G.B., Industrial age anthropogenic inputs of heavy metals into the pedosphere, p 500, Copyright (2002), with kind permission of Springer Science and Business Media)

humans during earlier historic periods. It is estimated that the pre-industrial cumulative Pb production is as large as industrial age Pb production (Settle and Patterson, 1980). This is supported by the observation that cumulative Pb deposition from anthropogenic sources in pre-industrial times (600 BC to 1800 AD) was similar to cumulative deposition during the industrial period (1800 AD to the present) (Renberg et al., 1994). Since the mobility of Pb in soils and sediments is very low, Pb in remote lakes and peat cores is derived from direct atmospheric precipitation, which, in turn, may be directly related to cumulative Pb production (Renberg et al., 1994).

The cumulative production of Cu, Zn and Cd dramatically increased after 1960–1970, as did the cumulative Cr production after the 1980's

(Fig. 9.5). Mercury and Ni relatively slowly increased during this period (Figs. 9.5–9.6). In 2000, industrial age cumulative world Cu, Zn and Pb production reached 451, 354 and 235 million metric tons, respectively (Table 9.2). The cumulative world Cr and Ni production was 105 and 36 million metric tons, respectively, while Cd and Hg production was 1.1 and 0.64 million metric tons, respectively (Table 9.2). Based on Settle and Patterson (1980), we estimate the cumulative pre-industrial (before 1850) Pb production was 227 million metric tons. The cumulative Hg production from mining as well as from world coal industry rapidly increased after the 1950s, resulting in the rapid increase in cumulative gross Hg production (Fig. 9.6). The overall contribution of Hg from the coal industry to total world Hg production ranged from 3.5% in 1900 to 6.7% in 2000 (Fig. 9.6, the inset).

Table 9.2. Global industrial age cumulative production of selected trace elements and heavy metals in 1880, 1900, 1950, 1990 and 2000 (million tons)[a]

Year	Cd	Cr	Cu	Hg	Ni	Pb	Zn	As
1880	0.02	0.03	6.58	0.04	0.09	10.7	5.97	0.04
1900	0.06	0.16	17.9	0.09	0.32	25.3	15.9	0.2
1950	0.31	4.99	100.9	0.31	3.92	94.2	84.7	2.0
1990	0.87	67.3	336.2	0.61	25.3	205.7	267.7	3.97
2000	1.1	105.4	451.1	0.64	35.7	234.6	354.1	4.5
1900/1880	2.7	4.7	2.7	2.4	3.5	2.4	2.7	3.7
1950/1900	5.3	30	5.6	3.6	12	3.7	5.3	13.2
2000/1900	18	643	25	7.3	110	9.3	22	30
2000/1950	3.4	21.1	4.5	2.0	9.1	2.5	4.2	2.3

[a]Data from Han et al., 2002a and 2003b. With kind permission of Springer Science and Business Media.

In 1900, industrial age gross cumulative world As production was 0.17 million tons, while in 2000 it reached 4.9 million tons (Fig. 9.7, Table 9.2). The cumulative world As mining production slightly increased before the 1920s, but then linearly grew (Fig. 9.7). The rate of the increase in the cumulative world As mining production is around 37,121 tons per year. In 2000, world cumulative As production from mining was 3.3 million metric tons, which was 55.1 times that in 1900 (Table 9.2). Based on the estimated annual As from world-wide coal and petroleum industries, the cumulative world As generated from coal and petroleum was calculated (Fig. 9.7). Since 1925, world As production from mining increased faster than that from the world coal and petroleum industries (Fig. 9.7). The cumulative global As production from coal and petroleum in 2000 was 1.24 million metric tons, 14 times that in 1900 (Fig. 9.7). The contribution of As from world coal and petroleum to the cumulative As gross production decreased from 85% in the 1850s to 25% in 1935–1975. Since then, the contribution from the world coal and petroleum industries has stabilized (around 30%) (Fig. 9.2). Industrial age world As cumulative production in 2000 was 22 times that in 1900 (Table 9.2). However, the actual gross cumulative anthropogenic As production is greater than the estimates presented here because (1) arsenic had been utilized by humans during earlier historic periods for pigment materials and medicine and (2) the high uncertainty associated with As concentration variation in coal and petroleum. Therefore, actual total anthropogenic cumulative As production may be greater than is estimated here.

Until 2000, industrial age cumulative world trace element production was in the following order: Cu > Zn > Pb >> Cr > Ni >> As > Cd > Hg. However, Pb has the highest total cumulative production if the pre-industrial output is included. Compared to the cumulative world trace element production in 1900, Cr and Ni had the highest increase, while Hg had the lowest increase. The ratios of cumulative production in 2000 to that in 1900 for the eight trace elements followed the trend: Cr (643) > Ni (110) > As (30) > Cu (25) > Zn (22) > Cd (18) > Pb (9.3) > Hg (7.3) (Table 9.2).

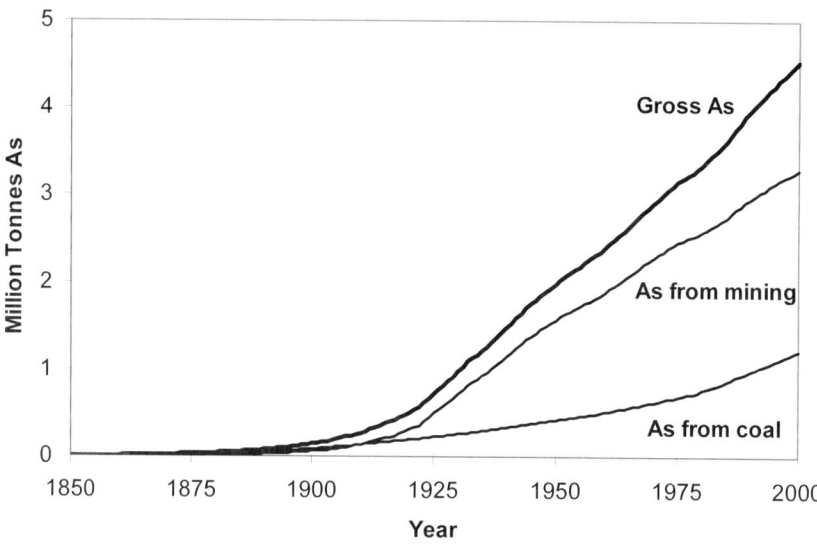

Figure 9.7. Global cumulative industrial age As production from mining, As generated from coal and petroleum, and cumulative gross As production since 1850 (after Han et al., 2003b. Reprinted from Naturwissenschaften, 90, Han F.X., Su Y., Monts D.L., Plodinec M.J., Banin A., Triplett G.B., Assessment of global industrial-age anthropogenic arsenic contamination, p 397, Copyright (2003), with kind permission of Springer Science and Business Media)

9.3 LIFE CYCLE OF TRACE ELEMENT PRODUCTS

The recovery rate of trace metals from both mining and processing (primary and secondary) is estimated at 85% (Rudawsky, 1986). The recycling efficiency with modern technology varies with the metal. Lead has the highest recycling efficiency (95%), followed by Cr (87%) and Zn the lowest (19%) (U.S. Dept. of Interior and U.S. Geol. Survey, 2000). However, the recovery rate of metals from mining highly depends on the technology used and the economy of the regions. In some undeveloped areas of China, the recovery rate of Hg and Pb-Zn from mining is 30–45% (Chen et al., 1999). Some metal (like Zn) is partly dispersed into the environment during product use as coatings, even before recycling. The recovery of trace elements in the form of first-generation products is approximately 80% for Pb, 74% for Cr and 16% for Zn. Most products containing trace elements as a minor component often last from 20 to 50 years, in fewer cases from 10 to 100 years. After two-three recycling processes, metal recoveries like Zn (possibly Cd and Ni) may be less than 10%. Non-recovered metals are mainly dispersed into the pedosphere as pollutants and wastes. The pedosphere is a major sink for trace elements compared to the atmosphere. Trace elements are almost completely dispersed into the pedosphere after about four generations of products for most metals, lasting for 80 to 200 years (complete-dispersion period); in fewer cases from 40 to 400 years. However, As-containing products have relatively shorter life periods. Wood preservatives are generally guaranteed to prevent lumber decay for a period of 30 years and other products last from five to 20 years (Loebenstein, 1994). Most agricultural products containing As have even shorter useful lives of less than one year and are dissipative (Loebenstein, 1994). Compared to the complete dissipation periods of heavy metal-containing products (Han et al., 2002a), As-containing products have much shorter useful lives (the short-life cycle), which makes the hypothesis (discussed above) and the source-based approach more reasonable and rational. The complete-dispersion period has decreased in recent years since the life of each generation of trace element-containing products has shortened. Trace elements enter the pedosphere through direct dispersion during primary/secondary (mining, smelting, and processing) and recycling processes (as air pollutants, liquid, and solid wastes), and the disposals of various outdated products and wastes, in which most trace elements are relatively concentrated. Further, agricultural practices including application of trace element-containing pesticides (As), fertilizers (Cd) and wastes are also important pathways (Han et al., 2000; Han et al., 2001a, 2001b; 2003b; 2004a ; Loebenstein, 1994).

Mercury, As, Cd and others are emitted into the atmosphere through human activities, such as combustion of coal/petroleum and during mining and processing (e.g., As during extractive metallurgical processing of nonferrous metals). In earlier centuries as today, most As was a byproduct of mining other metals (Loebenstein, 1994). It is globally transported and finally deposited by wet and dry processes into the pedosphere (Nriagu and Pacyna, 1988; Nriagu, 1979), a major sink for most trace elements. Since early in the past century, As compounds have been used as herbicides, pesticides, wood preservatives, animal feed additives, corrosion inhibitors, semi-conductors, metal alloys, and in glass manufacture. Of these, agricultural uses accounted for 90% of As utilization in the 1940s and then declined, while wood preservatives are the major current application of As (Loebenstein, 1994). However, as of January 1, 2004, USEPA will not allow chromated copper arsenate products to be used to treat wood intended for any of residential uses. This decision will decrease As dissipation in both the manufacturing and retail sectors (USEPA, 2002b).

9.4 POTENTIAL ANTHROPOGENIC TRACE ELEMENT INPUTS IN THE WORLD ARABLE SURFACE

Based on cumulative element production, the following indices are developed to estimate the potential environmental risks of accumulation of trace elements in the pedosphere: the potential world cultivated-land loading and potential global metal burden per capita.

Most human activities and industrial processes occur on or adjacent to arable land. World civilization-relevant land and arable land areas are used to calculate potential global cumulative trace element loading in the pedosphere by assuming trace elements excavated from the earth are eventually dispersed on the world land area (Table 9.3). World civilization-relevant land includes cultivated land, lakes and streams, swamps and marshes and human habitation areas, totaling about 23.8×10^6 km^2 (Banin et al., 1984). The global arable land accounts for about 10.7% of the world total land area, amounting to 16×10^6 km^2 (Banin et al., 1984). The potential cumulative trace element loading on global arable land is an indicator of the potential maximum average trace element loading over global arable land. The potential trace element cumulative loadings on world arable land in 2000 were 28190 for Cu, 22132 for Zn, 14661 for Pb, 6584 for Cr, 2231 for Ni, 283 for As, 66 for Cd and 39 kg km^{-2} for Hg. The potential anthropogenic trace element inputs in the world arable surface soil (0–10 cm) range from 0.15 mg Cd kg^{-1} soil, 0.15 mg Hg kg^{-1} soil to 49 mg

Cu kg^{-1} soil, and increase to 0.31 mg Hg kg^{-1} soil and 217 mg Cu kg^{-1} soil, respectively, if the cumulative elements extracted before 1950 and 2000 are completely dispersed in the pedosphere (Table 9.3).

As pointed out by Settle and Patterson (1980), most parts of the earth have been contaminated by trace elements. The trace element concentrations in world soils (as means) range from 0.035 mg kg^{-1} for Hg to 125 mg kg^{-1} for Cr (Table 9.4). Concentrations of these trace elements in the U.S. agricultural soils are similar to soils of the rest of the world (Holmgren et al. 1993). Lead and Cd concentrations in world soils are about 2–4 and 5 times that in the lithosphere, respectively (Table 9.4). This indicates that world soils may be contaminated by both Pb and Cd. Zinc concentration in the U.S., Chinese, and Indian uncontaminated soils is close to or slightly greater than that in the lithosphere, implying a slight contamination in world soils (Table 9.4). The total As concentration in the lithosphere and world soil averages 1.88 and 6.64 mg/kg, respectively (Table 9.4). However, the actual values of anthropogenic trace element inputs vary greatly depending on local industry and deposition properties. Potential anthropogenic As, Cd, Cu, Hg, Pb and Zn inputs (complete dispersion of cumulative metal produced before 2000) in the world surface arable soils are 1–10 times natural trace element concentrations in the lithosphere and 0.3–9 times current trace element concentrations in world soils (Table 9.3). Lead, Hg, and Cd have the highest potential anthropogenic inputs in the pedosphere compared to the lithosphere, while Hg, Cu and Pb have the highest inputs compared to world soils. This implies that Pb, Hg, Cd, Cu and Zn might comprise the most serious pollutants in the pedosphere. There may be much more environmentally benign capacity for newly used trace elements, such as Cr and Ni in the pedosphere. However, urban soils, mine and industrially contaminated soils, as well as river and lake sediments contain higher metal concentrations. In alluvial soils in England and Wales contaminated by base metal mining in the nineteenth and twentieth centuries, Cd, Cu, Zn and Pb concentrations were 0.6–15 (mean: 3.0), 5–1260 (376), 18–5544 (815) and 76–12600 (1492) mg kg^{-1}, respectively (recalculated from Davies 1980). These trace element concentrations are far greater than the potential average global anthropogenic trace element inputs due to concentrations from other human activities. Even agricultural soils accumulate trace elements such as Pb, Cd, Zn and Cu from normal agricultural practices (Holmgren et al., 1993; Han et al., 2000; Han et al., 2001a; 2001b). In addition to the dominant dispersion in arable lands, a portion of trace elements is dispersed into river, lake, and marine environments, especially as airborne pollutants. Arsenic inputs from the atmosphere into the world's oceans are relatively small compared to that into the world's soils (Nriagu and Pacyna, 1988). The retention time of As in the atmosphere averages from 7–10 days (Matschullat, 2000), while As in the pedosphere in moderate climates is around 1000–3000 years (Bowen, 1979).

Table 9.3. Potential anthropogenic inputs in world soils and global metal burden per capita[a]

Index	Year	Cd	Cr	Cu	Hg	Ni	Pb	Zn	As
Potential anthropogenic metal inputs (mg/kg) (civilization-relevant land surface soil)	1900	0.02	0.05	5.8	0.03	0.10	8.18	5.14	0.05
	1950	0.10	1.61	32.6	0.10	1.3	30.4	27.4	0.64
	1990	0.28	21.8	109	0.20	8.2	66.5	86.5	1.28
	2000	0.34	34.1	146	0.21	12	75.8	114	1.47
Potential anthropogenic metal inputs (mg/kg) (Arable surface soil)	1900	0.03	0.1	8.6	0.04	0.2	12.2	7.6	0.07
	1950	0.15	2.4	48.5	0.15	1.9	45.3	40.7	0.95
	1990	0.42	32.4	162	0.29	12.2	98.9	129	1.91
	2000	0.51	50.7	217	0.31	17.2	113	170	2.18
Ratio of anthropogenic cumulative input to metal content in world soil before	1900	0.06	0.001	0.22	1.20	0.01	0.45	0.16	0.01
	1950	0.30	0.02	1.24	4.33	0.08	1.68	0.85	0.14
	1990	0.83	0.26	4.14	8.33	0.49	3.66	2.68	0.29
	2000	1.02	0.41	5.56	8.73	0.69	4.18	3.55	0.33
Ratio of anthropogenic cumulative input to metal content in lithosphere before	1900	0.29	0.0004	0.10	0.84	0.002	1.04	0.12	0.04
	1950	1.54	0.01	0.55	3.03	0.02	3.85	0.63	0.51
	1990	4.26	0.17	1.85	5.83	0.12	8.42	1.98	1.02
	2000	5.23	0.26	2.48	6.11	0.17	9.60	2.62	1.16
Global metal burden per capita (kg cumulative metal per capita)	1900	0.04	0.1	10.8	0.05	0.2	15.3	9.6	0.1
	1950	0.12	2.0	40.1	0.13	1.6	37.5	33.7	0.8
	1990	0.16	12.7	63.5	0.11	4.8	38.9	50.6	0.8
	2000	0.18	17.3	74.2	0.10	5.9	38.6	58.2	0.7
	2000/1900	4.9	174	6.9	2.0	30	2.5	6.0	7.0

[a] Data from Han et al. 2002a and 2003b, with kind permission of Springer Science and Business Media.

Table 9.4. Trace element contents of the lithosphere and world soils (mg/kg) (Data from Han et al. 2002a, with kind permission of Springer Science and Business Media)

Lithosphere/ soils		Elements						
		Cd	Cr	Cu	Hg	Ni	Pb	Zn
Lithosphere			200[a]	100[a]	0.05[b]	100[a]	15[a]	50[a]
		0.098[c]	185[c]	75[c]		105[c]	8.5[c]	80[c]
World soil		0.62[d]	50[a]	59.8[d]	0.01[e]	25.8[d]	33.7[d]	29.2[d]
		0.30[f]	100-300[g]	15-40[g]	0.06[b]	20-30[g]	15-25[g]	50-100[g]
				30[b]		25[a]		40[a]
						40[h]		50[h]
								90[i]
	Average	0.5	125	39	0.035	25	27	48
US agricultural soil[j]	Range	<0.005-2.0		0.3-495		0.7-269	0.5-135	1.5-264
	95% sample	0.78		95		57	23	126
England contaminated soil[k]	Range	1.5-5.7		222-733	0.5-20		4563	277-2182
	Average	3.2		463	1.78		1935	906
China Uncontaminated soil	Average	0.097[l]	61[l]	22[m]	0.065[l]	23.4[l]	26[l]	100[m]
Contaminated soil[n]	Range	0.67-5		383-2081	55.64			117-1286
India Uncontaminated soil		0.1	114	51		63	144	114
Contaminated soil[o]		0.4	1221	77		113	155	173

[a]Berrow and Reaves, 1984
[b]Bowen, 1979
[c]Taylor and McLennan, 1985
[d]Ure and Berrow, 1982
[e]Brooks, 1977
[f]Lee and Keenev, 1975
[g]Aubert and Pinta, 1977
[h]Vinogradov, 1959
[i]Swaine, 1955
[j]Holmgren et al., 1993
[k]Recalculated from Davies, 1980
[l]China Environmental Monitor Station, 1990
[m]Liu et al., 1996
[n]Chen et al., 1999
[o]Ansari et al., 1999

The annual global trace element production in 2000 was 367, 66, 31, 17, 15, 9 and 3 times that in 1900 for Cr, Ni, As, Cu, Zn, Cd and Pb, respectively. However, the annual global Hg production in 2000 was only 0.4 times that in 1900. Compared to 1900, the increases in the potential global civilization-relevant land trace element load and potential global arable land load in 2000 were similar to the cumulative global element production. Han et al. (2002a) pointed out that the potential global arable land loading is a potential average loading. Thus, there is much fluctuation around this mean.

The most developed and industrialized areas (like Europe) with intensive human activities and relatively short recycling periods of the man-made products may be more highly contaminated, and trace element concentrations in these areas may be much higher than the average values (Tables 9.4–5). Many soils in Japan and central Europe either have become or will soon become overloaded with toxic trace elements (Nriagu and Pacyna, 1988). In undeveloped areas with less industrial activity and longer recycling periods of products, however, trace element concentrations may be much lower. But the newly industrialized countries like China and India are facing increasing trace element pollution in their regions (Tables 9.4–5). The main sources of trace element pollution in China are sewage irrigation, sludge application, and mining and smelting operations. Trace element polluted agricultural land due to point source pollution, such as industrial and mining activities are estimated 20.33×10^5 ha in China (recalculated from Chen et al., 1999). Cadmium and Hg polluted arable land due to sewage irrigation are 1.3×10^4 and 3.2×10^4 ha, respectively (Chen et al., 1999). In soils of China, trace element concentrations in uncontaminated soils are in the range of those in world soils except for Zn. Nickel, Zn, Cu and Pb in uncontaminated Indian soils are higher than those in world soils (Table 9.4). Clearly, trace element concentrations in some contaminated soils of these two countries are far higher than world soils. Copper and Hg in mining polluted soils of China were even higher than those in typical contaminated England soils (Table 9.4).

In southern U.S. agricultural soils, As concentration is significantly higher (8.09 mg As kg^{-1}) than uncultivated soils (3.63 mg As kg^{-1}) due to agricultural As additions, such as application of herbicides and pesticides (Pettry and Switzer, 2000). In urban soils of the northeastern U.S. affected by historical smelter emissions containing As, soil As concentration is around 26.4 mg kg^{-1} and some are as high as 163 mg kg^{-1} (Golding, 2001). However, in soils near smelter sites, As concentration may approach levels as high as 3000 mg kg^{-1}. Arsenic concentrations in streambed sediments across the continental United States range from 1.0 mg As kg^{-1} to 200 mg As kg^{-1}, with an average of 8.9 mg As kg^{-1} (Rice, 1999). Arsenic concentrations in these contaminated soils and sediments are far greater than the potential average global anthropogenic As inputs due to concentrations from human activities. In U.S. poultry wastes, As ranges from 10–40 mg kg^{-1} with high solubility (with water soluble As representing 40–80% of total As) (Han et al., 2004a). In 25 year-old poultry waste-amended soils, total As concentrations average 8.4 mg kg^{-1}, higher than in non-amended soils (average 2.1 mg kg^{-1}) (Han et al., 2004a). In addition, much waste in some industrial nations is currently sequestered in landfills, where trace elements are relatively immobilized. This greatly decreases the risks and rates of trace element dispersion in the pedosphere.

Table 9.5. Arsenic contents of the lithosphere and world soils (Data from Han et al., 2003b, with kind permission of Springer Science and Business Media)

Lithosphere/ soils		Concentration (mg kg^{-1})
Lithosphere	Mean	1.0[a], 1.5[b], 1.7[c], 1.8[d], 3.4[e]
World soil	Mean	5[e], 6[b], 7.2[f], 7.5[g], 5-10[h]
US agricultural soils[i]	Range	<0.1 - 93
	Mean	<10
Canada soils[j]		
Unpolluted soils	Range	4.8 -13.6
Polluted soils with smeltering	Range	50-100
Polluted soils with pesticides	Mean	54
Polluted soils with wood preservatives	Mean	6000
Sediments		
Ocean sediments[k]	Range	5-40
River and lake sediments[j]	Mean	< 20
US river and lake sediments[l]	Range	1.0 - 200
	Mean	8.9
Canada polluted sediments with mining[j]	Range	100-5000
Coal	Mean	5[b]
	Range	0.3-93[b], 0.34-130[m], 5-45[n]
Petroleum	Mean	0.26[b]
	Range	0.0024-1.63[n], 0.005-0.14[h]

[a]Taylor and McLenna, 1995
[b]Bowen, 1979
[c]Wedepohl, 1995
[d]Lide, 1996
[e]Bockris, 1977
[f]Koljonen, 1992
[g]Allard, 1995
[h]Baur and Onishi, 1969
[i]Kabata-Pendias and Pendias, 1992
[j]Environment Canada, 1993
[k]Neff, 1997
[l]Rice, 1999
[m]Piver, 1983
[n]Pacyna, 1987

9.5 POTENTIAL GLOBAL TRACE ELEMENT BURDENS PER CAPITA

Global population in 1900, 1950, 1980, and 2000 was 1.608×10^9, 2.502×10^9, 4.456×10^9 and 6.08×10^9, respectively (Woytinsky and Woytinsky, 1953; United Nations, 1962; U.S. Bureau of the Census, 1999).

The world population has exponentially increased during the 20th century (data not shown). The potential global trace element burden per capita for the eight trace elements has been calculated (Table 9.3). The potential global Cu and Zn burdens per capita greatly increased from early in the 20th century and continued to increase through the 1980s. The global Pb and Cd burdens per capita increased from the early 20th century, but stabilized after 1945–1950, while the global Cr burden increased rapidly after the 1960s. The global Hg, Cd and Ni burdens per capita were relatively low.

Compared to 1900, global trace element burdens per capita in 2000 are 174 times that in 1900 for Cr, 30 times for Ni and 2–7 times for Hg, Cd, Pb, Zn and Cu (Table 9.3). On an absolute basis, trace element burdens per capita in 2000 are from 0.10 kg for Hg, 0.18 kg for Cd to 58 kg for Zn and 74 kg for Cu (Table 9.3). The potential global As burden per capita greatly increased from the early 20th century and continued to increase through the 1950s. From the 1950s through 2000, the global As burden per capita stabilized around 0.8 kg cumulative As per capita, mainly due to the fast growth of the world population. Compared to 1900, global As burden per capita in 2000 was 0.7 kg, 8.2 times that in 1900 (Table 9.3). However, potential global trace element burden per capita also varies based on locality and historical human activities. Further, the potential threat of trace elements to humans in developed areas may be higher than that in undeveloped areas.

9.6 BIOAVAILABILITY, TOXICITY AND MOBILITY OF TRACE ELEMENTS

Bioavailability, toxicity and mobility of trace elements in the pedosphere highly depend upon environmental acidity, content of complexing ligands, and soil properties as discussed in the previous chapters (Han et al., 2001c). Acidification increases mobility, solubility, and toxicity of trace metals in the pedosphere (Huchinson and Whilby, 1977). Global acid rain, deposit of dry gaseous SO_2 and NO_x from the atmosphere into the pedosphere, and intensive application of inorganic fertilizers are the most important acidification sources (Paces, 1985). Currently, even though deposition of sulfur compounds has been reduced, N emission continues to increase (Wright and Schindler, 1995). Most of Northern Europe and Eastern North America are impacted by acid deposition (Baldwin, 1986). Various natural organic ligands are available for mobilization of trace elements (Almas et al., 2000a, b). Our current environment is tending to increase bioavailability and toxicity of trace elements in the pedosphere in the long-term.

9.7 CONCLUSIONS

The inventories of trace elements estimated here since the industrial age clearly indicate a large amount of trace element sources produced and dispersed into the pedosphere. In 2000, the cumulative industrial age anthropogenic global production of As, Cd, Cr, Cu, Hg, Ni, Pb and Zn was 4.53, 1.1, 105, 451, 0.64, 36, 235 and 354 million metric tons, respectively. The ratios of the potential anthropogenic trace element inputs to its content in world soil in 2000 were 0.3, 1.0, 0.4, 5.6, 8.7, 0.7, 4.2 and 3.5 for As, Cd, Cr, Cu, Hg, Ni, Pb, and Zn, respectively. Global trace element burdens per capita were 0.7, 0.18, 17.3, 74.2, 0.10, 5.9, 38.6 and 58.2 kg for As, Cd, Cr, Cu, Hg, Ni, Pb and Zn, respectively.

Unlike most organic pollutants, trace elements accumulate in the pedosphere once they enter the soil/water ecosystems. Remediation of contaminated soils is technically difficult, costly, and time-consuming. The most efficient approach to control trace element pollution in the pedosphere is to control anthropogenic pollution sources during trace element mining, processing, manufacturing, recycling, and the disposal of wastes. The technology of industrial processing needs to be continuously improved to reduce the trace element dispersion rates into the environment. Outdated trace element-containing products, by-products, and wastes need to be continually recycled. Finally, new substitute materials for trace elements are expected to be developed to eventually replace and limit usage of trace elements in our civilization. To develop substitute application materials to replace As in agriculture and forestry industries and to control Hg and As emissions from the coal industry would significantly decrease As and Hg pollution sources and dissipation rates into the environment. In addition, for trace element-polluted soils and water as a result of thousands of years of human civilization, comprehensive chemical remediation, phytoremediation, engineering remediation technologies, and modern genetics are essential to reduce the bioavailability, mobility, and toxicity of trace elements in ecosystems.

References

Adriano D.C. *Trace Elements in the Terrestrial Environment*. New York: Springer, 1986 and 2001.
Adriano D.L., Murphy L.S. Effects of ammonium polyphosphates on yield and chemical composition of irrigated corn. Agron J 1970; 62: 561–567.
Agbenin J.O. Zinc fractions and solubility in a tropical semi-arid soil under long-term cultivation. Bio Fertil Soils 2003; 37: 83–89.
Ahnstrom Z.A.S., Parker D.R. Cadmium Reactivity in Metal-Contaminated Soils Using a Coupled Stable Isotope Dilution-Sequential Extraction Procedure. Environ Sci Technol 2001; 35: 121–126.
Ainsworth C.C., Pilon J.L., Gassman P.L., Van Der Sluys W.G. Cobalt, cadmium, and lead sorption to hydrous iron oxide: Residence time effect. Soil Sci Soc Am J 1994; 58: 1615–1623.
Allard B. *"Groundwater."* In: *Trace Elements in Natural Waters,* B. Salbu, E. Steinnes, eds. Boca Raton, FL: CRC Press, 1995.
Allen K.N. Seasonal variation of selenium in outdoor experimental stream-wetland systems. J Environ Qual 1991; 20: 865–868.
Almas A.R., McBride M.B., Singh B.R. Changes in partitioning of cadmium-109 and zinc-65 in soils as affected by organic matter addition and temperature. Soil Sci Soc Am J 2000b; 64: 1951–1958.
Almas A.R., McBride M.B., Singh B.R. Solubility and lability of cadmium and zinc in two soils treated with organic matter. Soil Sci 2000a; 165: 250–259.
Amini M., Khademi H., Afyuni M., Abbaspour K.C. Variability of available cadmium in relation to soil properties and landuse in an arid region in central Iran. Water Air Soil Pollut 2005; 162: 205–218.
Amit R., Gerson R., Yaalon D.H. Stages and rate of the gravel shattering process by salts in desert Reg soils. Geoderma 1993; 57: 295–324.
Amrhein C., Mosher P.A., Brown A.D. The effects of redox on Mo, U, B, V, and As solubility in evaporation pond soils. Soil Sci 1993; 155: 249–255.
Ansari A.A., Singh I.B., Tobschall H.J. Status of anthropogenically induced metal pollution in the Kanpur-Unnao industrial region of the Ganga Plain, India. Environ Geology 1999; 38: 25–33.
Aubert H., Pinta M. *Trace Elements in Soils*. Elsevier, Amsterdam, 1977.
Avnimelech Y. Irrigation with sewage effluents: The Israeli experience, Environ Sci Technol 1993; 27: 1278–1281.
Awad F, Romheld V. Mobilization of heavy metals from contaminated calcareous soils by plant born, microbial and systhetic chelators and their uptake by wheat plants. J Plant Nutrit 2000a; 13: 1847–1855.
Awad F, Romheld V. Significant of root exudates in acquisition of heavy metals from a contaminated calcareous soil by graminaceous species. J Plant Nutrit 2000b; 23: 1857–1866.
Backes C.A., McLaren R.G., Rate A.W., Swift R.S. Kinetics of cadmium and cobalt desoprtion from iron and manganese oxides. Soil Sci Soc Am J 1995; 59: 778–785.
Badalucco L., Kuilman P. J. *"*Mineralization and immobilization in the rhizosphere.*"* In *The Rhizosphere*, Pinton Roberto, Varanini Zeno, Nannipieri Paolo, eds. New York, NY: Marcel Dekker, Inc., 2001.

Badawy S.H., Helal M.I.D., Chaudri A.M., Lawlor K., McGrath S.P. Soil solid-phase controls lead activity in soil solution. J Environ Qual 2002; 31: 162–167.

Badri M.A., Aston S.R. A comparative study of sequential extraction procedures in the geochemical fractionation of heavy metals in estuarine sediments. International Conference; Amsterdam. pp 705–708. 1981.

Bailey G.W., Akim L.G., Shevchenko S.M. "Predicting chemical reactivity of humic substances for minerals and xenobiotics: use of computational chemistry, scanning probe microscopy, and virtual reality." In *Humic Substances and Chemical Contaminants*, C.E. Clapp, M.H.B. Hayes, N. Senesi, P.R. Bloom, P.M. Jardine, eds. Madison, WI: Soil Science Society of America, Inc., 2001.

Baker A.J.M., Walker P.L. "Ecophysiology of metal uptake by tolerant plants." In *Heavy Metal Tolerance in Plants: Evolutionary Aspects*, A.J. Shaw, ed. Boca Raton, FL: CRC Press. 1989.

Baldwin J.H. Acid rain: A global perspective. Int J Ecol Environ Sci 1986; 12: 35–57.

Bandyopadhyay B.K., Bandyopadhyay A.K. Transformation of iron and manganese on coastal saline soil. J Indian Soc Soil Sci 1984; 32: 51–62.

Banin A., Gerstl Z., Fine P., Metzger Z., Newrzella D. Minimizing soil contamination through control of sludge transformations in soil. Joint German-Israel research projects. Final report. No. of Project: Wt 8678/458, 1990.

Banin A., Han F.X., Carina C. Total contents and solid-phase distribution of trace elements in 45 Israeli soils. Unpublished data. Hebrew University of Jerusalem, Rehonot, 76100. Israel. 1997a.

Banin A., Han F.X., Serban C., Ben-Dor E., Schachar Y. The dynamics of heavy metals partitioning and transformations in arid-zone soils. The Proceedings of The Fourth International Conference on Biogeochemistry of Trace Elements; 1997b; Berkeley, CA, USA.

Banin A., Lawless J.G., Whitten R.C. Global N_2O cycles - Terrestrial emissions, atmospheric accumulation and biospheric effects. Adv Space Res 1984; 4: 207–216.

Banin A., Navrot J. Origin of Life: Clues from relations between chemical compositions of living organisms and natural environments. Science 1975; 189: 550–551.

Banin A., Navrot J., Noi Y., Yoles D. Accumulation of heavy metals in arid-zone soils irrigated with treated sewage effluents and their uptake by rhodes grass. J Environ Qual 1981; 10: 536–540.

Banin A., Navrot J., Perl A. Thin-horizon sampling reveals highly localized concentrations of atmophile heavy metals in a forest soil. Sci Total Environ 1987; 61: 145–152.

Banin A., Nir S., Brummer G.W., Han F.X., Serban C., Krumnohler J. Cd pollution in soils: Long-term processes in the solid phase, there characterization and models for their prediction. Joint Israel-Commission of the European Communities research projects. Final report. No. of Project: ISC-8911-ISR (ENV), 1995.

Banuelos G.S. "Factors Influencing Field Phytoremediaton of Selenium-laden Soils." In *Phytoremediation of Contaminated Soil and Water*, N. Terry, G. Banuelos, eds. Boca Raton, FL: Lewis Publishers, 2000.

Barbarick K.A., Ippolito J.A. Termination of sewage biosolids application affects wheat yield and other agronomic characteristics. Agron J 2003; 95: 1288–1294.

Barbarick K.A., Ippolito J.A., Westfall D.G. Biosolids effect on P, Cu, Zn, Ni. and Mo concentrations in dryland wheat. J Environ Qual 1995; 24: 208–611.

Barbarick K.A., Ippolito J.A., Westfall D.G. Sewage biosolids cumulative effects on extractable-soil and grain elemental concentrations. J Environ Qual 1997; 26: 1696–1702.

Barbarick K.A., Workman S.M. Ammonium bicarbonate-DTPA and DTPA extractions of sludge-amended soil. J Environ Qual 1987; 16: 125–130.

References

Barbarick, K.A., Ippolito, J.A., Westfall, D.G. Extractable trace elements in the soil profile after years of biosolids application. J Environ Qual 1998; 27: 801–805.

Barber, Stanley A. *Soil Nutrient Bioavailability, A Mechanistic Approach*. New York: John Wiley and Sons, Inc., 1984.

Barrow N.J., Gerth J., Brummer G.W. Reaction kinetics of the adsorption and desorption of nickel, zinc and cadmium by goethite. II Modelling the extent and rate of reaction. J Soil Sci 1989; 40: 437–450.

Bartlett R.J, Kimble J.M. Behavior of chromium in soils. II. Hexavalent forms. J Environ Qual 1976; 5: 383–386.

Bartlett R.J. "Soil redox behavior" In *Soil Physical Chemistry*, Sparks D.L., ed, Boca Raton, FL: CRC Press, Inc., 1986.

Baur W.H., Onishi H. "Arsenic." In *Handbook of Geochemistry* II (3): 33., Wedepohl K.H., ed. Berlin: Springer, 1969.

Beckett P.H.T. The use of extractants in studies on trace metals in soils, sewage sludges and sludge-treated soils. Adv Soil Sci 1989; 9: 143–176.

Beckwith R.S., Tiller K.G., Suwadji E. "The effects of flooding on the availability of trace metals to rice in soils of differing organic matter status" In *Trace Elements in Soil-Plant-Animal Systems*, Nicholas DJD, Egan AR. eds. New York, NY: Academic Press, Inc., 1975.

Bell F.B., James B.R., Chaney R.L. Heavy metal extractability in long-term sewage sludge and metal salt-amended soils J Environ Qual 1991; 20: 481–486.

Belzile N., Lecomte P., Tessier A. Testing readsorption of trace elements during partial chemical extractions of bottom sediments. Environ Sci Technol 1989; 23: 1015–1020.

Bermond A., Benzineb K. The localization of heavy metals in sewage treated soils: Comparison of thermodynamics and experimental results. Water Air Soil Pollut 1991; 57–58: 883–890.

Berrow M.L., Burridge J.C., Reith J.W.S. Soil drainage conditions and related plant trace element contents. J Sci Food Agric 1983; 34: 53–54.

Berrow M.L., Reaves G.A. Proc Intl Conf On Environ Contamination. Edinburg, UK: GEP Consultants Ltd., 1984.

Bewers J.M., Barry P.J., MacGregor D.J. "Distribution and Cycling of Cadmium in the Environment." In *Cadmium in the Aquatic Environment*, J.O. Nriagu, J.B. Sprague, eds. New York: John Wiley & Sons, 1987.

Bewley R.J.F., Stotzky G. Effects of cadmium and zinc on microbial activity in soil; influence of clay minerals. Part II. Metals added simultaneously. Sci Total Environ 1983; 31: 57–69.

Bidwel A. M., Dowdy R.H. Cadmium and zinc availability to corn following termination of sewage sludge application. J Environ Qual 1987; 16: 438–442.

Biester H., Scholz C. Determination of mercury binding forms in contaminated soils: Mercury pyrolysis versus sequential extractions. Environ Sci Technol 1997; 31: 233–239.

Bilitewski, B., Härdtle G., Marek K. *Waste Management*, Springer, New York, 1994.

Bingham F.T., Page A.L., Mahler R.J., Ganji T.J. Yield and cadmium accumulation of forage species in relation to cadmium content of sludge-amended soil. J Environ Qual 1976; 5: 57–60.

Bingham F.T., Page A.L., Mahler R.J., Ganji T.J. Growth and cadmium accumulation of plants grown on a soil treated with a cadmium-enriched sewage sludge. J Environ Qual 1975; 4: 207–211.

Bingham F.T., Strong J.E., Sposito G. Influence of chloride salinity on cadmium uptake by Swiss chard. Soil Sci 1983; 83: 160–165.

Birkeland P. *Soils and Geomorphology*. Oxford, UK: Oxford University Press. 1984.

Bizily S.P., Rugh C.L., Meagher R.B. Phytodetoxification of hazardous organomercurials by genetically engineered plants. Nature Biotechnol 2000; 18: 213–217.

Blaylock A.D. Navy bean yield and maturity response to nitrogen and zinc. J Plant Nutri 1995; 18: 163–178.

Blaylock M.J., Huang J.W. "Phytoextraction of metals." In *Phytoremediation of Toxic Metals: Using Plants to Clean Up the Environment*, Raskin I., Ensley B.D., eds. New York, NY: John Wiley & Sons, Inc., 2000.

Bloom P.R., Bleam W.F., Xia K. "X-ray spectroscopy applications for the study of humic substances," In *Humic Substances and Chemical Contaminants*, C.E. Clapp, M.H.B. Hayes, N. Senesi, P.R. Bloom, P.M. Jardine, eds. Madison, WI: Soil Science Society of America, Inc., 2001.

Bloomfield C., Pruden G. The effects of aerobic and anaerobic incubation on the extractabilities of heavy metals in digested sewage sludge. Envion Pollut 1975; 8: 214–232.

Boawn L.C., Viets F.G. Jr., Crawford C.L. Plant utilization of zinc from various types of zinc compounds and fertilizer materials. Soil Sci 1957; 83: 219–227.

Bockris J.O.M. *Environmental Chemistry*. New York: Plenum Press, 1977.

Bolan N.S., Khan M.A., Donaldson J., Adriano D.C., Matthew C. Distribution and bioavailability of copper in farm effluent. Sci. Total Environ 2003; 309: 225–236.

Boon D.Y. "Potential selenium problems in great plains soils." In *Selenium in Agriculture and the Environment*, L.M. Jacobs, ed. Madison, IL: Soil Science of America, Inc. 1989.

Boon D.Y., Soltanpour P.N. Lead, cadmium, and zinc contamination of Aspen garden soils and vegetation. J Environ Qual 1992; 21: 82–86.

Borggaard O.K. Influence of iron oxides on cobalt adsorption by soils. J Soil Sci 1987; 38: 229–238.

Bowen H.J.M. *Environmental Chemistry of the Elements*. New York: Academic Press, 1979.

Bradford G.R., Chang A.C., Page A.L., Bakhtar D. Frampton JA, Wright H. Background Concentrations of Trace and Major Elements in California Soils. Kearney Foundation of Soil Science Special Report. University of California, 1996.

Bradford G.R., Page A.L., Lund L.J., Olmstead W. Trace element concentrations of sewage treatment plant effluents and sludges: Their interactions with soils and uptake by plants. J Environ Qual 1975; 4: 123–127.

Bragg L.J., Oman J.K., Tewalt S.J., Oman C.L., Rega N.H., Washington P.M., Finkelman R.B. The U.S. geological survey coal quality (COALQUAL) database, Version 2.0. U.S. Geol. Survey Open-File Report 97-134, 1998.

Brendecke J.W., Axelson R.D., Pepper I.L. Soil microbial activity as an indicator of soil fertility: long-term effects of municipal sewage sludge on an arid soil. Soil Bio Biochem 1993; 25: 751–758.

Brimecombe Melissa J, Leij Frans AD, Lynch James M. "The effect of root exudates on rhizosphere microbial populations." In *The Thizosphere*, Pinton Roberto, Varanini Zeno, Nannipieri Paolo, eds. New York, NY: Marcel Dekker, Inc., 2001.

Brookes P.C. The use of microbial parameters in monitoring soil pollution by heavy metals. Bio Fertil Soils 1995; 19: 269–279.

Brooks R.R. "Pollution through trace elements." In *Environmental Chemistry*, Bockris J.O.M., ed. New York: Plenum Press, 1977.

Brown G.E. Jr., Foster A.L., Ostergren J.D. Mineral surfaces and bioavailability of heavy metals: A molecular-scale perspective. Proc Natl Acad Sci 1999; 96: 3388–3395.

Brümmer G.W., Gerth J., Tiller K.G. Reaction kinetics of the adsorption and desorption of nickel, zinc and cadmium by goethite. I. Adsorption and diffusion of metals. J Soil Sci 1988; 39: 37–52.

References

Brummer G.W., Tiller K.G., Herms U., Clayton P.M. Adsorption-desorption and/or precipitation-dissolution processes of zinc in soils. Geoderma 1983; 31: 337–354.

Burns R.G. The uptake of cobalt into ferro-manganese nodules, soils, and synthetic manganese (IV) oxides. Geochim Cosmochim Acta. 1976; 40: 95–102.

Cai Z., Liu Z. Studies on oxidation states of cobalt extracted from soil with EDTA+HOAc+NH_4OAc. Pedosphere 1991; 1: 109.

Candelaria L.M., Chang A.C. Cadmium activities, solution speciation, and solid phase distribution of Cd in cadmium nitrate and sewage sludge-treated soil systems. Soil Sci 1997; 162 (10): 722–732.

Cartwright B., Merry R.H., Tiller K.G. Heavy metal contamination of soils around a lead smelter at Port Pirie, South Australia. Aust J Soil Res. 1976; 15: 69–81.

Cary E.E., Allaway W.H., Olson O.E. Control of chromium concentrations in food plants, I. Absorption and translocation of chromium by plants. J Agric Food Chem 1997; 25: 300–304.

Cavallaro N., McBride M.B. Effect of selective dissolution on pH-dependent charge in acid soil clays. Clays Clay Miner. 1984b; 32: 283–290.

Cavallaro N., McBride M.B. Zinc and copper sorption and fxation by an acid soil clay: effect of selective dissolutions. Soil Sci Soc Am J 1984a; 48: 1050–1054.

Chairidchai P., Ritchie G.S.P. Zinc absorption by a lateritic soil in the presence of organic ligands. Soil Sci Soc Am J 1990; 54: 1242–1248.

Chaney R.L. "Potential effects of waste constituents on the food chain." In *Land Treatment of Hazardous Wastes*, Parr J.F., ed. Park Ridge, New Jersey: Noyes Data Corp, 1983.

Chang A.C., Hyun H., Page A.L. Cadmium uptake for Swiss chard grown in composted sewage sludge treated field plots: Plateau or time bomb? J Environ Qual 1997; 26: 11–19.

Chang A.C., Page A.L., Warneke J.E., Grgurevic E. Sequential extraction of soil heavy metals following a sludge applications. J Environ Qual 1984; 13: 33–38.

Chang A.C., Pan G., Page A.L., Asano T. 2002. Developing human health-related chemical guidelines for reclaimed water and sewage sludge applications in agriculture. The Report to World Health Organization, University of California, Riverside, CA

Chao T. T. Selective dissolution of manganese oxides from soils and sediments with acidified hydroxylamine hydrochloride. Soil Sci Soc Am Proc 1972; 36: 704–768.

Chao T.T., Sanzolone R.F. Fractionation of soil selenium by sequential Partial dissolution. Soil Sci Soc Am J 1989; 53: 385–392.

Chao T.T., Zhou L Extraction techniques for selective dissolution of amorphous iron oxides from soils and sediments. Soil Sci Soc Am J 1983; 47: 225–232.

Charlatchka R., Cambier P. Influence of reducing conditions on solubility of trace metals in contaminated soils. Water Air Soil Pollut 2000; 118: 143–167.

Chen H., Zheng C., Tu C., Zhu Y. Heavy metal pollution in soils in China: Status and countermeasures. AMBIO 1999; 8: 130–134.

Chen X., Wright J.V., Conca J.L., Peurrung L.M. Evaluation of heavy metal remeidaiton using mineral apatite. Water Air Soil Pollut 1997; 98: 57–78.

Cherrey A., Chaignon V., Hinsinger P. Bioavailability of copper in thizosphere of rape and ryegrass cropped in vineyard soils. Proceedings of 5th International Conference on the Biogeochemistry of Trace Elements. Vienna, Austria, 1999.

Chilvers D.C., Peterson P.J. "Global cycling of arsenic." In *Lead, Mercury and Arsenic in the Environment*, T.C. Hutchinson, K.M. Meema, eds. New York: John Wiley. 1987.

China Environmental Monitor Station. Background concentrations of elements in soils of China. Beijing, China: Chinese Environ Science Press, 1990.

Chlopecka A., Adriano D.C. Mimicked in-situ stabilization of metals in a cropped soil: Bioavailability and chemical from of zinc. Environ Sci Technol 1996; 30: 3294–3303.

Chopin E. I. B., Black S., Hodson M.E., Coleman M.L., Alloway B.J. A preliminary investigation into mining and smelting impacts on trace element concentrations in the soils and vegetation around Tharsis, SW Spain. Mineral Soc Great Britain Ireland 2003; 67: 279–288.

Chung J., Burau R.G., Zasoski R.J. Chromate generation by chromate depleted subsurface materials. Water Air Soil Pollut 2001; 128: 407–417.

Cifuentes F.R., Lindemann W.C., Barton L.L. Chromium sorption and reduction in soil with implications to bioremediation. Soil Sci 1996; 161: 233–241.

Clarke L.B., Sloss L.L. Trace Elements – emissions from coal combustion and gasification. London: IEA Coal Research, 1992.

Clevenger T.E. Use of sequential extraction to evaluate the heavy metals in mining wastes. Water Air Soil Pollut 1990; 50: 241–254.

CNN. Toxin warnings grow for U.S. fish. Atlanta, GA, USA: CNN.com, Aug. 25, 2004.

Collins Y.E., Stotzky G. "Factors affecting the toxicity of heavy metals to microbes." In *Metal Ions and Bacteria*, Beveridge T.J, Doyle R.J., eds. New York, NY: Wiley, 1989.

Courchesne F., Gobran G.R. Mineralogical variations of bulk and rhizosphere soils from a Norway spruce stand, southwestern Sweden. Soil Sci Soc Am J 1997; 61: 1245–1249.

Courchesne F., Seguin V., Dufresne A. Solid phase speciation of metals in the rhizosphere of forest and industrial soils. Proceedings of the 5th International Conference on Biogeochemistry of Trace Elements, 1999, Austria.

Curtin D., Smillie G.W. Soil solution composition as affected by liming and incubation. Soil Sci Soc Am J 1983; 47: 701–707.

D'Amore J.J.D., Al-Abed S.R., Scheckel K.G., Ryan J.A. Methods for speciation of metals in soils: A Review. J Environ Qual 2005; 34: 1707–1745.

Dan J. The effect of dust deposition on the soils of the land of Israel. Quaternary Int 1990; 5: 107–113.

Dang Y.P., Danal R.C., Edwards D.G., Tiller K.G. Kinetics of zinc desorption from vertisols. Soil Sci Soc Am J 1994; 58: 1392–1399.

Danin A., Ganor E. Trapping of airborne dust by mosses in the Negev Desert, Israel. Earth Surf Proc Landf 1991; 16: 153–162.

Dar G. H. Effects of cadmium and sewage-sludge on soil microbial biomass and enzyme activities. Bioresour Technol 1996; 563: 141–145.

Dar G.H., Mishra M.M. Influence of cadmium on carbon and nitrogen mineralization in sewage sludge amended soils. Environ Pollut 1994; 84: 285–290.

Davies G., Ghabbour E.A., Cherkasskiy A. "Tight metal binding by solid phase peat and soil humic acids." In *Humic Substances and Chemical Contaminants,* C.E. Clapp, M.H.B. Hayes, N. Senesi, P.R. Bloom, and P.M. Jardine, eds. Madison, WI: Soil Science Society of America, Inc., 2001.

Davies, Brian E. *Applied Soil Trace Elements*. Chichester: John Wiley & Sons, 1980.

Davis J.A., Fuller C.C., Cook A.D. A model for trace metal sorption processes th the calcite surface: Adosprtion of Cd^{2+} and subsequent solid solution formation. Geochim Cosmochim Acta 1987; 51: 1477–1490.

Dean J.A. *LANGE'S Handbook of Chemistry*. (11th edition). New York: McGraw-Hill Co., 1973.

Dinkelaker B., Hahn G., Römheld V., Wolf G.A, Marschner H. Non-destructive methods for demonstrating chemical changes in the rhizosphere 1. Description of method. Plant Soil 1993; 155/156: 67–70.

Dregne H.E. Soils of Arid Regions. Developments in Soil Science 6. Amsterdam, Oxford: Elsevier, 1976.

Dudley L.M., McNeal B.L., Baham J.E., Coray C.S., Cheng H.H. Characterization of soluble organic compounds and complexation of copper, nickel, and zinc in extracts of sludge-amended soils. J Environ Qual 1987; 16: 341–348.
Dufey J.E., Genon J.G., Rufyikiri G., Delvaux B. Cation exchange properties of roots: Experimental and Modelling. Proceeding of 5th International Conference on the Biogeochemistry of Trace elements, Vienna, Austria, 1999.
Dungan R.S., Frankenberger W.T. Jr. Factors affecting the volatilization of dimethylselenide by *enterobacter cloacae* SLD1a-1. Soil Biol Biochem 2000; 1353–1358.
Eaton F.M., Harding R.B., Ganje T.J. Soil solution extractions at tenth-bar moisture percentages. Soil Sc 1960; 90: 253–258.
Ebbs S.D., Kochian L.V. Phytoextraction of zinc by oat (aAvena sativa), barley (Hordeum vulgare), and Indian mustard (Brassica juncea). Environ Sc Technol 1998; 32: 802–806.
Ebinghaus R., Tripathi R.M., Wallschlager, Lindberg S. E. "Natural and anthropogenic mercury sources and their impact on the air-surface exchange of mercury on regional and global scales." In *Mercury Contaminated Sites: Characterization, Risk Assessment and Remediation*, R. Ebinghaus, R.R. Turner, L.D. Lacerda, O. Vasiliev, W. Salomons, eds. Berlin: Springer-Verlag, 1999.
Edelstein D. Arsenic. In: *Mineral Facts and Problems*, 1985 Edition. BuMines B., 1985.
Elliott H.A., Liberat M.R., Huang C.F. Effect of iron oxide removal on heavy metal sorption by acid subsoils. Water Air Soil Pollut 1986; 27: 379–389.
Elliott H.A., Sparks D.L. Electrokinetic behavior of a paleudult profile in relation to mineralogical composition. Soil Sci 1981; 132: 402–409.
Elrashidi, M.A., Adriano D.C., Lindsay W.L. "Solubility, Speciation, and Transformations of Selenium in Soils." In *Selenium in Agriculture and the Environment*, L.M. Jacobs, ed. Madison, IL: Soil Science of America, Inc. 1989.
Emmerich W.E., Lund L.J., Page A.L., Chang A.C. Solid phase forms of heavy metals in sewage sludge-treated soils. J Environ Qual 1982; 11: 178–181.
Energy Information Administration (2001) International energy database. U.S. Department of Energy: Washington, DC.
Energy Information Administration. International energy annual 2000. U.S. Department of Energy, Washington, DC. 2002.
Environment Canada. Priority substances list assessment report, Arsenic and its compounds. Environment Canada, 1993.
Ericsson T.J., Linares, Lotse E.A. Mossbauer study of the effect of dithionite-citrate-bicarbonate treatment on a vermiculite, a smectite and a soil. Clay Miner 1984; 19: 85–91.
Essington M.E., Mattigod S.V. Element partitioning in size- and density-fractionated sewage sludge and sludge-amended soil. Soil Sci Soc Am J 1990; 54: 385–394.
Essington M.E., Mattigod S.V. Trace element solid-phase associations in sewage sludge and sludge-amended soil. Soil Sci Soc Am J 1991; 55: 350–356.
Evenari M., Shanan L., Tadmor N. The Negev – the Challenge of a Desert. Second edition, Cambridge and London: Harvard University Press, 1982.
FAO. "Lecture notes on the major soils of the world." In *World Soil Resources Reports 94*, P. Driessen, J. Deckers, O. Spaargaren, F. Nachtergaele, eds. Rome, 2001.
Feigin A., Ravina I., Shalhevet J. *Irrigation with Treated Sewage Effluent*. Springer-Verlag, 1991.
Feijtel T.C., Delaune R.D., Patrick Jr. W.H. Biogeochemical control on metal distribution and accumulation in Louisiana sediments. J Environ Qual 1988;17: 88–94.
Fenn L.B., Assadian N. Can rhizosphere chemical changes enhance heavy metal absorptuin by plants growing in calcareous soil? Proceedings of the 5th International Conference on the Biogeochemistry of Trace Elements; 1999 July11–July

15;Vienna, Austria: International Society for Trace Element Research, Vienna, Austria, 1999.
Fenton T.E. "Mollisols." In *Pedogenesis and Soil Taxonomy, II. The soil orders*, L.P. Wilding, N.E. Smeck, G.F. Hall, eds. Developments in Soil Science 11B, Elsevier, 1983.
Feth J.H., Roberson G.E., Polzer W.L. Sources of mineral constituents in water from granitic rocks, Sierra Nevada, California and Nevada. US Geol Surv Water Supply Pap 1964, 1535-M.
Fischer L., Muhlen E.Z., Brummer G.W., Niehus H. Atomic force microscopy investigations of the surface topography of a multidomain porous goethite. European J Soil Sci 1996; 47: 329–334.
Fitzgerald W.F. Is mercury increasing in the atmosphere, the need for an atmospheric mercury network. Water Air Soil Pollut 1995; 80: 245–254.
Fotovat A., Naidu R., Sumner M.E. Water: Soil ratio influences aqueous phase chemistry of indigenous copper and zinc in soils. Aust J Soil Res 1997; 35: 687–710.
Frankenberger W.T. Jr. Effects of trace elements on arsenic volatilization. Soil Bio Biochem 1997; 30: 269–274.
Franklin M.L, Morse J.W. The interaction of manganese(II) with the surface if aclciate in dilute solutions and seawater. Mar Chem 1983; 12: 241–254.
Fresquez P.R., Francis R.E., Dennis G.L. Sewage sludge effects on soil and plant quality in a degraded, semiarid grassland. J Environ Qual 1990; 19: 324–329.
Friesen D.K., Juo A.S.R., Miller M.H. Liming and lime-phosphorus-zinc interactions in two Nigerian Ultisols. I. Interactions in the soil. Soil Sci Soc Am J 1980; 44 (6): 1221–1226.
Frost R.R., Griffin R.A. Effect of pH on adsorption of arsenic and selenium from landfill leachate by clay minerals. Soil Sci Soc Am J 1977; 41: 53–57.
Galbreath K.C., Zygarlicke C.J. Mercy speciation in coal combustion and gasification flue gases. Environ Sc Technol 1996; 30: 2421–2426.
Gallagher P.J., Murphy L.S., Ellis R.J. Effects of temperature and soil pH on effectiveness of four zinc fertilizers. Commu Soil Sci Plant Anal. 1978; 9: 115–126.
Gerard E., Echevarria G., Sterckeman T., Morel J.L. Phytoavailability of cadmium in soils assessed by isotopic methods. Proceedings of 5th International Conference on the Biogeochemistry of Trace Elements. July 11–15, 1999, Vienna, Austria.
Gerth J., Brummer G. Adsorpiton festlegung von nickel, zinc and cadmium durch geothit. Anal Chem 1981; 316: 616–620.
Ghabru S.K., Arnaud R.J. St., Mermut A.R. Association of DCP-extractable iron with minerals in coarse soil clays. Soil Sci 1990; 149: 112–119.
Glasner A., Weiss D. The crystallization of calcite from aqueous solutions and the role of zinc and magnesium ions. I. Precipitation of calcite in the presence of Zn^{2+} ions. J Inorg Nucl Chem 1980; 42: 655–663.
Goenaga R., Chardon U. Growth, yield and nutrient uptake of Taro grown under upland conditions. J Plant Nutri 1995; 18: 1037–1048.
Goldberg S., Forster H.S., Heick E.L. Boron adsorption mechanisms on oxides, clay minerals and soils inferred from ionic strength effects. Soil Sci Soc Am J 1993; 57:704–708.
Goldberg S., Kapoor B.S., Rhoades J.D. Effect of aluminum and iron oxides and organic matter on flocculation and dispersion of arid zone soils. Soil Sci 1990; 150: 588–593.
Goldberg, S. Reanalysis of boron adsorption on soils and soil minerals using the constant capacitance model. Soil Sci Soc Am J 1999; 63: 823–829.
Golding S. Survey of typical soil arsenic concentrations in residential areas of the city of University Place. Washington State Department of Ecology, Olympia, Washington, 2001.

Goodson C.C., Parker D.R., Amrhein C., Zhang Y. Soil selenium uptake and root system development in plant taxa differing in Se-accumulating capability. New Phytol 2003; 159: 391–401.
Goyer R.A., Klaassen C.D., Waalkes M.P. Metal toxicology. San Diego, CA: Academic Press, 1995.
Grant C.A., Bailey L.D. Effects of phosphorus and zinc fertiliser management on cadmium accumulation in flaxseed. J Sci Food Agri 1997; 73: 307–314.
Grant C.A., Bailey L.D. Fertility management in canola production. Can J Plant Sci 1993; 73: 651–670.
Grant C.A., Peterson G.A., Campbell C.A. Nutrient Considerations for Diversified Cropping Systems in the Northern Great Plains. Agron J 2002; 94: 186–198.
Griffiths B.S., Diaz-Ravina M., Ritz K., McNicol J.W., Ebblewhite N., Baath E. Community DNA hybridization and %G+C profiles of microbial communities from heavy metal polluted soils. FEMS Microbiol. Ecol. 1997; 24: 103–112.
Grossman R.B., Millet J.C. Carbonate removal from soils by a modification of the acetate buffer method. Soil Science Soc Am Proc 1961; 25: 325–326.
Gupta S.K., Chen K.Y. Partitioning of trace metals in selective chemical fractions of nearshore sediments. Environ Lett 1975; 10: 129–158.
Gworek B. Inactivation of cadmium in contaminated soils using synthetic zeolites. Environ Pollut 1992a; 75: 269–171.
Gworek B. Lead inactivation in soils by zeolites. Plant Soil 1992b; 143: 71–74.
Haghiri F. Cadmium uptake by plants. J Environ Qual 1973; 2: 93–96.
Hall G.E.M., Gauthier G., Pelchat J.C., Pelchate P., Vaive J.E. Application of a sequential extraction scheme to ten geological certified reference materials for the determination of 20 elements. J Anal At Spectrom 1996; 11: 787–796.
Hamel S.C., Buckley B., Lioy P.J. Bioaccessibility of metals in soils for different liquid to solid ratios in synthetic gastric fluid. Environ Sci Technol 1998; 32: 358–362.
Han F.X., Banin A. Long-term transformations and redistribution of potentially toxic heavy metals in arid-zone soils. I: Incubation under saturated conditions. Water Air Soil Pollut 1997; 95: 399–423.
Han F.X., Banin A. Long-term transformations and redistribution of potentially toxic heavy metals in arid-zone soils. II: Incubation under field capacity conditions Water Air Soil Pollut 1999; 114: 221–250.
Han F.X., Banin A. Long-term transformations of Cd, Co, Cu, Ni, Zn, V, Mn and Fe in arid-zone soils under saturated conditions. Commun Soil Sci Plant Anal 2000; 31: 943–957.
Han F.X., Banin A. Selective sequential dissolution techniques for trace metals in arid-zone soils: The carbonate dissolution step, Commun Soil Sci Plant Anal 1995; 26: 553–576.
Han F.X., Banin A. Solid-phase manganese fractionation changes in saturated arid-zone soils: Pathways and kinetics, Soil Sci Soc Am J 1996; 60: 1072–1080.
Han F.X., Banin A. The fractional loading isotherm of heavy metals in an arid-zone soil. Commun Soil Sci Plant Anal 2001; 32 (17&18): 2691–2708.
Han F.X., Banin A., Kingery W.L., Li Z.P. Pathways and kinetics of transformation of cobalt among solid-phase components in arid-zone soils. J Environ Sci Health, Part A 2002b; 137: 175–194.
Han F.X., Banin A., Kingery W.L., Triplett G.B., Zhou L.X., Zheng S.J., Ding W.X. New approach to studies of redistribution of heavy metals in soils. Adv Environ Res 2003a; 8: 113–120.
Han F.X., Banin A., Su Y., Monts D.L., Plodinec M.J., Kingery W.L., Triplett G.B. Industrial age anthropogenic inputs of heavy metals into the pedosphere. Naturwissenschaften 2002a; 89: 497–504.

Han F.X., Banin A., Triplett G.B. Redistribution of heavy metals in arid-zone soils under a wetting-drying soil moisture regime. Soil Sci 2001a; 166: 18–28.

Han F.X., *Binding, distribution and transformations of polluting trace elements in soils of arid and semi-arid regions receiving waste inputs*. Ph.D. Dissertation, Hebrew University of Jerusalem. 1998.

Han F.X., Hargreaves J., Kingery W.L., Huggett D.B., Schlenk D.K. Accumulation, distribution and toxicity of copper in soils of catfish pond receiving periodic copper sulfate applications. J Environ Qual 2001b; 30: 912–919.

Han F.X., Hu A.T., Qin H.Y. Transformation and distribution of forms of zinc in acid, neutral and calcareous soils of China. Geoderma 1995; 66: 121–135.

Han F.X., Hu A.T., Qin H.Y., Fractionation and availability of added Cadmium in soil environment, *Environ. Chem.* (in Chinese) 1990a; 9: 49–53.

Han F.X., Hu A.T., Qin H.Y., Fractionation of zinc bound to organic matter in soil, J Nanjing Agric Univ 1990b; 13: 68–74.

Han F.X., Hu A.T., Qin H.Y., Shi R.H. Enrichment capability of the native zinc by some components of soils in China. Acta Pedological Sinica 1991; 28: 327–333.

Han F.X., Hu A.T., Qin H.Y., Shi R.H. Fractionation and availability of added soluble zinc in various soil environments. (In Chinese, with English abstract). China Environ Sci 1992; 12: 108–112.

Han F.X., Hu A.T., Qin H.Y., Shi R.H. Study on mechanism of zinc deficiency in calcareous soils. (In Chinese, with English abstract). Environ Chem 1993; 12: 36–41.

Han F.X., Kingery W.L., Selim H.M. "Accumulation, redistribution, transport and bioavailability of heavy metals in waste-amended soils." In *Trace Elements in Soil: Bioavailability, Fluxes and Transfer*, Iskander IK, Kirkham MB, eds. Boca Raton, FL: CRC Press, 2001c.

Han F.X., Kingery W.L., Selim H.M., Gerald P. Accumulation of heavy metals in a long-term poultry waste-amended soil. Soil Sci 2000; 165: 260–268.

Han F.X., Kingery W.L., Selim H.M., Gerard P.D., Cox M.S. Arsenic solubility and distribution in poultry-waste and long-term amended soil. Sci Total Environ 2004a; 320: 51–61.

Han F.X., Sridhar B.B.M., Monts D.L., Su Y. Phytoavailability and toxicity of trivalent and hexavalent chromium to *Brassica juncea L. Czern.* New Phytologist 2004c; 162 (2): 489–499.

Han F.X., Su Y., Monts D.L., Plodinec M.J., Banin A., Triplett G.B. Assessment of global industrial-age anthropogenic arsenic contamination. Naturwissenschaften 2003b; 90: 395–401.

Han F.X., Su Y., Monts D.L., Sridhar B.B.M. Distribution, transformation and bioavailability of trivalent and hexavalent chromium in contaminated soil. Plant Soil 2004b; 265: 243–252.

Han F.X., Su Y., Monts D.L., Waggoner C. Bioavailability of mercury in Oak Ridge site. United State Department of Energy's Mercury Workshop, Oak Ridge, TN, Sept. 2005.

Han F.X., Zhu Q.X. Fractionation and availability of zinc in paddy soils of China. Pedosphere 1992; 2: 283–288.

Hansel C.M., Force M.J.L., Fendorf S., Sutton S. Spatial and temporal association of As and Fe species on aquatic plant roots. Environ Sci Technol 2002; 36: 1988–1994.

Hansen D., Duda P.J., Zayed A. Terry N. Selenium removal by constructed wetlands: role of biological volatilization. Environ Sc Technol 1998; 32: 591–597.

Hardiman R.T., Banin A., Jacoby B. The effect of soil type and degree of metal contamination upon uptake of Cd, Pb and Cu in bush beans. Plant Soil 1984b; 81: 3–15.

References

Hardiman R.T., Jacoby B., Banin A. Factors affecting the distribution of cadmium, copper and lead and their effect upon yield and zinc content in bush beans (Phaseolus vulgaris L.). Plant Soil 1984a; 81: 17–27.

Harriss R.C., Hohenemser C. Mercury: Measuring and managing the risk. Environ 1978; 20: 25–36.

Harter R.D. Effect of soil pH on adsorption of lead, copper, zinc, and nickel. Soil Sci Soc Am J 1983; 47: 47–51.

Hassen A., Jedidi N., Cherif M., M'Hiri A., Boudabous A., Van Cleemput O. Mineralization of nitrogen in a clayey loamy soil amended with organics wastes enriched with Zn, Cu, and Cd. Bioresour Technol 1998; 64: 39–45.

Hattori H. Influence of cadmium on decomposition of sewage-sludge and microbial activities in soils. Soil Sc Plant Nutr 1989; 35: 289–299.

Hayes K.F., Traina S.J. "Metal ion speciation and its significance in ecosystem health." In *Soil Chemistry and Ecosystem Health*, Huang P.M., Adriano D.C., Logan T.J., Checkai R.T., eds. Madison, WI: Soil Sci Soc Am Inc, 1998.

Heinrichs H., Mayer R. The role of forest vegetation in the biogeochemical cycle of heavy metals. J Environ Qual 1980; 9: 111–118.

Helal H.M., Haque S.A., Rmadan A.B., Schnug E. Salinity-heavy metal interaction as evaluated by soil extraction and plant analysis. Commun Soil Sci Plant Anal 1996; 57: 1355–1361.

Hickey M.G., Kittrick J.A. Chemical partitioning of cadmium, copper, nickel, and zinc in soils and sediments containing high levels of heavy metals. J Environ Qual 1984; 13: 372–376.

Hinkley T.K., Lamother P.J., Meeker G.P., Jiang X., Miller M.E., Fulton R. Trace elements deposited with dusts in Southwestern U.S. – enrichments, fluxes, comparison with records from elsewhere. Proceedings of the ICAR5/GCTE-SEN Joint Conference, International Center for Arid and Semiarid Lands Studies; 2002; Lubbock, Texas: Texas Tech University, 2002.

Hinsinger P. Bioavailability of trace element as related to root-induced chemical changes in the rhizosphere. Proceeding of 5th International Conference on the Biogeochemistry of Trace elements, Vienna, Austria, 1999.

Hinz C., Selim H.M. Kinetics of Zn sorption-desorption using a thin disk flow method. Soil Sci 1999; 164: 92–100.

Hirsh D., Banin A. Cadmium speciation in soil solutions. J Environ Qual 1990; 19: 366–372.

Ho M.D., Evans G.J. Operational speciation of cadmium, copper, lead and zinc in the NIST standard reference materials 2710 and 2711 (Montana soil) by the BCR sequential extraction procedure and flame atomic absorption spectrometry. Anal Commun 1997; 34: 353–364.

Hodgson J.F., Lindsay W.L., Trierweiler J.F. Micronutrient cation complexing in soil solution.: II. Complexing of zinc and copper in displaced solution from calcareous soils. Soil Sci Soc A Proc 1966; 30: 723–726.

Holmgren G.G.S., Meyer M.W., Chaney R.L., Daniels R.B. Cadmium, lead, zinc, copper and nickel in agricultural soils of United States of America. J Environ Qual 1993; 22: 335–348.

Hong S., Candelone J.P., Patterson C.C., Boutron C.F. Greenland ice evidence of hemispheric lead pollution two millennia ago by Greek and Roman civilization. Science 1994; 265: 1841–1843.

Hopkins B.G., Whitney D.A., Lamond R.E., Jolley V.D. Phytosiderophore release by sorghum, wheat, and corn under zinc deficiency. J Plant Nutri 1998; 21: 2623–2637.

Horne A.J., "Phytoremediation by constructed wetlands." In *Phytoremediation of Contaminated Soil and Water*, N. Terry, G. Banuelos, eds. Boca Raton, FL: Lewis Publishers, 2000.

Howari F. Heavy metal speciation and mobility assessment of arid soils in the vicinity of Al Ain landfill, United Arab Emirates Int J Environ Pollut 2004; 22: 721–731.

Hsu J.H., Lo S.L. Charaterization and extractability of copper, manganese, and zinc in swine manure composts. J Environ Qual 2000; 29: 447–453.

Huang J.W., Cunningham S.D. Lead phytoextraction: Species variation in lead uptake and translocation. New Phytol 1996; 134: 75–84.

Huchinson T.C., Whilby L.M. The effects of acid rainfall and heavy metals particulates on a boreal forest ecosystem near the Sudbury smelting region of Canada. Water Air Soil Pollut 1977; 7: 421–438.

Hutchinson T.C., Meema K.M., *Lead, Mercury, Cadmium and Arsenic in the Environment*. New York: Wiley, 1987.

Hyun H., Chang A.C., Parker D.R., Page A.L. Cadmium solubility and phytoavailability in sludge-treated soils: Effects of soil organic carbon. J Environ Qual 1998; 27: 329–334.

Iyengar S.S., Martens D.C., Miller W.P. Distribution and availability of soil zinc fractions. Soil Sci Soc Am J 1981; 45: 735–739.

Jackson B.P., Miller W.P. Effectiveness of phosphate and hydroxide for desorption of arsnic and selenium species from iron oxides. Soil Sci Soc Am J 2000; 64: 1616–1622.

Jackson M. L. *Soil Chemical Analysis.* Englewood Cliffs NJ: Prentice-Hall Inc, 1958.

James B.R. Hexavalent chromium solubility and reduction in alkaline soils enriched with chromite ore processing residue. J Environ Qual 1994; 23: 227–233.

Jarvis S.C., Jones L.H.P., Hopper M.J. Cadmium uptake form solution by plants and its transport from roots to shoots. Plant Soil 1976; 44: 179–191.

Jarvis S.C. The association of cobalt with easily reducible manganese in some acidic permanent grassland soils. J Soil Sci 1984; 35: 431

Jeffery J.J., Uren N.C. Copper and zinc species in the soil solution and the effects of soil pH. Aust J Soil Res 1983; 21: 479–488.

Jiang T.H. Fractionation and availability of micronutrients in some soils of Jiangsu province. Master thesis, Nanjing Agricultural University, Nanjing, P.R. China, 1983.

Jiao Y., Bailey L.D., Grant C.A. Effects of phosphorus and zinc fertilizer on cadmium uptake and distribution in flax and durum wheat. J Sci Food Agri 2004; 84: 777–785.

Jin Q., Wang Z., Shan X., Tu Q., Wen B., Chen B. Evaluation of plant availability of soil trace metals by chemical fractionation and multiple regression analysis. Environ Pollut 1996; 91: 309–315.

Johnson C.D., Vance G.F. Long-term land application of biosolids: soil and plant trace element concentrations. University of Wyoming, Cooperative Extension Service, B-1062. 1998

Jugsujinda A., Patrick W.H.Jr. Growth and nutrient uptake by rice in a flooded soil under controlled aerobic-anaerobic and pH conditions. Agron J 1977; 69: 705–710.

Jump R.K., Sabey B.R. "Soil test extractants for predicting selenium in plants." In *Selenium in Agriculture and the Environment*, L.M. Jacobs, ed. Madison, IL: Soil Science of America, Inc. 1989.

Jung J., Logan T.J. Effects of sewage sludge cadmium concentration on chemical extractability and plant uptake. J Environ Qual 1992; 21: 73–81.

Kabata-Pendias A., Pendias H. Trace elements in soils and plants. Boca Raton, FL: Lewis Publishers, 1992.

Kadria E., Michel J. Effects of salinity on toxic element transfers in soils associated to agricultural wastewater reuse: mobility and bioavailability of zinc and lead for ryegrass in soils irrigated with saline water. 2004. http://Kuk.uni-

References

freiburg.de/hosted/eurosoil/2004/full-papers/Id1053_El-AZAB-full.pdf (2005 access).

Kanwar J.S., Singh S,S. Boron in normal and saline-alkali soils of the irrigated areas of the Punjab. Soil Sci 1961; 92: 207–211.

Kaushansky P.,Yariv S. The interactions between calcite particles and aqueous solutions of magnesium, barium or zinc chloride, Appl Goechem 1986; 1: 607–618.

Kelly T., Buckingham D., DiFrancesco C., Porter K., Goonan T., Sznopek J., Berry C., Cran M. Historical statistics for mineral commodities in the United States. U.S. Geological Survey, 2002.

Khalid R.A., Gambrell, R.P., Patrick, W.H.Jr., Chemical availability of cadmium in Mississippi River sediment. J Environ Qual 1981; 10: 523–528.

Khan A., Soltanpour P.N. Effect of wetting and drying on DTPA-extractable Fe, Zn, Mn and Cu in soils. Commu Soi Sci Plant Anal. 1978; 9: 193–202.

Kheboian C., Bauer C. Accuracy of selective extraction procedures for metal speciation in model aquatic sediments. Anal. Chem 1987; 59: 1417–1423.

Khoshgoftarmanesh A.H., Kalbasi M. Effect of municipal waste leachate on soil properties and growth and yield of rice. Commun Soil Sci Plant Anal 2002; 33: 2011–2020.

Kidron G., Barzilay E., Sachs E. Microclimate control upon sand microbiotic crusts, western Negev Desert, Israel. Geomorphology 2000; 36: 1–18.

Kim J.G., Dixon J.B., Chusuei C.C., Deng Y. Oxidation of chromium(III) to (VI) by manganese oxides. Soil Sci Soc Am J 2002; 66: 306–315.

Kim, N.D. and Fergusson, J.E., Effectiveness of a commonly used sequential extraction techniques in determining the speciation of cadmium in soils. Sci Total Environ 1991; 105: 191–209.

King P.M., Alston A.M. "Diagnosis of trace element deficiencies in wheat on Eyre Peninsula, South Australia" In *Trace Elements in Soil-Plant-Animal Systems*, D.J.D. Nicholas, A.R. Egan, eds. New York, NY: Academic Press, Inc., 1975.

Kingery W.L, Oppenheimer S.F., Han F.X., Selim H.M. Adsorption/desorption hysteresis of trace metals with soil components: A Dynamical systems approach. Proceedings of the Fifth International Conference on the Biogeochemistry of Trace Elements, Vienna, Austria. 1999.

Kinniburgh D.G., Jackson M.L., Syers J.K. Adsorption of alkaline earth, transition and heavy metal cations by hydrous gels of iron and aluminium. Soil Sci Soc Am J 1976; 40: 796–799.

Knudtsen K., O'Connor G.A. Characterization of iron and zinc in Albuquerque sewage sludge. J Environ Qual 1987; 16: 85–90.

Kochian L.V. "Zinc absorption from hydroponic solutions by plants roots." In *Zinc in Soils and Plants*, A.D. Robson, ed. Boston: Kluwer Academic Publishers, 1993.

Koljonen T. *Geochemical Atlas of Finland, Part 2*. Geological Survey of Finland, 1992.

Kongshaug G., Bockman O.C., Kaarstad O., Morka H. Inputs of trace elements to soils and plants. Proceeding of Chemical Climatology and Geomedical Problems. Norsk Hydro, Oslo, Norway, 1992.

Kornicker W.A., Morse J.W., Damascenos R.N. The chemistry of Co^{2+} interaction with calcite and aragonite surface. Chem Geol 1985; 53: 229–236.

Kubota J. "Regional distribution of trace element problems in North America" In *Applied Soil Trace Elements*, B.E. Davies, ed. New York: John Wiley and Sons, Ltd., 1980.

Lagerwerff J.V., Biersdorf G.T., Brower D.L. Retention of metals in sewage sludge: I. Constitutent heavy metals. J Environ Qual 1976; 5: 19–23.

Lake D.L., Kirk P.W.W., Lester J.N. Fractionation, characterization, and speciation of heavy metals in sewage sludge and sludge-amended soils: A review. J Environ Qual 1984; 13: 175–183.

Latterell J.J. Dowdy R.H., Larson W.E. Correaltion of extrctable metals and metal uptake of snap beans grown on soil amended with sewage sludge. J Environ Qual 1978; 7: 425–440.

LeClaire, J.P., Change A.C., Levesque C.S, Sposito G. Trace metal chemistry in arid-zone field soils amended with sewage sludge. IV: Correlations between zinc uptake and extracted soil zinc fractions. Soil Sci Soc Am J 1984; 48: 509–513.

Lee B.D., Graham R.C., Laurent T.E., Amrhein C., Creasy R.M. Spatial distributions of soil chemical conditions in a serpentinitic wetland and surrounding landscape. Soil Sic Soc Am J 2001; 65: 1183–1196.

Lee K.W., Keeney D.R. Cadmium and zinc additions to Wisconsin soils by commercial fertilizers and wastewater sludge application. Water Air Soil Pollut 1975; 5: 109–112.

Lerch R.N., Barbarick K.A., Westfallm D.G., Follett R.H., McBride T.M., Owen W.F. Sustainable rates of sewage sludge for dryland winter wheat production 1. Soil nitrogen and heavy metals. J Prod Agri 1990; 3: 60–65.

Levy, D.B., Barbarick A., Siemer E.G., Sommers L.E. Distribution and partitioning of trace metals in contaminated soils near leadville, colorado, J Environ Qual 1992; 21: 185–195.

Li M., Hue N.V., Hussain S.K.G. Changes of metal forms by organic amendments to Hawaii soils. Commun. Soil Sci Plant Anal 1997; 28: 381–394.

Li Z., Ryan J.A., Chen J., Al-Abed S.R. Adsorption of cadmium on biosolids-amended soils, J Environ Qual 2001; 30: 903–911.

Liang C.A, Tabatabai M.A. Effect of trace elements on nitrification in soils. J Environ Qual 1978; 7: 291–293.

Lide D.R. CRC handbook of chemistry and physics. 77th ed. Boca Raton, USA: CRC Press, 1996.

Lide D.R. *Handbook of Chemistry and Physics*. 82nd edition. Boca Raton: CRC Press, 2001.

Lindau C.W., Hossner L.R. Sediment fractionation of Cu, Ni, Zn, Cr, Mn, and Fe in one experimental and three natural marshes, J Environ Qual 1982; 11: 540–545.

Lindqvist O., Johansson K., Aastrup M., Andersson A., Bringmark L., Hovsenius G., Hakanson L., Iverfeldt A., Meili M., Timm B. Mercy in the Swedish environment-recent research on causes, consequences and corrective methods. Water Air Soil Pollut 1991; 55: 1–251.

Lindsay W.L. *Chemical Equilibria in Soils*. New York: John Wiley & Sons, 1979.

Lindsay W.L., Norvell W.A. Development of a DTPA soil test for zinc, iron, manganese, and copper. Soil Sci Soc Am J 1978; 42: 421–428.

Lindsay W.L., Norvell W.A. Equilibrium relationships of Zn^{2+}, Fe^{3+}, Ca^{2+}, and H^+ with EDTA and DTPA in soils. Soil Sci Soc Am Proc 1969; 33: 62–68.

Liu, Z. *Micronutrients in Soils of China*. Nanjing: Jiangsu Sceince and Technology Publishing House, 1996.

Loebenstein J.R. The materials flow of arsenic in the United States. U.S. Bureau of Mines Information, 1994.

Loganathan P., Burau R.G., Fuerstenau D.W. Influence of pH on the sorptionof Co^{2+}, Zn^{2+} and Ca^{2+} by a hydrous manganese oxide. Soil Sci Soc Am J 1977; 41: 57–62.

Lombi E., Wenzel W.W., Gobran G.R., Adriano D.C. Rhizosphere-contaminant interaction and its role in phytoremediation: A review. Proceedings of the 5th International Conference on Biogeochemistry of Trace Elements, Austria, 1999.

Losi M.E., Amrhein C., Frankenberger W.T. Factors affecting chemical and biological reduction of Cr(VI) in soil. Environ. Toxicol Chem 1994; 13: 1727–1735.

Losi M.E., Frankenberger W.T. Microbial oxidation and solubilization of precipitated elemental selenium in soil. J Environ Qual 1998; 27: 836–843.

Ma L.Q. Factor influencing the effectiveness and stability of aqueous lead immobilization by hydroxyapatite. J Environ Qual 1996; 25: 1420–1429.
Ma L.Q., Choate A.L., Rao G.N. Effects of incubation and phosphate rock on lead extractability and speciation in contaminated soils. J Environ Qual 1997a; 26: 801–807.
Ma L.Q., Komart K.M., Tu C., Zhang W., Cai Y., Kennelly E.D. A fern that hyperaccumulates arsenic. Nature 2001; 409: 579.
Ma L.Q., Lindsay W.L. Divalent zinc activity in arid-zone soils by chelation. Soil Sci Soc Am J 1990; 54: 719–722.
Ma L.Q., Rao G.N. Effects of phosphate rock on sequential chemical extraction of lead in contaminated soils. J Environ Qual 1997b; 26: 788–794.
Ma Q.Y., Logan T.J., Traina S.J. Lead immobilization from aqueous solutions and contaminated soils using phosphate rocks. Environ Sci Technol 1995; 29: 1118–1126.
Ma, L.Q. and Rao, G.N., Chemical fractionation of cadmium, copper, nickel and zinc in contaminated soils. J Environ Qual 1997a; 26: 259–264.
Mahan K.I., Foderaro T.A., Garza T.L., Martinez R.M., Maroney G.A., Trivisonne M.R., Willging E.M. Microwave digestion techniques in the sequential extraction of cadmium, iron, chromium, manganese, lead and zinc in sediments. Anal Chem 1987; 59: 938–945.
Mahler R.J., Bingham F.T., Page A.L. Cadmium-enriched sewage sludge application to acid and calcareous soils: Effect on yield and cadmium uptake by lettuce and chard. J Environ Qual 1978; 7: 274–281.
Mahler R.J., Bingham F.T., Sposito G., Page A.L. Cadmium-enriched sewage sludge application to acid and calcareous soils: Relation between treatment, cadmium in saturation extracts, and cadmium uptake. J Environ Qual 1980; 9: 359–364.
Mandal B., Chatterjee J., Hazra T.L., Mandal L.N. Effect of preflooding on transformation of applied zinc and its uptake by rice in lateric soil. Soil Sci 1992; 153: 250–257.
Mandal L.N., Mandal B., Zinc fractions in soils in relation to zinc nutrition of lowland rice. Soil Sci. 1986;142: 141–148.
Mandal L.N., Mitra R.R. Transformation of iron and manganese in rice soils under different moisture regimes and organic matter applications. Pland Soil 1982; 69: 45–56.
Manecki M., Maurice P.A., Traina S.J. Kinetics of aqueous Pb reaction with apatites. Soil Sci 2000; 165: 920–933.
Mangaroo A.S., Himes F.L., McLean E.O. The adsorption of zinc by some soils after various pre-extraction treatments. Soil Sci Soc Am Proc 1965; 29: 242–245.
Mann S.S., Ritchie G.S.P., Changes in the form of cadmium with time in western Australian soils. Aust J Soil Res 1994; 32: 241–250.
Marrett D.J., Page A.L., Bradford G.R., Cardenas R., Graham R.C., Chang A.C. Background levels of soil trace elements in southern California soils, Annual report submitted to Southern California Edison Co., Rosemead, CA, University of California, Riverside, CA, 1992.
Marschner H. *Mineral Nutrition of Higher Plants*. London: Academic Press, 1986.
Marschner H., Romheld V. Root induced changes in the availability of micronutrients in the rhizosphere. In *Plants Roots: The Hidden Half*, Y. Waisel, E. Amram, U. Kafkafi, eds. New York: Marcel Decker Inc., 1996.
Martin J.M., Nirel P., Thomas A.J., Sequential extraction techniques: Promises and Problems, Mar Chem 1987; 22: 313–341.
Martinez C.E., McBride M.B. Aging of coprecipitated Cu in alumina: Changes in structureal location, chemical form, and solubility. Geochim Cosmochim Acta 2000; 64: 1729–17369.

Martinez C.E., Motto H.L. Solubility of lead, zinc and copper added to mineral soils. Environ Pollut 2000; 107: 153–158.

Martinez Garcia M.J., Moreno-Grau S., Martinez Garcia J.J., Moreno J., Bayo J., Guillen Perez J.J., Moreno-Clavel J. Distribution of the metals lead, cadmium, copper and zinc in the top soil of Cartagena, Spain. Water Soil Soil Pollut 2001; 131: 329–347.

Matschullat J. Arsenic in the geosphere: a Review. Sci Total Environ 2000; 249: 297–312.

McBride M.B. "Forms and distribution of copper in solid and solution phases of soils." In *Copper in Soils and Plants*, J.F. Loneragan, A.D. Robson, R.D. Graham, eds. Sydney: Academic Press, 1981.

McBride M.B. Chemisorption and precipitation of Mn^{2+} at $CaCO_3$ surfaces. Soil Sci Soc Am J 1979; 41: 693–698.

McBride M.B. Chemisorption of Cd^{2+} on calcite surfaces. Soil Sci Soc Am J 1980; 42: 26–28.

McBride M.B. Processes of heavy and transition metal sorption by soil mineral. In *Interactions at the Soil Colloid-Soil Solution Interface*, G.H. Bolt, M.F. De Boodt, M.H.B.Hayes, M.B. McBride, eds. NATO ASI Series (Series E: Applied Sciences-Vol 190). Dordrecht, Netherlands: Kluwer Academic Publishers, 1991.

McBride M.B. Toxic metal accumulation from agricultural use of sludge: Are USEPA regulations protective? J Environ Qual 1995; 24:5–18.

McBride M.B., Blasiak J.J. Zinc and copper solubility as a function of pH in an acid soil. Soil Sci Soc Am J 1979; 43: 866–870.

McBride M.B., Richards B.K., Steenhuis T., Russo J.J., Sauve S. Mobility and solubility of toxic metals and nutrients in soil fifteen years after sludge application. Soil Sci 1997; 162: 487–500.

McBride. Reactions controlling heavy metal solubility in soils. Adv. Soil Sci 1989; 10: 1–56.

McFadden L.D., McDonald E.V., Wells S.G., Anderson K., Quade J., Forman S.L. The vesicular layer and carbonate collars of desert soils and pavements: formation, age and relation to climate change. Geomorphology 1998; 24: 101–145.

McGowen S.L., Basta N.T., Brown G.O. Use of diammonium phosphate to reduce heavy metal solubility and transport in smelter-contaminated soil. J Environ Qual 2001; 30: 493–500.

McGrath S. P., Cegarra J. Chemical extractability of heavy metals during and after long-term applications of sewage sludge to soil. J Soil Sci 1992; 43: 313–321.

McGrath S.P. "Behaviour of trace elements in terrestrial ecosystems." In *Contaminated Soils*, Prost R., ed. Paris: Editions INRA, Les Colloques no. 85, 1997.

McGrath S.P. "Adverse effects of cadmium on soil microflora and fauna." In *Cadmium in Soils and Plant*, McLaughlin M.J., Singh B.R., eds. Netherlands: Kluwer Academic, 1999.

McIntosh A.W., Shephard B.K., Mayes R.A., Atchison G.J., Nelson D.W. Some aspects of sediment distribution and macrophyte cycling of heavy metals in a contaminated lake. J Environ Qual 1978; 7: 301–305.

McKenzie R.M. "The mineralogy and chemistry of soil cobalt." In *Trace Element in Soil-Plant-Animal Systems,* Nicholas, D.J.D., Egan, A.R., eds. New York: Academic Press Inc.,1975.

McKenzie R.M. The adsorption of lead and other heavy metals on oxides of manganese and iron. Austral J Soil Res 1980; 18: 61–73.

Mclaren R.G., Crawford D.V. Studies on soil copper: I. The fractionation of copper in soils. J Soil Sci 1973; 24: 172–181.

McLaren R.G., Lawson D.M., Swift R.S. Sorption and desorption of cobalt by soils and soil components. J Soil Sci 1986b; 37: 413–426.

McLaren R.G., Lawson D.M., Swift R.S. The forms of cobalt in some Scottish soils as determined by extraction and isotopic exchange. J Soil Sci 1986a; 37: 223–243.

McLaren R.G., Lawson D.M., Swift R.S., Purves D. The effects of cobalt additions on soil and herbage cobalt concentrations in some S. E. Scotland pastures. J Agri Sci 1985; 105, 347–363.

McLaren R.G., Ritchie G.S.P. The long-term fate of copper fertilizer applied to a lateristic sandy soil in Western Australia. Aust J Soil Res 1993; 93: 39.

McLaughlin M.J., Andrew S.J., Smart M.K., Smolders E. Effects of sulfate on cadmium uptake by Swiss chard: I. Effects of complexation and calcium competition in nutrient solutions. Plant Soil 1998a: 202: 211–216.

McLaughlin M.J., Lambrechts R.M., Smolders E., Smart M.K. Effects of sulfate on cadmium uptake by Swiss chard: II. Effects due to sulfate addition to soil. Plant Soil 1998b; 202: 217–222.

McLaughlin M.J., Tiller K.G., Beech T.A., Smart M.K. Soil salinity causes elevated cadmium concentrations in field-grown potato tubers. J Environ Qual 1994: 23: 1013–1018.

McNeal J.M., Balistrieri L.S. "Geochemistry and occurrence of selenium: An overview." In *Selenium in Agriculture and the Environment*, L.M. Jacobs, ed. Madison, IL: Soil Science of America, Inc. 1989.

Meagher RB. Phytoremediation of toxic elemental and organic pollutants. Curr Opin Plant Biol 2000; 3: 153–62.

Megharaj K.V.M., Sethunathan N., Naidu R. Bioavailability and toxicity of cadmium to microorganisms and their activities in soil: a review. Adv Environ Res 2003; 8:121–135.

Meigs P. *Review of Research on Arid Zone Hydrology*. Paris: UNESCO, 1953.

Mench M.J., Didier V.L., Loffler M., Gomez A., Masson P. A mimicked in-situ remediation study of metal-contaminated soils with emphasis on cadmium and lead. J Environ Qual 1994; 23: 58–63.

Merritts D., De Wet A., Menking K. Environmental Geology – An earth systems approach. New York: W.H. Freeman & Co, 1997.

Miller W.P., Martens D.C., Zelazny L.W. Effect of sequence in extraction of trace metals from soils. Soil Sci Soc Am J 1986; 50: 598–601.

Miller W.P., Mcfee W.W. Distribution of Cd, Zn, Cu, and Pb in soils of industrial northwestern Indiana. J Environ Qual 1983; 12: 29–33.

Mingelgrin U., Biggar J.W. Copper species in aqueous sewage sludge extract. Water Air Soil Pollut 1986; 28: 351–359.

Misra A., Sarkunan K.V., Das M., Nayar P.K. Transformation of added heavy metals in soils under flooded condition. J India Soc Soil Sci 1990; 38: 416–418.

Misra S.G. Pandey G. Evaluation of a suitable extractant for available nickel in soils. Plant Soil 1974; 41: 697–700.

Misra S.G., Pandey G. Evaluation of suitable extractant for available lead in soils. Plant Soil 1976; 45: 693–696.

Mitchell G.A., Bingham F.T., Page A.L. Yield and metal composition of lettuce and wheat grown on soils amended with sewage sludge enriched with cadmium, copper, nickel and zinc. J Environ Qual 1978; 7: 165–171.

Mitchell L.G., Grant C.A., Racz G.J. Effect of nitrogen application on concentration of cadmium and nutrient ions in soil solution and in durum wheat. Can J Soil Sci 2000; 80: 107–115.

Mitchell R.L. In *Chemsitry of the Soil*, Bear F.E., ed. New York: Reinhold, 1964.

Moghe V.B., Mathur G.M. Status of boron in soil arid soils of western Rajasthan. Soil Sc iPlant Nutr 1966; 12: 11–14.

Monger H.C. "Arid Soils." In *Encyclopedia of Soil Science*, H. Tan, ed. New York: Marcel Dekker, 2002.

Monger H.C., Martinez-Rios J., Khresat S.A. "Tropical soils: Arid and Semi-arid." In *Encyclopedia of Soils in the Environment*, D. Hillel, ed. Amsterdam: Elsevier, 2004.

Moraghan J.T. Manganese nutrition of flax as affected by FeEDDHA and soil air drying. Soil Sci Soc Am J 1985a; 49: 668–671.

Moraghan J.T., Freeman T.J. Influence of FeEDDHA on growth and manganese accumulation in flax. Soil Sci Soc Am J 1978; 42: 455–460.

Moraghan J.T.. Manganese deficiency in soybeans as affected by FeEDDHA and low soil temperature. Soil Sci Soc Am J 1985b; 49: 1584–1586.

Moral R., Gilkes R.J., Jordan M.M. Distribution of heavy metals in calcareous and non-calcareous soils in Spain. Water Air Soil Pollut 2005; 162: 127–142.

Moreno J.L, Garcia C., Landi L., Falchini L., Pietramellara G., Nannipieri P. The ecological dose value for assessing Cd toxicity on ATP content and DHA and urease activities of soil. Soil Biol Biochem 2001; 33: 483–489.

Moreno J.L., Hernandez T., Garcia C. Effect of a cadmium-contaminated sewage sludge compost on dynamics of organic matter and microbial activity in an arid soil. Bio Ferti Soils 1999; 28: 230–237.

Mortvedt J.J. "Heavy metal contaminants in inorganic and organic fertilizers". In *Fertiliers and Environment*, C. Rodriguez-Barrueco, ed. Netherlands: Kluwer Academic Publishers, 1996.

Mortvedt J.J., Cadmium levels in soils and plants from some long-term soil fertility experiments in the United States of America. J Environ Qual 1987; 16: 137–142.

Mortvedt J.J., Mays D.A., Osborn G. Uptake by wheat of cadmium and other heavy metal contaminants in phosphate fertilizers. J Environ Qual 1981; 10: 193–197.

Motaium E., Badawy S.H. Effect of irrigation using sewage water on the distribution of some heavy metals in bulk and rhizopshere soils and different plant species: Cabbage plants (Brassica Oleracea L.) and organge trees (Citrus sinensis L). Proceeding of the 5th International Conference on the Biogeochemistry of Trace Elements, Volume I, 1999, Vienna, Austria.

Mullins G.L., Sommers L.E. Characterization of cadmium and zinc in four soils treated with sewage sludge. J Environ Qual 1986; 15: 328–387.

Mullins, C.L., Martens D.C., Miller W.P., Kornegay E.T., Hallock D.L. Copper availability, form and mobility in soils from three annual copper-enriched hog manure applications. J Environ Qual. 1982; 11: 316–320.

Murray J.W., Dillard J.G. The oxidation of cobalt (II) adsorbed on manganese oxide. Geochim Cosmochim Acta 1979; 43: 781–787.

Murthy A.S.P., Zinc fractions in wetland rice soils and their availability to rice. Soil Sci 1982; 133: 150–154.

Myneni S.C.B., Tokunaga T.K., Brown G.E.Jr. Abiotic selenium redox transformations in the presence of Fe(II, III) hydroxides. Science 1997; 278: 1106–1109.

Myttenaere C., Mousny J.M. The distribution of chromium-51 in lowland rice in relation to the chemical form and to the amount of stable chromium in the nutrient solution. Plant Soil 1974; 41: 65–72.

Nannipieri P. "The potential use of soil enzyme as indicators of productivity, sustainability and pollution." In *Soil Biota, Management in Sustainable Farming Systems*, C.E. Pankhurst, B.M. Doube, V.V.S.R. Gupta, P.R. Grace, eds. Australia: CSIRO Publications, 1994.

Narwal R.P., Singh B.R. Effect of organic materials on partitioning, extractability and plant uptake of metals in an alum shale soil. Water Air Soil Pollut 1998; 103: 405–421.

National Research Council. *Use of Reclaimed Water and Sludge in Food Crop Production*, National Academy of Press, Washington, D.C., 1996.

Neff J.M. Ecotoxicology of arsenic in the marine environment. Environ Toxicol Chem 1997; 16: 917–927.

References

Nettleton W.D., Peterson F.F. "Aridisols." In *Pedogenesis and Soil Taxonomy. II. The Soil Orders,* P. Wilding, N.E. Smeck, G.F. Hall, eds. Amsterdam: Elsevier, 1983.

Nirel P.M.V., Morel F.M.M. Pitfalls of sequential extractions. Water Res 1990; 24: 1055–1056.

Nissenbaum A. Distribution of several metals in chemical fractions of sediment core from the sea of Okhotsk. Israel J Earth Sci 1972; 21: 143–154.

Nolan A.L., McLaughlin M.J., Mason S.D. Chemical speciation of Zn, Cd, Cu and Pb in pore waters of agricultural and contaminated soils using donnan dialysis. Environ Sci Technol 2003; 37: 90–98.

Norvell W.A., Lindsay W.L. Reactions of EDTA complexes of Fe, Zn, Mn, and Cu with soils. Soil Sci Soc Am Proc. 1969; 33: 86–91.

Nriagu J.O. Global inventory of natural and anthropogenic emissions of trace metals to the atmosphere. Nature 1979; 279: 409–411.

Nriagu J.O., Pacyna J.M. Quantitative assessment of worldwide contamination of air, water and soils by trace metals. Nature 1988; 333: 134–139.

Nriagu J.O. *The Biogeochemistry of Mercury in the Environment.* New York: Elsevier/North-Holland Biomed Press, 1991.

Oliver D.D., Brockman F.J., Bowman R.S., Kieft T.L. Microbial reduction of hexavalent chromium under vadose zone conditions. J Environ Qual 2003; 32: 317–324.

Onken B.M., Hossner L.R. Plant uptake and determination of Arsenic species in soil solution under flooded conditions. J Environ Qual 1995; 24: 373–381.

Oremland R.S, Hollibaugh J.T., Maest A.S., Presser T.S., Miller L.G., Culbertson C.W. Appl Environ Microbiol 1989; 55: 2333–2343.

Oscarson D.W., Huang P.M., Liaw W.K., Hammer U.T. Kinetics of oxidation of arsenite by various manganese dioxides. Soil Sci Soc Am J 1983; 47: 644–648.

Paces T. Sources of acidification in central Europe estimated from elemental budgets in small basins. Nature 1985; 315: 31–36

Pacyna J.M. Atmospheric emissions of arsenic, cadmium, lead and mercury from high temperature processes in power generation and industry. In: *Lead, Mercury, Cadmium and Arsenic in the Environment,* T.C. Hutchinson, K.M. Meema, eds. SCOPE 1987; 31: 69–87.

Page A.L. Fate and effects of trace elements in sewage sludge when applied to agricultural lands. US Environmental Protection Agency Report No. EPA670/2–74–005, 1974.

Page A.L., Bingham, F.T., Nelson C. Cadmium absorption and growth of various plant species as influenced by solution cadmium concentration. J Environ Qual 1972; 1: 288–291.

Page A.L., Ganje T.J., Joshi M.S. Lead quantities in plants, soils and air near some major highways in southern California. Hilgardia 1971; 41: 1–31.

Palmer C.D., Puls R.W. Natural attenuation of hexavalent chromium in ground water and soils. EPA/540/S-94/505, 1994.

Papadopoulos P., Rowell D.L. The reaction of cadmium with calcium carbonate surfaces. J Soil Sci 1988; 39: 23–36 .

Pascual J.A., Garcia C., Hernandez T., Ayuso M. Changes in the microbial activity of an arid soil amended with urban organic wastes. Bio Fertil Soils 1997; 24: 429–434.

Pascual J.A., Hernandez T., Garcia C., Ayuso M. Carbon mineralization in an arid soil amended with organic wastes of varying degrees of stability. Commun Soil Sci Plant Anal 1998; 29: 835–846.

Pasricha N.S., Ponnamperuma F.N. Influence of salt and alkali on ionic equalibria in submerged soils. Soil Sci Soc Am J 1976; 40: 374–376.

Pavaleyev T. Boron in chernozems and gray forest soils of northern Bulgaria. Sov Soil Sci 1958; 9: 1042–1048.

Payne G.G., Martens D.C., Winarko C., Perera N.F. Availability and form of copper in three soils following eight annual applications of copper-enriched swine manure. J Environ Qual 1988; 17: 740–746.

Pepper I.L., Bezdicek D.F., Baker A.S., Sims J.M. Silage corn uptake of sludge-applied zinc and cadmium as affected by soil pH. J Environ Qual 1983; 12 (2): 270–275.

Petruzelli G., Guidi G., Lubrano L. Influence of organic matter on lead adsorption by soil. Z. Pflanzenernahr. Dung Bodenkd 1981; 144: 74–77.

Pettry D.E., Switzer R.E. Arsenic concentrations in selected soils and parent materials in Mississippi. Mississippi State University: Mississippi State, MS 39762, USA. 2000.

Pickering I.J., Brown G.E.Jr., Tokunaga T. X-ray absorption spectroscopy of selenium transformations in Kesterson Reservoir soils. Environ Sci Technol 1995; 29: 2456–2459.

Pierzynski, G.M. and Schwab, A.P. Bioavailability of zinc, cadmium, and lead in a metal-contaminated alluvial soil. J Environ Qual 1993; 22: 247–254.

Pingitore N.E., Eastman M.P. The experimental partitioning of Ca^{2+} into calcite. Chem Geol 1986; 45: 113–120.

Pinton R., Varanini Z., Nannipieri P. "The rhizosphere as a site of biochemical interactions monag soil scompoents, plants and microorganisms." In *The Rhizosphere*, Pinton Roberto, Varanini Zeno, Nannipieri Paolo, eds. New York, NY: Marcel Dekker, Inc., 2001.

Piver W.T. "Mobilization of arsenic by natural and industrial processes." In *Biological and Environmental Efforts of Arsenic,* B.A. Fowler ed. Amsterdam: Elsevier, 1983.

Pokrovsky O., Schott J., Thomas F. Dolomite surface speciation and reactivity in aqutic systems. Geochim Cosmochim Acta 1999; 63: 3133–3143.

Prasad B., Mehta A.K., Singh M.K. Zinc fractions and availability of applied zinc in calcareous soil treated with organic matter. J Indian Soc Soil Sci 1990; 38: 248–253.

Prasad R.B., Basavaiah S., Rao A.S., Rao I.S. Forms of copper in soils of grape orchards. J India Soc Soil Sci 1984: 318–322.

Rabinowitz M.B., Wetherill G.W. Identified sources of lead contamination by stable isotope techniques. Environ Sci Technol 1972; 16: 705–709.

Rai D., Eary L.E., Zachara J.M. Environmental chemistry of chromium. Sci Total Environ 1989; 86: 15–23.

Ramos L., Hernandez L.M., Gonzalez, M.J. Sequential fractionation of copper, lead, cadmium and zinc in soils from or near Donana national park. J Environ Qual 1994; 23: 50–57.

Ramos M.C., Lopez-Acevedo M. Zinc levels in vineyard soils from the Alt Penedes-Anoia region (NE Spain) after compost application. Adv Environ Res 2004; 8: 687–696.

Rapin F., Tessier A., Campbell P.G.C., Carignan R. Potential artifacts in the determination of metal partitioning in sediments by a sequential extraction procedure. Environ Sci Technol 1986; 20: 836–840.

Raskin I., Ensley B.D. *Phytoremediation of Toxic Metals: Using Plants to Clean the Environment.* New York: John Wiley & Sons Inc, 2000.

Ravikovitch S., Margolin M., Navrot J. Micronutrients in soils of Israel. Soil Sci 1961; 92: 85–89.

Ravikovitch S., Bidner-Bar Hava N. Saline soils in the Zevulum valley. Bull Agri Exp Sta, Rehoboth. No. 49, 1948.

Raychaudhuri S.P., Datta B.N.R. Trace element status of India soils. J Indian Soc. Soil Sci 1964; 12: 207–214.

Reed S.T., Martens D.C. "Copper and zinc". In *Methods of soil analysis.* Part 3. Chemical methods. Sparks D.L., ed. Madison, WI: SSSA Book Ser. 5, 1996.

References

Reeder R., Prosky J.L. Composiitonal sector zoning in dolomite. J Sediment Petrol 1986; 56: 237–247.

Reeder R.J., Grans J.C. Sector zoning on calcite cement crystals: Implications for trace element fistrubtions in carbonates. Geochim Cosmochim Acta 1987; 51: 187–194.

Reeves R.D., Baker A.J.M. "Metal-accumulating Plants." In *Phytoremediation of Toxic Metals: Using Plants to Clean Up the Environment*, I. Raskin, B.D. Ensley, eds. New York, NY: John Wiley & Sons, Inc., 2000.

Rehm G.W., Wiese R.A., Hergert G.W. Response of corn to zinc source and rate of zinc band applied with either orthophosphate or polyphosphate. Soil Sci 1980; 129: 36–41.

Renberg I., Persson M.W., Emteryd O. Pre-industrial atmospheric lead contamination detected in Swedish lake sediments. Science 1994; 368: 323–326.

Rendell P.S., Batley G.E., Cameron A.J. Adsorption as a control of metal concentrations in sediment extracts, Environ Sci Technol 1980; 14: 314–318.

Renella G., Landi L., Nannipieri P. Degradation of low molecular weight organic acids complxed with heavy metals in soil. Geoderma 2004; 122: 311–315.

Reuter D.J. "The recognition and correction of trace element deficiencies" In *Trace Elements in Soil-Plant-Animal Systems*, D.J.D. Nicholas, A.R. Egan, eds. New York, NY: Academic Press, Inc., 1975.

Rice K.C. Trace element concentrations in streambed sediment across the continental United States. Environ Sci Technol 1999; 33: 2499–2504.

Robarge W.P. Precipitation/dissolution reactions in soils. In *Soil Physical Chemistry*, 2nd Sparks D.L., ed. Boca Raton, FL: CRC Press, 1999.

Robert M., Terce M. "Effect of gel and coatings on clay mineral chemical properties." In *Inorganic Contaminants in the Vadose Zone*, B. Bar-yosef, N.J. Barrow, J. Goldsmith, eds. Berlin, 1989.

Romheld V. The role of phytosiderophores in acquisition of iron and other micronutrients in graminaceous species: An ecological approach. Plant Soil 1991; 130: 127–134.

Ross S.M. *Toxic Metals in Soil-Plant Systems*. Chichester, UK: John Wiley & Sons, 1994.

Ruby M.V., Davis A., Nicholson A. In situ formation of lead phosphates in soils as a method to immobilize lead. Environ Sci Technol 1994; 14: 877–880.

Rudawsky O. *Mineral Economics*. Amsterdam: Elsevier, 1986.

Ryan J, Masri S., Garabet S. Geographical distribution of soil test values in Syria and their relationship with crop response. Commun Soil Sci Plant Anal. 1996; 27: 1579–1593.

Sadana U.S., Takkar P.N. Effect of sodality and zinc on soil solution chemistry of manganese under submergerged conditions. J Agr Sci 1988; 111: 51–55.

Sadig M. Solubility and speciation of zinc in calcareous soils. Water Air Soil Pollut 1991; 57–58: 411–421.

Saha U.K., Taniguchi S., Sakurai K. Adsorption behavior of cadmium, zinc and lead on hydroxyaluminum- and hydroxyaluminosilicate-montmorillonite complexes. Soil Sc Soc Am J 2001; 65: 694–703.

Sajwan K.S., Lindsay W.L. Effect of redox, zinc fertilization and incubation time on DTPA-extractable zinc, iron and manganese. Commun Soil Sci Plant Anal 1988; 19: 1–11.

Salt D.E., Kramer U. "Mechanisms of metal hyperaccumulation in plants." In *Phytoremediation of Toxic Metals: Using Plants to Clean Up the Environment*, I. Raskin, B.D. Ensley, eds. New York, NY: John Wiley & Sons, Inc., 2000.

Salt D.S., Prince R.C., Pickering I.J., Raskin I. Mechanisms of cadmium mobility and accumulation in Indian mustard. Plant Physiol 1995; 109: 1427–1433.

Sanders J.R. The effect of pH on the total and free ionic concentrations of manganese, zinc and cobalt in soil solutions. J Soil Sci 1983; 34: 315–323.

Sauve S., McBride M.B., Norvell W.A., Hendershot W.H. Copper solubility and speciation of in situ contaminated soils: Effects of copper level, pH and organic matter. Water Air Soil Pollut 1997; 100: 133–149.

Schaffer B., Larson K.D., Snyder G.H., Sanchez C.A. Identification of mineral deficiencies associated with mango decline by DRIS. Hort Sci 1988; 23: 617–619.

Schnitzer M., Skinner S.I.M. Organo-metallic interactions in soils. 7. Stability constants of Pb^{2+}, Ni^{2+}, Mn^{2+}, Co^{2+}, Cu^{2+}, and Mg^{2+} fulvic acid complexes. Soil Sci 1967; 103: 247–252.

Schwab A.P., Lindsay W.L. The effect of redox on the solubility and availability of manganese in a calcareous soil Rice plants. Soil Sci Soc Am J 1983; 47: 217–220.

Sedbery J. E., Reddy C.N. The distribution of zinc in selected soils in Louisiana. Commun in Soil Sci and Plant Anal 1976; 7: 787–795.

Sequi P., Aringhieri R. Destruction of organic matter by hydrogen peroxide in the presence of pyrophosphate and its effect on soil specific surface area. Soil Sci Soc Am J 1977; 41: 340–343.

Settle D.M., Patterson C.C. Lead in Albacore: Guide to lead pollution in Americans. Science 1980; 207: 1167–1176.

Shacklette H.T., Boerngen J.G. Element concentrations in soils and other surficial materials of the conterminous United States: an account of the concentrations of 50 chemical elements in samples of soils and other regoliths. Washington, 1984.

Shahandeh H., Hossner L.R. Plant screening for chromium phytoremediation. Int. J. Phytoremediat 2000; 2: 31–51.

Shainberg I. "Chemical and mineralogical components of crusting." In *Soil Crusting, Chemical and Physical Processes*, M.E. Sumner, B.A. Stewart, eds. Advances in Soil Science, Boca Raton: Lewis Publishers, 1990.

Shannon R.D., White J.R. The selectivity of a sequential extraction procedure for the determination of iron oxyhydroxides and iron sulfides in lake sediments. Biogeochem 1991; 14: 193–208.

Shenker M., Fan T.W.M., Crowley D.E. Phytosiderophores influence on cadmium mobilization and uptake by wheat and barley plants. J Environ Qual 2001;30: 2091–2098.

Sheppard J.C., Funk W.H. Trees as environmental sensors monitoring long-term heavy metal contamination of Spokane River, Idaho. Environ Sci Technol 1975; 9: 638–643.

Shuman L.M. "Chemical forms of micronutrients in soils." In *Micronutrients in Agriculture* (2nd Edition), Mortvedt, J.J., Cox, F.R., Shuman, L.M., Welch, R.M., eds. Madison, WI.: Soil Sci Soc Am Inc., 1991.

Shuman L.M. Effect of removal of organic matter and iron or manganese-oxides on zinc adsorption by soil. Soil Sci 1988; 146: 248–254.

Shuman L.M. Effect of roganic waste amendments on zinc adsorption by two soils. Soil Sci 1999; 164: 197–205.

Shuman L.M. Effects of tillage on the distribution of manganese, copper, iron, and zinc in soil fractions. Soil Sci Soc Am J 1985b; 49: 1117–1122.

Shuman L.M. Fractionation method for soil microelements. Soil Sci 1985a; 140: 11–22.

Shuman L.M. Zinc adsorption isotherms for soil clays with and without iron oxides removed. Soil Sci Soc Am J 1976; 40: 349–352.

Shuman L.M., Separating soil iron- and manganese-oxide fractions for microelement analysis, Soil Sci Soc Am J 1982; 46: 1099–1102.

Shuman L.M., Wang J. Effect of rice variety on zinc, cadmium, iron and manganese content in rhizosphere and non-rhizosphere soils fractions. Commun. Soil Sci Plant Anal 1997; 28: 23–36.

Sillanpaa M. *Micronutrients and the Nutrient Status of Soils: A Global Study*. Rome: Food and Agricultural Organization of the United Nations, 1982.

References

Simeoni L.A., Barbarick K.A., Sabey B.R. Effect of small-scale composting of sewage sludge on heavy metal availability to plants. J Environ Qual 1984; 13: 264–268.

Simyu G.M., Tole M.P., Davies T.C. Trace elements concentrations in Hell's Gate National Park and health implicatins for zebra (Equus burchelli), Kenya. Retrieved Jan 2005 from: www.unep.org/gef/content/pdf/34-Gelas.pdf.

Singer A. The Soils of the Land of Israel. 2007. To be published by Springer.

Singer A., Dultz S., Argaman E. Properties of the non-soluble fractions of suspended dust over the Dead Sea. Atmos. Environ. 2004; 38: 1745–1753.

Singh B., Gilkes R.J. Properties and distribution of iron oxides and their association with minor elements in the soils of south-western Australia. J Soil Sci 1992; 43: 77–98.

Singh B., Singh M., Dang Y.P. Distribution of forms of zinc in some soils of Haryana. J. Indian Soc. Soil Sci 1987; 35: 217–224.

Singh B.R., Almas A., Jeng A.S., Narwal R.P. Crop uptake and extractability of cadmium in soils naturally high in metals at different pH levels. Commun Soil Sci Plant Anal 1996; 26: 2123–2142.

Singh R.R., Parsad B., Sinha. Forms of copper in calcareous soils. J Indian Soc Soil Soc 1989; 37: 45–51.

Singh, M.V., Abrol, I.P., Transformation and movement of zinc in an alkali soil and their influence on the yield and uptake of zinc by rice and wheat crops. Plant Soil. 1986; 94: 445–449.

Smolders E., McLaughlin M.J. Effect of soil solution chloride on cadmium availability to Swiss chard. Soil Sci Soc Am J 1996: 60: 1443–1447.

Smolders E., Tiller K.G., McLaughlin M.J. Lambregts R.M. Effect of soil solution chloride on cadmium availability to Swiss chard. J Environ Qual 1998; 27: 426–431.

Soil Survey Staff. Soil taxonomy, a basic system of soil classification for making and interpreting soil surveys. U.S. Dept. Agric. Handbook, 436. Washington D.C.: Soil Conserv Serv, 1975.

Soltanpour P.N, Workman S.M. Use of NH_4HCO_3-DTPA soil test to assess the availability and toxicity of selenium to alfalfa palnts. Commun Soil Sci Plant Anal 1980; 11: 1147–1156.

Soltanpour P.N., Olsen S.R., Goos R. J. Effects of nitrogen fertilization of dryland wheat on grain selenium concentration. Soil Sci Soc Am J 1982; 46: 430–433.

Soltanpour P.N., Schwab A. P. A new soil test for simultaneous extraction of macro- amd micro-nutrients in alkaline soils. Commun Soil Sci Plant Anal, 1977; 8: 195–207.

Sposito G., LeVesque C.S., LeClaire J.P., Chang A.C. Trace elements chemistry in arid-zone field soils amended with sewage sludge: III. Effect of the time on the extraction of trace metals. Soil Sci Soc Am J 1983; 47: 898–902.

Sposito G., Lund L.J., Chang A.C. Trace metal chemistry in arid-zone field soils amended with sewage sludge: I. Fractionation of Ni, Cu, Zn, Cd, and Pb in solid phases,. Soil Sci Soc Am J 1982; 46: 260–264.

Sposito G., Page A.L. In Circulation of Metal Ions in the Environment: Metal Ions in Biological Systems. New York: Marcel Dekker, 1984.

Sposito, G. "Distinguish adsorption from surface precipitation." In *Geochemical Processes at Mineral Surfaces*. J.A. Davis, K.F. Hayes, eds. Washington D.C.: Americal Chemical Society Symposium Series No. 323, 1986.

Stahl R.S., James B.R. Zinc sorption by manganese-oxide-coated sand as a function of pH. Soil Sci Soc Am J 1991; 55: 1291–1294.

Steinhilber P., Boswell F.C. Fractionation and characterization of two acrobic sewage sludge. J Environ Qual 1983; 12: 529–534.

Steveninck R.F.M. Van, Babare A., Fernando D.R., Steveninck M.E.Van. The binding of zinc, but not cadmium, by phytic acid in roots of crop plants. Plant Soil 1994; 167: 157–164.

Stevenson F.J. :Organic matter reactions involving pesticides in soil." In *Bound and Conjugated Pesticide Residues*, D.D. Kaufman, G.G. Still, G.D. Paulson, S.K. Bandal, eds. Washington: American Chemical Society, 1976.

Stevenson F.J. Nature of divalent transition metal complexes of humic acids as revealed by a modified potentiometric titration method. Soil Sci 1977; 123: 10–17.

Stevenson F.J., Fitch A. "Reactions with organic matter". In *Copper in Soils and Plants*, J.F. Loneragan, A.D. Robson, R.D. Graham, eds. Sydney: Academic Press. 1981.

Stewart M.A., Jardine P.M., Barnett M.O., Mehlhorn T.L., Hyder L.K., McKay L.D. Influence of soil geochemical and physical properties on the sorption and bioaccessibility of chromium(III). J Environ Qual 2003; 32: 129–137.

Stipp S.L., Hochella M.F.Jr., Parks G.A., Leckie J.O. Cd^{2+} uptake by calcite, solid-state diffusion, and the formation of solid-solution: Interface processes observed with near-surface sensitive techniques (XPS, LEED, and AES). Geochim Cosmochim Acta 1992; 56: 1941–1954.

Street J.J., Lindsay W.L., Sabey B.R. Solubility and plant uptake of cadmium in soils amended with cadmium and sewage sludge. J Environ Qual 1977; 6: 72–77.

Stumm W. Morgan J.J. *Aquatic Chemistry*, 2nd ed, New York: John Wiley and Sons, 1982.

Sutherland R.A., Tack F.M.G. Fractionation of Cu, Pb and ZSn in certifed reference soils SRM 2710 and SRM 2711 using the optimized BCR sequential extraction procedures. Adv Environ Res 2003; 8: 37–50.

Swaine D.J. The trace element content of soils. Common Bur. Soil Sci Tech Commun 48, Baltimore, M.D.: Lord Baltimore Press, 1955.

Swarup A., Anand S. Transformation and availability of iron and manganese in submerged sodic soils in relation to yield and nutrition of rice. Fertilizer News 1989; 34: 21–23.

Taylor S.R., McLennan S.M. *The continental crust: Its composition and evolution*, Oxford: Blackwell, 1985.

Taylor S.R., McLennan S.M. The geochemical evolution of continental crust. Rev Geophy 1995; 33: 241–265.

Tessier, A., Campell, P.G.C., and Bisson, M., Sequential extraction procedure for the speciation of particulate trace metals. Anal Chem 1979; 51: 844

Thompson A., Parker D.R., Amrhein C. Selenate partitioning in field-situated constructed wetland mesocosms. Ecol Eng 2003; 20: 17–30.

Thornthwaite C.W. An approach toward a rational classification of climate. Geogr Rev 1948; 38: 55–94.

Tiller K.G. "Micronutrients." In *Soils: An Australian Viewpoint*, CSIRO, Melbourne. London: Acad Press, 1983.

Tiller K.G., Merry R.H. "Copper Pollution of Agricultural Soils." In *Copper in Soils and Plants*, J.F. Loneragan, A.D. Robson, R.D. Graham, eds. Sydney: Academic Press, 1981.

Tokunaga T., Brown G.E. Jr., Pickering I.J., Sutton S.R., Bajt S. Selenium transport between ponded waters and sediments. Environ Sci Technol 1997; 31: 1419–1425.

Tokunaga T., Pickering I.J., Brown G.E. Jr. X-ray absorption spectroscopy studies of selenium transformations in ponded sediments. Soil Sci Soc Am J 1996; 60: 781–790.

Tokunaga T.K., Wan J., Firestone M.K., Hazen T.C., Olson K.R., Herman D.J., Sutton S.R., Lanzirotti A. In situ reduction of chromium (VI) in heavily contaminated soils through organic carbon amendment. J Environ Qual 2003; 32: 1641–1649.

Traina S.J., Doner H.E. Heavy metal induced released of manganese (II) from a hydrous manganese dioxide. Soil Sci Soc Am J 1985; 49: 317–321.

Traina S.J., Laperche A. Contaminant bioavailaiblity in soils, sediments, and aquatic environments. Proc Natl Acad Sci USA 1999; 96: 3365–3371.

References

Treeby M., Marschner H., Römheld V. Mobilisation of iron and other micronutrient cations from a calcareous soil by plant borne, microbial and synthetic chelators. Plant Soil 1989; 114: 217–226.

U.S. Bureau of the Census. *World Population Profile: 1999*. U.S. Government Printing Office, Washington, DC., 1999.

U.S. Environmental Protection Agency (USEPA) Manufacturers to Use New Wood Preservatives, Replacing Most Residential Uses of CCA, http://www.epa.gov/pesticides/factsheets/chemicals/cca_transition.htm (verified 7 May 2003), 2002b.

U.S. Environmental Protection Agency (USEPA). Implementation guidance for the arsenic rule. U.S. Environmental Protection Agency; U.S. Government Printing Office: Washington, DC., 2002a.

UNEP. *World Atlas of Desertification*. London, New York: Edward Arnold, 1993.

United Nations. *Demographic Yearbook*, 14 Issue. New York: United Nation, 1962.

United Nations. *World Energy Supplies 1950-1974*. New York: United Nations, 1976.

United Nations. *World Energy Supplies 1972-1976*. New York :United Nations, 1978.

Ure A.M, Berrow M.L. The elemental constituents of soils. In: Bowen HJM (ed) Environmental chemistry. London: Royal Soc Chem, 1982.

US Bureau of Mines. *Minerals Yearbook. Metals and Minerals*. Washington, D.C.: US Dept of Interior, 1978–1980.

US Dept. of Interior, US Geol. *Survey Minerals yearbook. Metals and minerals*. Washington, D.C.: US Gov Printing Office, 1994–2000.

US Dept. of Interior, US Geol. Survey. Mercury in U.S. coal – Abundance, distribution, and modes of occurrence, http://pubs.usgs.gov/factsheet/fs095-01/fs095-01.html (Verified 28 Aug. 2002), 2002.

USEPA. Lead in paint, dust and soil. http://www.epa.gov/lead/. 2006.

USGS. Commodity Statistics and Information. http://minerals.usgs.gov/minerals/pubs/commodity/. 2006.

USGS. Mercury in the Environment, Fact Sheet 146-00. http://minerals.usgs.gov/mercury, 2000.

Van der Lelie D., Schwitzguebel J.P., Glass D.J., Vangronsveld J., Baker A. Assessing phytoremediation's progress in the United States and Europe. Environ Sci Technol 2001; 35: 447A–452A.

Van H. W. Materials Balance and Technology Assessment of Mercury and its Compound on National and Regional Bases. EPA 560/3-75-007. U.S. Environmental Protection Agency, Washington, DC, 1975.

Verheye W. "Soils of arid and semi-arid areas." In *UNESCO Encyclopedia of Life Support Systems*. 2006. Submitted for publication.

Vig K., Megharaj M., Sethunathan N., Naidu R. Bioavailability and toxicity of cadmium to microorganisms and their activities in soil: a review. Adv Environ Res 2003; 8: 121–135.

Villarroel, De. Jr., Chang, A.C., Amrhein, C. Cd and Zn phytoavailability of a field-stabilized sludge-treated soil. Soil Sci 1993; 155: 197–205.

Vinogradov A.P. *The Geochemistry of Rare and Dispersed Chemical Elements in Soils*. New York: Consultants Bureau Inc, 1959.

Vulkan T., Mingelgrin V., Ben-Asher J., Frenkel H. Copper and zinc speciation in the solution of a soil-sludge mixture. J Environ Qual 2002; 31: 193–203.

Wakefield Z.T. Distribution of Cadmium and Selected Heavy Metals in Phosphate Fertilizer Processing (Bulletin Y-159), National Fertilizer Development Center, TVA, 1980.

Wang P.X., Qu E.F. A preliminary study on the chemical forms and availability of Nickel in manured loessial soil. Proceedings of the international symposium on the role of sulphur, magnesium and micronutrients in balanced plant nutrition, S. Porth, ed. Washington, 1992.

Warden B.T., Reisenauer H.M. Fractionation of soil manganese forms important to plant availability. Soil Sci Soc Am J 1991; 55: 345–349.

Webber J.H., Beauchamp E.G. Cadmium concentration and distribution in corn (Zea mays L) growing on a calcareous soil for three annual sludge applications. J. Environ Sc Health B 1979; 14: 459–474.

Wedepohl K.H. The composition of the continental crust. Geochim Cosmochim Acta 1995; 59: 1217–1232.

Weggler-Beaton K., McLaughlin M.J., Graham R.D. Salinity increased cadmium uptke by wheat and Swiss chard from soil amended with biosolids. Aust J Soil Res 2000; 38: 37–45.

Welch A.H., Watkins S.A., Helsel D.R., Focazio M.F. Arsenic in groundwater resources of the United States: U.S. Geological Survey Fact Sheet 063-00, On line at: http://co.water.usgs.gov/trace/pubs/fs-063-00/. US Geological Survey, 2000.

Wenzel W.W., Kirchbaumer N., Prohaska T., Stingeder G., Lombi E., Adriano D.C. Arsenic fractionation in soils using an improved sequential extraction procedure. Anal Chimica Acta 2001; 436: 309–323.

Whiting S.N., Souza M.D.m Terry N. Rhizosphere bacteria mobilize Zn for hyperaccumulation by Thlaspi caerulescens Environ Sc Technol 2001; 35: 3144–3150.

Williamson A., Johnson M.S. "Reclamation of metalliferous mine wastes." In *Effect of Heavy Metal Pollution on Plants. Vol. 2 Metals in Environment*, N.W. Lepp, ed. London, UK: Applied Science Publishers, 1981.

Wolt, Jeff, *Soil Solution Chemistry. Applications to Environmental Science and Agriculture.* New York: John Wiley and Sons, Inc., 1994.

Woolson E.A., Axley J.H., Kearney P.C. The chemistry and phytotoxicI. Contaminated field soils. Soil Sci Soc Am J 1971; 35: 938–943.

Woytinsky W.S., Woytinsky E.S. World population and production: Trends and outlook. Baltimore MD: Lord Baltimore Press, 1953.

Wright R.F., Schindler D.W. Interaction of acid rain and global changes: Effects on terrestrial and aquatic ecosystems. Water Air Soil Pollut 1995; 85: 89–99.

Wu J., Laird D.A., Thompson M.L. Sorption and desorption of copper on soil clay components. J Environ Qual 1999; 28: 334–338.

Xu Y., Schwartz F.W. Sorption of Zn^{2+} and Cd^{2+} on hydroxyapatite surfaces. Environ Sci Technol 1994; 28: 1472–1480.

Yin, Y. Allen H.E., Li Y., Huang C.P., Sanders P.F. Adsorption of mercury (II) by soil: effect of pH, chloride, and organic matter. J Environ Qual 1996; 25: 837–844.

Zachara J.M., Cowan C.E., Resch C.T. Sorption of divalent metals on calcite. Geochim Cosmochim Acta 1991; 55: 1549–1564.

Zachara J.M., Kittrick J.A, Harsh J.B. Solubility and surface spectroscopy of zinc precipitates on calcite. Geochim Cosmochim Acta 1989; 53: 9–19.

Zachara J.M., Kittrick J.A, Harsh J.B. The mechanism of Zn^{2+} adsorption on calcite. Geochim Cosmochim Acta 1988; 52: 2281–2291.

Zasoski R. J., Burau R.G. Sorption and sorptive interaction of cadmium and zinc on hydrous manganese oxide. Soil Sci Soc Am J 1988; 52: 81–87.

Zeien H., Brummer G.W. Determination of the mobility and binding of heavy metals in soil by sequential extraction. Mitteilungen der deutschen bodenkundlichen gesellschaft. 1991; 66: 397–400.

Zhang F., Romheld V., Marschner H. Diurnal rhythm of release of phytosiderophores and uptake rate of zinc in iron-deficient wheat. Soil Sci Plant Nutri 1991; 37: 671–678.

References

Zhang P., Ryan J.A. Formation of chloropyromorphite from galena (PbS) in the presence of hydroxyapatite. Environ Sci Technol 1999; 33: 618–624.

Zhao F.J., Dunham S.J., and McGrath S.P. Arsenic hyperaccumulation by different fern species. New Phytol 2002; 156: 27–31.

Zhu Q. " Zinc in soils." In *Micronutrients in Soils of China*, Liu Z. ed., Nanjing: Jiangsu Science and Technology Publishing House, 1996.

Index

A
Abrasion and erosion, 29
Adsorption, Cd, Ni, Pb, Zn, 102
Adsorption/desorption process, 101–102
Adsorption kinetics, 140–144
Aeolian Material Derived Soil, 28–34
Aeolian sands, 12
Africa, 19–20
African vertisols, Cr, Zn, Mn, Zn, Cu, 57
Aging of oxides, 138
Agricultural management, 284–286
Alfisols, 40–42
Alluvial sands, 11
Analytical techniques for solid-phase speciation, 148–149
Animal waste, 286
Annual transfer fluxes, 198–199
Antarctica, 19
Argilluviation, 40–42
Arid northeastern Brazil, 17
Aridisols, 20, 42–43
Aridity index, 3, 5
Arizona, Cd, Zn, Cu, Ni, Pb, 63
As
   California soils, 64
   China soil, 60
   coal, 307
   earth's crust, rocks, 56
      lithosphere, 321
   problems in the world, 304
   reduction/oxidation, 103
   soil, 56
   sources, 304–314
   volatilization, 301–302
   world soils, 56, 325
As adsorption
   Fe, Mn, Al oxides, 139
   kaolinite, 146
   montmorillonite, 146
As fractionation, 130
Atmospheric inputs, 281–284
Atomic properties of elements, 160

Australia, Co, Mn, Mo, Zn, 60
Australia, Ni, 61
Australian soils, Zn solution. 83

B
Ba selectivity on calcite, 148
Binding of elements in solid-phase, 131–167
Bioavailability
   complexed ions in solution, 234
   rhizosphere, 227–229
   solid-phase, 236
   solution, 233–234
   trace elements, 221–266
Biological remediation, 297–301
Biological surface crusts, 32–34
Biosolid, 133
Bonn procedures, 121–124
Boron
   adsorption, 138–139
   adsorption on montmorillonite, 146
   China soil, 59
   earth's crust, rocks, 56
   Israeli soil, 66
   Serbia, 62
   soil, 56
   Uzbekistan, 61
   world soils, 56
Bulgaria, Co, Cu, Zn, 61

C
Calcite selectivity sequence
   Cd, Zn, Mn, Co, Ni, Ba, Sr, 148
California soil
   Cd fractionation, 151–152
   Cd speciation, 86
   Cr fractionation, 152
   Cu fractionation, 153
   Cu speciation, 85
   Mn, Pb, Co, Cu, As, Ni, Zn, Cd, Hg, Mo, Cr, 64
   Ni fractionation, 153–154
   P levels on Cd and Zn speciation, 94
   Pb fractionation, 154

Zn fractionation, 155
Zn speciation, 84
Carbonate dissolution, 110–120
Carbonate dissolution/precipitation, 95–100
Cation exchange capacity, 140–141
Cd
   affinity on amorphous Fe oxide, 137
   affinity on humic acid, 132
   African soil, 56
   Arizona, 63
   California soil, 64, 152
   China soil, 60
   distribution in solid-phase, 151–152
   earth's crust, rocks and, world soils, 55
   interaction with Fe oxide, 136
   Israeli soil, 65
   Kenya, 56
   reaction with organic matter, 133
   reactivity in sludge-amended soils, 132–133
   Rocky Mountains, 62
   soil, 55
   solid-phase control, 163
   Spain, 62
   speciation, California, 88
   speciation, Israeli soil, 87
   speciation, South Australia, 87
   Syria, 66
   United Arab Emirates, 56–57
   US soils, 55, 62–64
   waste-amended soil, 152
Cd fractionation
   California soil, 151–152
   Israeli soil, 151–152, 181–182
   sludge-amended soil, 150–152
   Spain soil, 151–152
Cd, Cr, Cu, Hg, Ni, Pb, Zn in lithosphere, 323
Cd, Cr, Cu, Hg, Ni, Pb, Zn in world soils, 323
Cd, the initial fast process, 172–176
Cd, Zn, Pb, Ni adsorption, 102
Central African Republic, Cu, 57
Central Asian Plains, 18

Chad, Cr, Mn, Co, Pb, Cu and Ni, 57
Chernozems
   Bulgaria, Co, Cu, Zn, 61
   Romania, Cu, Ni, Zn, 62
China soil
   Co fractionation, 157
   Cu fractionation, 153
   Mn, 58
   Mo, 59–60
   Ni fractionation, 154
   Se, Cd, As, Hg, Pb, Cr, 60
   Zn, Cu, Ni, Co, B, 59
   Zn fractionation, 155–156
Chinese fern, 299
Classification of arid and semi-arid soils, 42–44
Co
   adsorption on Fe/Mn oxides, 139–140
   affinity on amorphous Fe oxide, 137
   affinity on geothite (FeOOH), 137
   affinity on haematite, 137
   African vertisols, 57
   Australia, 60
   Azerbaidzhan, 62
   Bulgaria, 61
   California soils, 64
   Chad, 57
   China soil, 59
   deficiency, 263
   distribution in solid-phase, 157–159
   earth's crust, rocks and world soils, 51–52
   fractionation and transformation, 215–220
   fractionation in China soil, 157
   fractionation in Israeli soil, 158
   Israeli soil, 66
   Russia, 61
   selectivity on calcite, 148
   soil, 51–52
   solid-phase control, 166–167
   Uzbekistan, 61
CO_2-H_2O System, 77–78
Coastal dunes, 12
Colorado soils

Cu fractionation, 153
Pb fractionation, 154
Zn solution, 83
Complexation of DOC and organics, 102–103
Concentrations of trace elements
  Egypt, 74
  Israeli, 74
  saturated paste extracts, 75
  sludge-amended soils, 72–74
  soil solution, 72–76
  solutions of California soils, 72–73
  South Australia, 74, 76
Contents of bioavailable trace elements
  Africa, 253
  Asia, 253–257
  China, 253–256
  East, 258–260
  Europe, 257
  India, 256
  Iraq, 258–259
  Iran, 259
  Israel, 259
  Lebanon, 259
  Mexico, 258
  North America, 258
  Pakistan, 257
  Syria, 260
  US, 258
Contents of trace elements in plants, 225
Correlation
  Co and Fe contents, 135
  Cu and MnO contents, 136
  stable isotope dilution-sequential extraction procedure, 132–133
Cr
  African soil, 57
  African vertisols, 57
  California soils, 64
  Chad, 57
  China soil, 60
  distribution in solid-phase, 152
  earth's crust, rocks and world soils, 50–51
  fractionation in California soil, 152
  fractionation in Israeli soil, 152, 183
  fractionation in sludge-amended soil, 152
  fractionation in Spain soil, 152
  interaction with organic matter, 131
  Israeli soil, 65
  Madagascar, 57
  reaction with organic matter, 133
  reduction/oxidation, 103–104
  soil, 50–51
  solid-phase control, 165–166
  United Arab Emirates, 56–57
Crescentic dunes, 12–13
Cu
  African Republic, 57
  African soil, 56
  African vertisols, 57
  Arizona, 63
  Bulgaria, 61
  California soils, 64
  Chad, 57
  China soil, 59
  deficiency, 262
  earth's crust, rocks and, world soils, 52–54
  humic acid, 132
  Indian soil, 57–58
  the initial fast process, 172–176
  Israeli soil, 65
  Romania, 62
  Russia, 61
  soil, 52–54
  solid-phase, 153
  Spain, 62
  speciation, California, 86
  speciation, Israeli soil, 86
  speciation, sludge-amended soil, 86
  Syria, 66
  United Arab Emirates, 56–57
  US, 62–64
  Uzbekistan, 61
Cu affinity
  amorphous Fe oxide, 137

geothite (FeOOH) and haematite, 137
Cu fractionation
  California soil, 153
  China soil, 153
  Colorado soil, 153
  Indian soil, 153
  Israeli soil, 183–186
  sludge-amended soil, 153
  Spain soil, 153
Cupric ferrite, 165
Cu solid-phase control, 163–165

D

Dasht-i-Lat and Dasht-i-Kavri deserts, 17
Deficiency, 260–263
Deflation, 28–29
Desert Pavement Formation, 25–28
Dissolution
  major elements by NaOAc-HOAc, 114–118
  trace elements by NaOAc-HOAc, 114–118
Dissolution capacity of NaOAc-HOAc, 111–114
Dissolution of Ca, 114
Dissolution/precipitation/co-precipitation, 95–100
Dissolved organic carbon on speciation, 102
Distribution
  arid and semi-arid areas, 15–20
  bioavailabile trace elements, 253–260
  trace elements among solid-phase components, 150–159
    Cd, 151–152
    Cr, 152
    Cu, 153
    Co, 157–159
    Mn, 156–157
    Ni, 153–154
    Pb, 154
    Zn, 155–156
  trace elements in leaves, 222
  trace elements in solid-phase, 131–167
Distribution of trace elements, 47–68
  Africa, 56–57
  Asia, 57–60
  Australia, 60–61
  Azerbaidzhan, 62
  Bulgaria, 61
  Central African Republic, 57
  Chad, 57
  China, 58–60
  Europe, 61–62
  Former Soviet Union, 61–62
  India, 57–58
  Kenya, 56
  Madagascar, 57
  Middle East, 65–66
  North America, 62–64
  Romania, 62
  Russia, 61
  Spain, 62
  United Arab Emirates, 56–57
  Uzbekistan, 61
Dry subhumid areas, 5
DTPA, DPTA-TEA, DTPA-AB, 229–232
Dust Accretion, 28–34

E

Earth's crust, B, As, Se, Cu, Co, Cr, Zn, Ni, Pb, Cd, Mn, Mo, Hg, 50–56
EDTA, 231–232
Effects of element inputs on bioavailability, 241–244
  element interaction, 250–251
  N and P fertilizers, 250–251
  organic matter, DOC, 244–246
  organic matter removal on heavy metal adsorption, 133–134
  pH of NaOAc-HOAc on the solid-phase components, 118–120
  pH on Zn binding by Fe oxide and hydroxides, 137
  plant species and growth, 251–252
  salt concentrations, 248–249

Index

sewage on contents of trace
    elements in plants, 225–227
soil eh and moisture on
    bioavailability, 239–241
soil pH on bioavailability, 237–239
sources of elements, 246–248
time on bioavailability, 241–244
Egypt soil, solution concentration, 74
Electron techniques, 149
Element burdens per capita, 325–326
Element in the earth's crust, 47–48
Entisols, 43
Ephemeral stream, 7
Europe, 20
Evapotranspiration, 5
Extractants for bioavailable trace
    elements, 229–233
Extractants for solid-phase fractions,
    124–125
  carbonate Fraction, 126
  exchangeable fraction, 124–125
  organically bound fraction,
    126–127
  readily reducible oxide fraction, 126
  reducible oxide fraction, 127–128
  residual fraction and total, 128

F
Factors affecting bioavailability of
    trace elements, 237–252
Factors affecting contents of trace
    elements, 66–68
  agricultural management/ industrial
    activities, 68
  climate effects, 66–67
  parent materials, 67
  soil processes and properties, 67–68
Factors affecting trace element
    speciation in soil solution,
    91–95
  dissolved organic matter, 92
  soil pH and Eh, 91–92
  total Inorganic ligands, 92–93
Fe, Mn, Al oxide with As, Co, 139–140
Fe oxide and hydroxide with Zn, 137
Fe oxide removal, 140–144

Field capacity regime, 169–220
First hydrolysis constants, 82
Fluvial landforms, 9–10
Forms of trace elements in sludge,
    276–278
Frankinite, 165

G
Geomorphic landforms, 7–15
Geomorphic processes, 7–15
Global perspective, 303–327
Great Indian Desert, 17
The Great Plains, 15–17

H
Hg
  California soils, 64
  China soil, 59
  earth's crust, rocks and, world
    soils, 55
  soil, 55
  sources, 307
  US soils, 55
High energy binding sites, 136
Humic acid reaction with Cu, Cd, Pb
    and Zn, 132
Hyper-arid, 5
Hyper-arid environments, 5
Hysteresis of Zn
    desorption/adsorption, 134

I
Incubation experiments, 169–221
India, Mn, Cu, Ni, Zn, Mo, 57–58
Indian mustard, 300
Indian soil
  Cu fractionation, 153
  Zn fractionation, 155
Industrial age annual production
  As, 304–307
  Cd, 311
  Cr, 311–313
  Cu, Ni, Zn, 313–314
  Hg, 307–310
  Pb, 310–311
  trace elements, 303–314

Industrial age cumulative production of trace elements, 314–318
Initial Fast Processes of Cu, Ni, Zn, Cd, Zn, 172–176
Inner sphere complexion, 135
Ion probes (proton induced X-ray emission), 149
Ionic potential definition, 49
Iron and Mn oxides and hydroxides with trace elements, 135–145
Irrigation with sewage, 75, 267–273
Israeli soil
   Cd fractionation, 151–152
   Cd speciation, 87
   Co fractionation, 158, 215–220
   Cr fractionation, 152
   Cr, Cr, Ni, Cu, Pb, Zn, Cd solid-phase, 181–189
   Cu, 169–220
   Cu, Cd, Cr, Ni, Zn, 65
   Cu fractionation, 153
   Cu speciation, 85
   Mn, Co, B, 66
   Mn fractionation, 156–157, 203–205
   Ni, 169–220
   Ni fractionation, 153–154
   Pb fractionation, 154
   residence time of metals, 201–202
   saturated paste extracts, 75
   solution concentration, 75
   transfer flux, 169–220
   Zn, 169–220
   Zn fractionation, 155

K
Kenya, Cd, Mo, Zn, 56
Kinetics
   solution of $CaCO_3$, 114
   transformation of Mn, 211–213
Kinetics of Co transformation among solid-phase fractions, 218–219

L
Lacustine Landforms, 10
Life cycle of products, 319–320
Linear or longitudinal dunes, 13

Loess, 14
Long-term application of sewage, 270

M
Madagascar, Cr, Zn, 57
Magnetic spectroscopy, 149
Major physico-chemical and biological processes, 95–105
Manganese Partitioning, 202–213
Martonne aridity index, 3, 5
Mass spectrometric techniques, 149
Mass-wasting processes, 8
Melanization, 36–38
Metal affinities for humic acid, 132
Metal affinities on Fe oxides, 137
Mineral surface crusts, 32
Mineralogical characteristics of trace elements, 160–167
Mining and industrial activities, 286–289
Mn
   affinity on geothite (FeOOH) and haematite, 137
   African vertisols, 57
   Australia, 60
   California soils, 64
   Chad, 57
   China soil, 58
   deficiency, 262–263
   distribution at the field capacity regime, 203–205
   distribution at the saturated paste regime, 203–205
   distribution in solid-phase, 156–157
   earth's crust, rocks and, world soils, 55
   fractionation in Israeli soil, 156–157
   Israeli soil, 66
   selectivity on calcite, 148
   soil, 55
   solid-phase control, 166–167
   Syria, 66
   Uzbekistan, 61
Mo
   African soil, 56
   Azerbaidzhan, 62
   California soils, 64

Index

China soil, 59–60
earth's crust, rocks and, world soils, 55
Indian soil, 57–58
Kenya, 56
Queensland, Australia, 60
Russia, 61
soil, 55
Mobilization of clay, 41
Mollisols, 36–38
Monitoring pollution, 290–293
enzymes, 290–293
microbes, 290–293
plants, 290

N
NaCl effect on Cd speciation, 89
Ni
Arizona, 63
Australia, 60
California soils, 64
Chad, 57
China soil, 59
distribution in solid-phase, 153–154
earth's crust, rocks and, world soils, 54
fractionation
California soil, 153–154
China soil, 154
Israeli soil, 153–154, 186–187
Indian soil, 57–58
the initial fast process, 172–176
Israeli sludge-amended soil, 192–194
Israeli soil, 65
Romania, 62
selectivity on calcite, 148
soil, 54
solid-phase control, 163–165
United Arab Emirates, 56–57
US, 62–64
Ni affinity
amorphous Fe oxide, 137
geothite (FeOOH) and haematite, 137
Nutrients in sewage, 268

O
Order of extraction steps in SSD, 128–129

P
Patagonian desert, 17
Pb
affinity on
amorphous Fe oxide, 137
geothite (FeOOH) and haematite, 137
humic acid, 131
Arizona, 63
California soils, 64
Chad, 57
China soil, 60
distribution in solid-phase, 154
earth's crust, rocks and, world soils, 54
fractionation in
California soil, 154
Colorado soil, 154
Israeli soil, 154, 189
Sludge-amended soil, 154
Spain soil, 154
the initial fast process, 172–176
Israeli sludge-amended soil, 192–194
Rocky Mountains, 62
soil, 54
solid-phase control, 163
sources, 310–311
Spain, 62
speciation, Southern Australia, 89
toxicity, 311
United Arab Emirates, 56–57
US, 62–64
Uzbekistan, 61
pe + pH on Mn redistribution, 203, 207–208
Pedoturbation, 38–40
pH effects on Zn adsorption kinetics, 142–143
pH effects on Zn binding, 137
Phosphate
on Cd and Zn solution speciation, 94
dissolution/precipitation, 100–101

Physical and chemical remediation, 293–296
Phytoremediation, 297–301
Plant species for phytoremediation, 298–300
Plant uptake, 221–227
Pollution, 267–302
Potential inputs of trace elements, 320–325
Precipitation and accumulation of clay, 42
Processes in Semi-Arid Areas, 36–42
Properties of arid soils, 20–21

R

Reaction of trace elements with organic matter, 131–134
Reagents for removing Fe oxides, 140
Redistribution index, 177–179
Reduced partitioning parameter, 180
Reduction-oxidation, 103–105
Reg Soils, 25–28
Relationship of Co and Mn transformation, 216–218
Remediation, 293–302
Remediation of Pb pollution, 295
Residence time
  Cu in soil solution, 201
  Ni in soil solution, 201
  Zn in soil solution, 201–202
Residual sands, 11
Rhizosphere chemistry, 227–229
Rocks and, B, As, Se, Cu, Co, Cr, Zn, Ni, Pb, Cd, Mn, Mo, Hg, 50–56
Rocky Mountains, 15–17
Rocky Mountains, Cd, Pb, Zn, 62–63
Romania, Cu, Zn, Ni, 62
Root uptake of trace elements, 222
Rub al Khali deserts, 17
Russia, Mo, Cu, Zn, Co, 61

S

Saline Soils, 34–36
Salt Accumulation and Distribution, 34–36
saturated paste extracts, 75
Saturated paste regime, 169–221
Saturation extraction, 232
Saudi Arabia, Zn speciation, 84
Se
  adsorption on kaolinite and montmorillonite, 146
  China soil, 60
  deficiency, 264
  earth's crust, rocks and, world soils, 56
  soil, 56
  volatilization, 301
  Western US, 289
Selective sequential dissolution, 107–130
Selenium with irrigation, 272
Semi-arid areas, 5
Sequential fractionation for As, 129–130
Serbia, B, 62
Serpentine and seleniferous soils, 289–290
Sewage sludge, 273–281
Sludge-amended soil, fractionation, 192–194
  Cd fractionation, 150–152
  Cr fractionation, 152
  Cu fractionation, 153
  Cu speciation, 86
  Pb fractionation, 154
  solution concentrations, 72–74
  Zn speciation, 84
Soil
  enzymes, 290–292
  forming factors and processes, 22–42
  organic matter removal, 132
  processes in arid (desert) areas, 25–36
  solution composition, 70–71
  solution definition, 69
Solid-phase component, 108
solubility
  equilibrium, 160–162
  solid phases, 160–167
  trace elements in sludge, 275

Index 365

Solubility product constants, 98–99
Solution concentration in Israeli soils with sewage irrigation, 73
Solution speciation, 76–91
  Ca and Mg, 78
  Cd, 86–89
  Cr(III) and Cr(VI), 89
  Cu, 85–86
  Hg, 90
  Ni, 86
  Pb, 89
  Se, 90–91
  Zn, 78–85
  Zn, pH effects, 83
Source of pollution, 267–290
  agricultural management, 284–286
  atmospheric inputs, 281–284
  irrigation with sewage, 267–273
  mining and industrial activities, 286–289
  serpentine and seleniferous soils, 289–290
  sewage sludge, 273–281
  traffic inputs, 281–284
Source trace elements in soils, 47–48
South America, 17
South Australia
  Cd speciation, 87
  Pb speciation, 89
  soil solution concentration, 76
Southern Queensland, Zn speciation, 83
Spain soil
  Cd fractionation, 151–152
  Cr fractionation, 152
  Cu fractionation, 153
  Cu, Pb, Cd and Zn, 62
  Pb fractionation, 154
Sr selectivity on calcite, 148
Steppe regions, 13–15
Storage of trace elements in plants, 221–227

T
Trace elements reactions, 135–148
  clay minerals, 145–146
  carbonate, 146–148
  iron, manganese, and aluminum oxides, 135–145
Syria, Zn, Cu, Mn, Cd, 66
Takla-Makan and Gobi deserts, 17
Tharsis mine, 287–288
Thin disk flow method, 134, 141
Thornthwaite method, 7
Toxicity, 264–266
Trace elements
  definition, 47
  fertilizers, 284–285
  lithosphere, 323
  sewage, 269
  sludge, 274
  soils and rocks, 48–56
  world soils, 323
Traffic inputs, 281–284
Transfer flux
  Cu, 169–221
  Ni, 169–221
  trace elements, 169–221
  Zn, 169–221
Translocation of trace elements in plants, 222–223
Transport of clay through the soil body, 41
Transport of Hg, 310
Turkestan desert, 17

U
United Arab Emirates, Cd, Cr, Ni, Pb, Cu, Zn, 56–57
United States, Cd, Zn, Cu, Ni, Pb, 62–64
uptake of trace elements by plants, 223–224
US soils, Zn, Ni, Pb, Cu, Hg, Cd, 54–55
USEPA limits of trace elements in sludge, 279–281
UV-visible / luminescence spectroscopy, 149
Uzbekistan, Co, Cu, Mn, B, Pb, Zn, 61

V
Vertisols, 38–40
Vibrational spectroscopy, 149
Volatilization, 301–302

W

Waste-amended soil, 152
Wetting-drying cycle regime, 169–221
Wind transport, 29–32
World soils, B, As, Se, Cu, Co, Cr, Zn, Ni, Pb, Cd, Mn, Mo, Hg, 50–56

X

X-ray diffractograms
   the carbonate fraction, 111–113
X-ray methods, 149

Z

Zn
   Arizona, 63
   Australia, 60
   Bulgaria, 61
   California soils, 64
   China soil, 59
   deficiency, 260–262
   distribution in solid-phase, 155–156
   earth's crust, rocks and, world soils, 54
   Indian soil, 57–58
   the initial fast process, 172–175
   interaction with Fe oxide, 136
   Israeli sludge-amended soil, 192–194
   Israeli soil, 65
   Kenya, 56
   Madagascar, 57
   reaction with organic matter, 133
   Rocky Mountains, 62
   Romania, 62
   Russia, 61
   selectivity on calcite, 148
   soil, 54–55
   solid-phase control, 163–165
   solution in Australian soils, 83
   solution, Colorado soil, 83
   Spain, 62
   Syria, 66
   United Arab Emirates, 56–57
   US soils, 54–55
   Uzbekistan, 61
Zn adsorption
   kinetics, 140–144
   soils with Fe removal, 142–143
Zn affinity
   amorphous Fe oxide, 137
   geothite (FeOOH) and haematite, 137
   humic acid, 132
Zn fractionation
   California soil, 155
   China soil, 155–156
   Indian soil, 155
   Israeli soil, 187–189
   African vertisols, 57
Zn speciation
   California, 84
   Saudi Arabia, 84
   sludge-amended soil, 84
   southern Queensland, 84

ENVIRONMENTAL POLLUTION

1. J. Fenger, O. Hertel and F. Palmgren (eds.): *Urban Air Pollution – European Aspects.* 1998   ISBN 0-7923-5502-4
2. D. Cormack: *Response to Marine Oil Pollution – Review and Assessment.* 1999   ISBN 0-7923-5674-8
3. S.J. Langan (ed.): *The Impact of Nitrogen Deposition on Natural and Semi-Natural Ecosystems.* 1999   ISBN 0-412-81040-9
4. C. Kennes and M.C. Veiga (eds.): *Bioreactors for Waste Gas Treatment.* 2001   ISBN 0-7923-7190-9
5. P.L. Younger, S.A. Banwart and R.S. Hedin: *Mine Water: Hydrology, Pollution, Remediation.* 2002   ISBN 1-4020-0137-1; Pb 1-4020-0138-X
6. K. Asante-Duah: *Public Health Risk Assessment for Human Exposure to Chemicals.* 2002   ISBN 1-4020-0920-8; Pb 1-4020-0921-6
7. R. Tykva and D. Berg (eds.): *Man-Made and Natural Radioactivity in Environmental Pollution and Radiochronology.* 2004   ISBN 1-4020-1860-6
8. L. Landner and R. Reuther (eds.): *Metals in Society and in the Environment. A Critical Review of Current Knowledge on Fluxes, Speciation, Bioavailability and Risk for Adverse Effects of Copper, Chromium, Nickel and Zinc.* 2004   ISBN 1-4020-2740-0; Pb 1-4020-2741-9
9. P.F. Ricci: *Environmental and Health Risk Assessment and Management. Principles and Practices.* 2006   ISBN 1-4020-3775-9
10. J. Davenport and J.L. Davenport: *The Ecology of Transportation: Managing Mobility for the Environment.* 2006   ISBN 1-4020-4503-4
11. M.B. Adams, D.R. DeWalle and J.L. Hom (eds.): *The Fernow Watershed Acidification Study.* 2006   ISBN 1-4020-4614-6
12. Y. Onishi, O.V. Voitsekhovich and M.J. Zheleznyak (eds.): *Chernobyl – What Have We Learned? The Successes and Failures to Mitigate Water Contamination Over 20 Years.* 2007   ISBN 1-4020-5348-7
13. F.X. Han: *Biogeochemistry of Trace Elements in Arid Environments.* 2007   ISBN 978-1-4020-6023-6